**Making Crystals by Design**

*Edited by*
*Dario Braga and Fabrizia Grepioni*

## 1807–2007 Knowledge for Generations

Each generation has its unique needs and aspirations. When Charles Wiley first opened his small printing shop in lower Manhattan in 1807, it was a generation of boundless potential searching for an identity. And we were there, helping to define a new American literary tradition. Over half a century later, in the midst of the Second Industrial Revolution, it was a generation focused on building the future. Once again, we were there, supplying the critical scientific, technical, and engineering knowledge that helped frame the world. Throughout the 20th Century, and into the new millennium, nations began to reach out beyond their own borders and a new international community was born. Wiley was there, expanding its operations around the world to enable a global exchange of ideas, opinions, and know-how.

For 200 years, Wiley has been an integral part of each generation's journey, enabling the flow of information and understanding necessary to meet their needs and fulfill their aspirations. Today, bold new technologies are changing the way we live and learn. Wiley will be there, providing you the must-have knowledge you need to imagine new worlds, new possibilities, and new opportunities.

Generations come and go, but you can always count on Wiley to provide you the knowledge you need, when and where you need it!

*William J. Pesce*
President and Chief Executive Officer

*Peter Booth Wiley*
Chairman of the Board

# Making Crystals by Design

## Methods, Techniques and Applications

*Edited by*
*Dario Braga and Fabrizia Grepioni*

WILEY-VCH Verlag GmbH & Co. KGaA

**The Editors**

*Prof. Dario Braga*
Department of Chemistry
University of Bologna
Via F. Selmi 2
40126 Bologna
Italy

*Prof. Fabrizia Grepioni*
Department of Chemistry
University of Bologna
Via F. Selmi 2
40126 Bologna
Italy

**Cover illustration**

The front cover picture is a photo taken by the editors in their lab. The heart was there as part of a solid batch of crystals prepared by a German visiting student.
Although this coincidence seems contradictory to the book title "Making Crystals by Design", it reminds us that the greatest designer is nature itself.

**Library of Congress Card No.:**
applied for

**British Library Cataloguing-in-Publication Data**
A catalogue record for this book is available from the British Library.

**Bibliographic information published by the Deutsche Bibliothek**
The Deutsche Bibliothek lists this publication in the Deutsche Nationalbibliografie; detailed biliographic data are available in the Internet at http://dnb.d-nb.de.

**Typesetting**   Dörr + Schiller GmbH, Stuttgart
**Printing**   betz-druck GmbH, Darmstadt
**Binding**   Litges & Dopf Buchbinderei GmbH, Heppenheim

Printed in the Federal Republic of Germany
Printed on acid-free paper

**ISBN:**   978-3-527-31506-2

# List of Contents

*Making Crystals by Design*. Edited by Dario Braga and Fabrizia Grepioni
Copyright © 2007 WILEY-VCH Verlag GmbH & Co. KGaA, Weinheim
ISBN: 978-3-527-31506-2

# Preface

"Organizing molecules into predictable arrays is the first step in a systematic approach to designing (...) solid-state materials."
(M. C. Etter et al., J. Am. Chem. Soc., **1987**, *109*, 7786.)

*Making crystals by design* is the paradigm of crystal engineering, the modern discipline at the intersection of supramolecular and materials chemistry. The *engineering* idea is that crystals with desired properties can be constructed from the convolution of the physico-chemical properties and intermolecular bonding capacity of the building blocks with the periodicity and symmetry of the crystal. Such a "bottom-up" approach from molecules and ions to aggregates requires crystal-oriented synthetic strategies.

Crystal syntheses do not differ, in their essence, from classical chemical experiments in which molecules are modelled, synthetic routes devised, products characterized and their properties evaluated. However, this typical chemical approach needs, in a sense, to be repeated twice: first, in order to prepare the building blocks (whether molecules or ions), and then to arrange the building blocks in a desired way via nucleation, precipitation and crystallization to attain and/or control crystal properties. Obviously, crystallization invariably implies the need to characterize the solid product, often in the form of a polycrystalline powder, for which routine analytical and spectroscopic laboratory tools are much less useful than in the case of solution chemistry.

The ultimate product of a crystal making exercise is a crystal, therefore understanding and reproducing the crystal-making process requires a good knowledge of solid-state techniques, such as differential scanning calorimetry, thermogravimetry, solid-state NMR, variable temperature X-ray powder diffraction, etc., which are not used routinely in traditional academic chemistry laboratories.

Crystal makers need to master both covalent chemistry in solution, in order to prepare building blocks, and supramolecular chemistry in the solid state, in order to assemble building blocks in crystals. This also applies to the study of coordination networks, where the knowledge of coordination chemistry and metal-ligand bonding is fundamental.

Furthermore, the intriguing possibility that the same building block may lead to different crystal structures, i.e. crystal polymorphs, with their different physico-

*Making Crystals by Design*. Edited by Dario Braga and Fabrizia Grepioni
Copyright © 2007 WILEY-VCH Verlag GmbH & Co. KGaA, Weinheim
ISBN: 978-3-527-31506-2

chemical properties, needs to be taken into account. Scientists engaging in crystal engineering endeavors must be well aware of this phenomenon and of its – at times dramatic – implications for the "making crystal by design" initial assumption. It is essential to understand why a crystallization process leads to one crystal form or another, and whether these crystal forms interconvert, in order to be able to reproduce the experimental conditions and product selection on a laboratory or plant scale.

When planning this book we were led by awareness that crystal makers come mainly from two traditionally distinct and often separate backgrounds. On the one hand, synthetic chemists, who know well how to make molecules, often possess a limited perception of how their molecules will behave once put together with billions of identical molecules in a crystal, and find problematic the characterization, handling, representation and modelling of crystalline solids. On the other hand, solid-state scientists, of whom crystallographers and solid-state theoreticians represent the most abundant populations, know how to use sophisticated techniques and computational methods for the investigation, characterization and evaluation of crystalline solids but may have limited experience in the handling of complex molecules in laboratory preparations.

With these ideas in mind we have asked a number of highly qualified researchers from different branches of the field to take part in the preparation of a book that could collect the experience of theoreticians, synthetic-solution and solid-state chemists, crystallographers and spectroscopists with a common focus on the project of *making crystals by design*.

This book should serve both as a state-of-art overview of "what's going on" in frontier areas (theoretical evaluation of noncovalent interactions, hydrogen-bonded crystals, coordination networks, solid-state reactivity and reactions taking place in the solid state, crystal polymorphism, etc.) and as an entry point to the fundamental methods and techniques required for a successful investigation of crystalline solids (crystallography, solid-state NMR spectroscopy, atomic force microscopy etc.).

The response has been extraordinary and we are proud to present to the readership a handbook, which is also a research book, written by a consortium of topnotch scientists but with the main objective of serving as an introductory overview, seen from its various angles, of this burgeoning field. Clearly, the book has no pretensions of exhaustiveness: our main objective has been that of providing an Arianna's thread to beginners and advanced levels students (as well as to lecturers for their teaching) to guide their way in this cross-disciplinary area of chemical science. If the reading of these chapters stimulates a few young scientists to choose and pursue their focused directions of research in the domain of *crystal making* we shall be successful. The increasing presence of *molecular materials, supramolecular solid-state chemistry* or *crystal engineering* topics in postgraduate high level courses (Masters and PhD), suggests that this book could also be a useful tool for teachers.

We would like to thank the authors of the chapters and acknowledge their patience with our insistence with details and deadlines, we are also grateful to Wiley-VCH people, in particular to Dr. Elke Maase, Dr. Bettina Bems and Dr. Tim Kersebohm for their guidance and assistance throughout this project.

D. Braga and F. Grepioni, University of Bologna, Italy                    August 2006

# List of Contributors

**Christer B. Aakeröy**
Department of Chemistry
Kansas State University
111 Willard Hall
Manhattan, KS 66506–3701
USA

**Joel Bernstein**
Department of Chemistry
Ben-Gurion University of the Negev
Beer Sheva
Israel 84105

**Dario Braga**
Department of Chemistry
University of Bologna
Via F. Selmi 2
40126 Bologna
Italy

**Lee Brammer**
Department of Chemistry
University of Sheffield
Brook Hill
Sheffield S3 7HF
UK

**Lucia Carlucci**
Dipartimento di Chimica Strutturale e
Stereochimica Inorganica
Via G. Venezian 21
20133 Milano
Italy

**Maria A. Carvajal**
Department de Química Física
Universitat de Barcelona,
Av. Diagonal 647
08028 Barcelona
Spain

**Neil R. Champness**
School of Chemistry
University of Nottingham
University Park
Nottingham NG7 2RD
UK

**Gianfranco Ciani**
Dipartimento di Chimica Strutturale e
Stereochimica Inorganica
Via G. Venezian 21
20133 Milano
Italy

**Tomislav Friščić**
Department of Chemistry
University of Iowa
305 Chemistry Building
Iowa City, IA 55242–1294
USA

*Making Crystals by Design*. Edited by Dario Braga and Fabrizia Grepioni
Copyright © 2007 WILEY-VCH Verlag GmbH & Co. KGaA, Weinheim
ISBN: 978-3-527-31506-2

**Angelo Gavezzotti**
Dipartimento di Chimica Strutturale e
Stereochimica Inorganica
University of Milano
Via Venezian 21
20133 Milano
Italy

**Leonard R. MacGillivray**
Department of Chemistry
University of Iowa
423B Chemistry Building
Iowa City, IA 55242–1294
USA

**Roberto Gobetto**
Department of Inorganic Chemistry
University of Torino
Via P. Giuria 7
10125 Torino
Italy

**Gerd Kaupp**
Department of Organic Chemistry
University of Oldenburg
Carl-von Ossietzky-Str. 9–11
26129 Oldenburg
Germany

**Guillermo Mínguez Espallargas**
Department of Chemistry
University of Sheffield
Brook Hill
Sheffield S3 7HF
UK

**Juan J. Novoa**
Department de Química Física
Universitat de Barcelona
Av. Diagonal 647
08028 Barcelona
Spain

**Emiliana D'Oria**
Department de Química Física
Universitat de Barcelona
Av. Diagonal 647
08028 Barcelona
Spain

**Davide M. Proserpio**
Dipartimento di Chimica Strutturale e
Stereochimica Inorganica
Via G. Venezian 21
20133 Milano
Italy

**Nate Schultheiss**
Department of Chemistry
Kansas State University
111 Willard Hall
Manhattan, KS 66506–3701
USA

**Satoshi Takamizawa**
International Graduate School of Arts
and Sciences
Yokohama City University
22–2 Seto
Kanazawa-ku, Yokohama
Kanagawa 236–0027
Japan

**Fumio Toda**
Department of Chemistry
Okayama University of Science
1–1 Ridai-cho
Okayama 700–0005
Japan

# 1
# Geometry and Energetics

## 1.1
## Supramolecular Interactions: Energetic Considerations
*Angelo Gavezzotti*

"God always geometrizes"
(attributed to Plato, showing, if confirmed, that even great philosophers can be badly mistaken)

### 1.1.1
### Introduction

In science as well as in everyday life, things are seldom as they appear to the unaided eye. Sight, that privileged sense so cherished by humans, leading as it does to the concepts of pleasure and beauty, is but of limited use in scientific proceedings, especially those connected with the behavior of matter at a molecular level. At a higher level of elaboration, geometry can be called for in the interpretation of objective phenomena, with some progress, especially in structural chemistry, where the geometrization of structure is of considerable help in the description of molecular-size objects. Chemistry, however, cannot be entirely built upon either sight or geometry. Electrons and nuclei act under the laws of quantum mechanics under well specified Hamiltonian operators, all around the key observable quantity, interelectronic and internuclear energy. Macroscopic systems span a space of dynamic variables, velocities and momenta, and evolve under the firm guidance of the two principles of thermodynamics, the most unnegotiable of all physical laws. There is very little solid chemistry without quantitative estimates of energy and entropy. This chapter is designed to outline the principles of the quantitative energetic evaluation of intermolecular interactions, the basis of nonreactive chemical phenomena occurring among organic molecules in condensed phases.

*Making Crystals by Design.* Edited by Dario Braga and Fabrizia Grepioni
Copyright © 2007 WILEY-VCH Verlag GmbH & Co. KGaA, Weinheim
ISBN: 978-3-527-31506-2

1.1.2
**Enthalpy**

### 1.1.2.1 The Quantistic Approach: Molecular Orbital (MO) Theory

The fundamental equation of quantum mechanics is the time-dependent Schrö-dinger equation:

$$H \Psi = -h/i (\partial \Psi / \partial t) \tag{1.1.1}$$

where $H$ is the hamiltonian operator, $h$ is Planck's constant, and i is the imaginary unit ($i = \sqrt{-1}$). When the potential energy is independent of time, the wavefunction $\Psi$ factorizes into a time-dependent part, $f(t)$, and a time-independent, position-dependent part, $\psi(x)$, so that:

$$H \psi(x) = E \psi(x) \tag{1.1.2}$$

$$\Psi(x, t) = f(t) \psi(x) \tag{1.1.3}$$

After normalization of the wavefunction, multiplying on the left by $\psi^*(x)$, and integrating over all space, one gets the expectation value of the electronic energy:

$$<E> = \int \psi^*(x) H \psi(x) \, d\tau \Big/ \int \psi^*(x) \psi(x) \, d\tau = \int \psi^*(x) H \psi(x) \, d\tau \tag{1.1.4}$$

A hydrogen atom consists of a nucleus of unit positive charge and one electron. The quantum mechanical equation for a hydrogen atom is [1]:

$$-h^2 /(2m_e) \nabla^2 \psi - (e^2/r)\psi = E \psi \tag{1.1.5}$$

where the first term corresponds to the kinetic energy part of the Hamiltonian, and the second represents the Coulombic potential between nucleus and electron. $\nabla^2$ is the second derivative operator. This equation is the only one that can be exactly solved in quantum chemistry. For helium, one gets:

$$-h^2/2m_e \nabla_1^2 \psi - (2e^2/r_1)\psi - (h^2/2m_e)\nabla_2^2 \psi - (2e^2/r_2)\psi + (e^2/r_{12})\psi$$
$$= H_1\psi + H_2\psi + (e^2/r_{12})\psi = E \psi \tag{1.1.6}$$

where now the equation contains one cross term for the Coulombic interaction between the two electrons. Because of this last term, the analytical solution of the above equation is impossible.

For a polyatomic molecule, nuclei are considered motionless, because the motion of electrons is on a much faster timescale than the motion of nuclei (the Born–Oppenheimer approximation). The corresponding Hamiltonian is:

$$H = \Sigma_k H(k) + e^2 \Sigma_{k>j} 1/r_{kj} \tag{1.1.7}$$

where $H(k)$ is the part for electron $k$, kinetic energy plus coulombic interactions with all the nuclei, and the second summation corresponds to the electron–electron Coulombic potential. Since an analytical solution of the Schrödinger equation is impossible even for a helium atom, the case with a hamiltonian like (1.1.7) is plainly hopeless.

Solution of the Schrödinger equation provides two main pieces of information, the total electronic energy of the system (the repulsion energy between motionless nuclei can be simply calculated by a Coulomb sum), and the wavefunction $\Psi(x, y, z, t)$, which in turn gives the electron density $\varrho(x,y,z)$ and a formal answer to the question 'where are the electrons?', because the quantity:

$$P(d\tau) = [\Psi\Psi^*] \, dx \, dy \, dz \tag{1.1.8}$$

is interpreted as the probability, $P$, that an electron be found in the infinitesimal volume $dV = dx \, dy \, dz$. $\Psi^*$ is the complex conjugate of $\Psi$.

In order to solve Eq. (1.1.2) with Hamiltonian (1.1.7), a large number of approximations are needed. Enter the favorite son of quantum chemistry, the molecular orbital (MO) method [1]. As appears from Eq. (1.1.4), a great many integrals are needed, so the wavefunction must be written in an easily integrable form. The nodal properties, crucial for the energetic considerations, just as the number of nodes relates to the energy of a vibrating string, must be taken into account: the spherical harmonics, $Y_{lm}(\theta, \phi)$, are used to take care of this. In MO theory, the radial part of atomic orbitals, $R_i(r)$, is written as a combination of gaussian basis functions $g(r)$, complete atomic orbitals $\chi_i$ are written as products of the radial and angular parts, and molecular orbitals $\varphi_i$ are then obtained as linear combinations of atomic orbitals (LCAO).

$$g(r) = N \, \alpha^{(2n+1)/4} \, r^{(n-1)} \, \exp(-\alpha r^2) \tag{1.1.9}$$

$$R_i(r) = \Sigma_j \, a_{ij} \, g_j \tag{1.1.10}$$

$$\chi_i = N' \, R_i(r) \, Y_{lm}(\theta, \phi) \tag{1.1.11}$$

$$\varphi_i = \Sigma_j \, c_{ij} \, \chi_j \tag{1.1.12}$$

in which $r$ is the distance of the electron from the nucleus, $\alpha$ is called the exponent of the gaussian, $n$ is 1, 2 or 3 for s, p or d functions respectively, and $N$ and $N'$ are normalization constants.

The total wavefunction $\Phi$ for a molecular system is then taken as a product of molecular orbitals, each of which contains the coordinates of one electron only, i.e., $\Phi$ is a product of one-electron orbitals. This is a very severe approximation, because this wavefunction cannot describe the effects of electron correlation, or the simultaneous displacement, or other simultaneous change in properties, of many electrons at a time.

Spin is a further complication. According to the Pauli principle, acceptable wavefunctions must change sign on formal exchange of the coordinates of two electrons [1]. This is taken care of by writing the total wavefunction in the form of a determinant (the Slater determinant), in which each MO is multiplied by an appropriate spin wavefunction.

The solution of Eq. (1.1.2) is done in an iterative, self-consistent way using the variational principle: whenever the true wavefunction is approximated by some incomplete function that depends on a number of parameters, the expectation value of the energy, Eq. (1.1.4), is higher than the expectation value that competes to the exact wavefunction. The procedure requires a lot of mathematical manipulation, but the problem has been solved once and for all and the task is nowadays performed by very efficient black-box computer packages.

An MO calculation provides, among many others, the following chemically useful items: (i) the energies of the molecular orbitals; these do not have a real physical meaning, but are sometimes useful, being an estimate of the energy levels of separate electron pairs in the molecule; (ii) the expectation value, or the total electronic energy of the molecule; (iii) the electron density, $\varrho(x, y, z)$. For the purposes of supramolecular chemistry, total energies can be used to evaluate the intermolecular interaction energies in pairs or clusters of molecules, while the electron density determines the molecular electrostatic potential and hence the cohesive properties of the molecule and its molecular recognition abilities:

$$\varrho(x, y, z) = \Sigma_r \Sigma_s P_{rs} \chi_r (x, y, z) \chi_s(x, y, z) \tag{1.1.13}$$

where $P_{rs}$ is a sum of products of LCAO coefficients in Eq. (1.1.12). Energies obtained by the methods so far described are called Hartree–Fock (HF) energies. They represent the kinetic energies of electrons and the electrostatic effects between electrons and nuclei, but cannot take into account electron correlation, because of the use of one-electron orbitals. This limitation forbids the treatment of dispersion energy, a crucial part of the intermolecular potential. To enchance the scope of MO methods beyond HF, configuration interaction can be used. In the basic HF treatment, the total wavefunction is represented by one Slater determinant, but one could simultaneously take into account all the possible electronic configurations, the ground state with no excitation, plus single and double or even triple excitations, taking the total wavefunction as a linear combination of all the resulting Slater determinants. The final equations become very complicated, but are still manageable.

The inclusion of excited states is one way of representing electron correlation. In the Moeller–Plesset (MP) approximation, at the $n$th level (MP$n$) all the $n$-fold excited configurations are taken into account, and the total energy results as:

$$E = E^0 + E^1 + E^2 + E^3 + E^4 + ... \tag{1.1.14}$$

The MP0 term is the sum of the energies of the molecular orbitals, the MP1 energy is the HF energy, and the energies that include upper terms in the expansion are called MP2, MP3, MP4... energies and represent electron correlation.

### 1.1.2.2 The Quantistic Approach: Density Functional Theory (DFT)

The foundation of DFT [2–4] is the Hohenberg–Kohn theorem: the energy of a system of electrons in an external potential $v(R)$ is a functional of the electron density (a functional is a mathematical entity that has a given value for each given value of another function), and the ground state energy of a many-electron system is uniquely determined by the ground-state density. In the Kohn–Sham assumption, a local potential, $V_{KS}(R)$, can be introduced for a molecular electron cloud in the field of the nuclei, such that a system of noninteracting electrons moving in the $V_{KS}(R)$ field will have the same density as the exact density of the interacting electron system. $V_{KS}$ has terms for the electron–nuclei and electron–electron coulombic interaction, and a further term which incorporates the exchange [3] and electron correlation effects:

$$V_{KS}(R) = V_{eZ}(R) + V_{ee}(R) + V_{exch}(R) \qquad (1.1.15)$$

At this point, the Kohn–Sham Hamiltonian operator can be written and the expectation value determined (compare with the above proceedings for the MO theory):

$$H_{KS} = -1/2\nabla^2 + V_{KS}(R) \qquad (1.1.16)$$

$$\int \Phi^* H \, \Phi \, d\tau = <E_{KS}> \qquad (1.1.17)$$

$$E_{KS} = T_{KS} + U_{eZ} + U_{ee} + E_{exch,KS} \qquad (1.1.18)$$

where $T_{KS}$ is the kinetic energy, $U_{eZ}$ is the electron-nuclei attraction, $U_{ee}$ is the electron–electron repulsion, and $E_{exch,KS}$ is the exchange energy over the Kohn–Sham orbitals. The crucial point is the following: the exchange energy $E_{exch,KS}$ is replaced by an effective contribution $E_{ex,eff}$ that also includes correlation effects as well as possible, and whose functional form $\varepsilon_{ex}(R)$ can be more or less empirically adapted to the purpose. This effective correction energy has a term in the difference between the true kinetic energy and the kinetic energy of the noninteracting electrons, and a term in the difference between the total electron interaction energy and the coulombic electron–electron repulsion. The latter term incorporates the correlation energy, as much as the functional $\varepsilon_{ex}(R)$ is indeed a realistic energy density. In the local density approximation, LDA, one takes the exchange-correlation energy of an electron gas of uniform density, equal to the local density in the molecular system at point R: a drastic approximation. Other more sophisticated approaches have been proposed, hence the various names (B, BP, B-LYP, etc.).

### 1.1.2.3 The Quantistic Approach: the Crystal Orbital Method

So far, only energies of isolated molecules have been considered. In a crystal, [5] the translational invariance of the potential energy $V$ at a generic point $r$ in real space is reflected in a translational invariance of the quantum problem, which is, as usual, the solution of a steady state Schrödinger equation:

$$V(r) = V(r - R) \tag{1.1.19}$$

$$H \psi (r) = E \psi (r) \text{ or } H \psi (r - R) = E \psi (r - R) \tag{1.1.20}$$

where $R$ is any translational vector. The eigenfunctions which are proper solutions of Eq. (1.1. 20) must be symmetry-adapted to translation; they are called Bloch functions:

$$\Phi(r + R, k) = \exp(ikR) \, \Phi(r, k) \tag{1.1.21}$$

$k$ is a vector whose components have the dimensions of a reciprocal length, and the dot product in the exponential is a pure number. Since $\exp(ikR)$ is a wave, $k$ expresses its wavelength (hence the name, wave vector).

A specification of the actual form of Bloch functions in terms of atomic orbitals is needed. They can be written as a product of a plane wave and a function, $\Omega(r, k)$, with the translational periodicity of the lattice. With the Schrödinger problem written in $k$-dependent form, the solutions take the form of a linear combination of Bloch functions, whose coefficients are to be determined. For all the atomic orbitals $\chi_j(r - r_j)$ in the unit cell, each centered at position $r_j$, in a crystal with $N$ unit cells, a combination which has the same translational periodicity as the lattice is the replacement of a single AO in the reference cell with a sum over all translationally equivalent AOs in the extended crystal. The final form of the crystal orbitals is then:

$$\Phi(r, k) = N^{-1/2} \Sigma_R \exp[ikR] \, \chi_{jR} \tag{1.1.22}$$

The solution of Eq. (1.1. 20) with the hamiltonian of Eq. (1.1.7) and under the Bloch conditions (1.1. 22) is conceptually similar to the solution under the LCAO-MO assumption (1.1.12). A major difference is that while for the molecular problem one solution is sufficient, the crystal problem is a function of $k$ and therefore the variational problem must be solved a number of times equal to the number of sampling points in the independent part of $k$-space (the first Brillouin zone). These computational difficulties limit the applicability of the crystal orbital method to rather small molecules and unit cells, the urea crystal being probably the limit. Moreover, the crystal orbital method cannot be applied with a proper consideration of correlation energy because of computational limitations.

For a crystal, orbital energies are a function of $k$ and hence $E(k)$ plots are obtained, called energy bands. Like molecular orbitals, bands are partly filled with electrons and partly empty. The energy difference between the Fermi level (highest occupied point of a band) and the energy of the lowest unoccupied band is called the band-gap, the equivalent of the HOMO–LUMO gap in an isolated molecule. Metals have half-filled electronic energy bands and a zero band-gap, so they are good conductors, while organic crystals have wide band-gaps and are therefore semiconductors or electric insulators.

The physical meaning of *k*-dependence can be further explained by comparing with the molecular case. In a molecule, atomic orbitals combine into molecular orbitals, whose energy increases with the number of nodes. In an infinite periodic system, like a crystal, one gets in principle, an infinite number of crystal orbitals, which merge into a continuum represented in the continuous nature of the energy bands (which however, in practice, are sampled at a finite number of points in the Brillouin zone). Thus, *k* labels the different crystal orbitals and counts their number of nodes. Since *k* is a vector in crystal reciprocal space (the same reciprocal space used by X-ray crystallographers), how bands evolve in k depends on the nature of the electron density along the corresponding directions in crystal space.

### 1.1.2.4 The Classical Approach: Vibrational and Nonbonded ("Force Field") Energies

The Born–Oppenheimer approximation uncouples electron and nuclear motion. The latter concerns massive (at least, relative to electrons) bodies, and much lower velocities: while the formal velocity of an electron may approach the speed of light, a molecule in the gas phase travels at about the speed of a supersonic jet plane. While electronic energies must be calculated by quantum mechanics, nuclear motions are more easily described in a classical framework.

For a molecule with $N$ nuclei, there are $3N$ degrees of freedom, with three ascribed to rigid-body translation, three to rotation, and $3N$-6 to vibration, or the change of the reciprocal positions of the nuclei. The classical potential energy of such vibrations can be written as:

$$E(\text{pot,vib})= \Sigma_i\, 1/2\ k_i\ (g_i - g_i°)^2 + \Sigma_{i,j}\ 1/2\ k_{ij}\ (g_i - g_i°)\ (g_j - g_j°) + \Sigma_n f(\tau_n) \qquad (1.1.23)$$

where the $k$s are the force constants, provided by molecular spectroscopy or by empirical fitting; the $g$s are the actual values of some internal bond stretching or bond bending coordinates, and $g°$ is the corresponding reference, or 'strainless' value, obtained usually by statistical studies of spectroscopic or diffraction experiments. $f(\tau)$ is some function of torsional angles. In addition, one must have some way of representing the soft attractive potential that arises from interaction between atoms not bound to one another and not bound to the same atom, as well as the sharp repulsion that arises when these atoms come into too close contact. This nonbonded potential energy is usually given the empirical form: [6–9]

$$E\ (\text{nonbonded}) = \Sigma_{i,j}\, f(R_{ij}) + E(\text{Coulombic}) \qquad (1.1.24)$$

$$f(R_{ij}) = -AR_{ij}^{-n} + BR_{ij}^{-m} + C\exp(-DR_{ij}) \qquad (1.1.25)$$

where $R_{ij}$ is the distance between two atomic nuclei, $A$, $B$, $C$, $D$..., $n$, $m$,... are empirical disposable parameters. Typical values are $n=6$ and $m=12$. The Coulombic term is supposed to represent the direct coulombic interactions between polarized regions of the electron distribution of the molecule; in the simplest case, it takes the point-charge form: [10]

$$E(\text{Coulombic}) = 1/(4\pi\varepsilon°) \; q_i q_j \; R_{ij}^{-1} \tag{1.1.26}$$

where the $q_i$s are point-charge parameters. In an immediate extension, these same functional forms can be used for intermolecular interactions, the idea being that the interaction energy between the two phenyl rings in 1,3-diphenylpropane has the same origin as that between two benzene rings in the 1,3-diphenylpropane crystal. Table 1.1.1 shows some typical stretching force constants, Table 1.1.2 some typical nonbonded functions. It may at once be seen that a 0.1 Å displacement in the region around the minimum causes an energy rise of 5–10 kJ mol$^{-1}$ for the former, and of 0.5–1 kJ mol$^{-1}$ for the latter. There is an order-of-magnitude difference between intra-and intermolecular stretchings.

**Table 1.1.1** Some typical force constants over internal molecular coordinates (units of N m$^{-1}$).

| | |
|---|---|
| C–C stretching, alkanes | 430–450 |
| $Csp^3$–$Csp^2$ | 500 |
| C–C stretching, benzene | 625 |
| C–C double bond stretching, olefins | 910 |
| C–H stretching, alkanes | 450–470 |
| C=O stretching | 800 |

**Table 1.1.2** UNI force field [8, 9] potential parameters: $A, B, C$ for $A \exp(-BR) - CR^{-6}$ Distances in Å, energies in kJ mol$^{-1}$.

| | | $A$ | $B$ | $C$ | $R°$ | Well depth |
|---|---|---|---|---|---|---|
| H | H | 24158.4 | 4.010 | 109.20 | 3.36 | −0.042 |
| H | C | 120792.1 | 4.100 | 472.79 | 3.29 | −0.205 |
| H | N | 228279.0 | 4.520 | 502.08 | 2.98 | −0.394 |
| H | O | 295432.3 | 4.820 | 439.32 | 2.80 | −0.505 |
| H | F | 64257.8 | 4.110 | 248.36 | 3.29 | −0.110 |
| H | S | 268571.0 | 4.030 | 1167.34 | 3.35 | −0.458 |
| H | CL | 292963.7 | 4.090 | 1167.34 | 3.30 | −0.501 |
| C | C | 226145.2 | 3.470 | 2418.35 | 3.89 | −0.387 |
| C | N | 491494.5 | 3.860 | 2790.73 | 3.49 | −0.851 |
| C | O | 393086.8 | 3.740 | 2681.94 | 3.61 | −0.674 |
| C | F | 196600.9 | 3.840 | 1168.75 | 3.50 | −0.350 |
| C | S | 529108.6 | 3.410 | 6292.74 | 3.96 | −0.909 |
| C | CL | 390660.1 | 3.520 | 3861.83 | 3.83 | −0.678 |

Table 1.1.2 (continued)

|   |    | A        | B     | C       | R°   | Well depth |
|---|----|----------|-------|---------|------|------------|
| N | N  | 365263.2 | 3.650 | 2891.14 | 3.70 | − 0.629    |
| N | O  | 268571.0 | 3.860 | 1522.98 | 3.49 | − 0.464    |
| N | F  | 249858.9 | 3.930 | 1277.90 | 3.43 | − 0.435    |
| N | S  | 630306.9 | 3.590 | 5576.76 | 3.75 | − 1.108    |
| O | O  | 195309.1 | 3.740 | 1334.70 | 3.61 | − 0.336    |
| O | F  | 182706.1 | 3.980 | 868.27  | 3.39 | − 0.320    |
| O | S  | 460909.4 | 3.630 | 3790.70 | 3.71 | − 0.801    |
| O | CL | 338297.3 | 3.630 | 2782.36 | 3.71 | − 0.588    |
| F | F  | 170916.4 | 4.220 | 564.84  | 3.20 | − 0.293    |

The conformation of a gas-phase molecule can be derived by minimizing the strain energy Eqs. (1.1.23)–(1.1.26). A powerful simulation tool for the collective behavior of a molecular ensemble is molecular dynamics (MD) [11]. In the classical approach, the total instantaneous energy of a molecular system, $E$(tot) is again a sum of terms as in Eqs. (1.1.23)–(1.1.26). Note that the above functional forms assume that the electronic structure is not significantly perturbed, that is, $E$(electronic) = constant.

Forces $F_{i,k}$ ($k$th cartesian component of the position vector of atom $i$) can be calculated easily, given the simple functional forms of vibrational and nonbonded potentials, as their first derivatives. The time evolution of such a system can be described using the equations of motion of classical Newtonian mechanics, $F_{i,k} = m_i(d^2x_{i,k}/dt^2)$. Given a reasonable starting configuration $(x_{i,k})°$, integration of the above differential equation gives the trajectories of all atoms in the molecule. Of course the integration cannot be carried out analytically, but is carried out by fast and efficient numerical iterative techniques.

The total kinetic energy of the multimolecular system is the sum of the kinetic energies of all atoms, and can be equated to the equipartition value, in the key link between molecular motion and temperature:

$$E(\text{kin}) = \Sigma_i 1/2m_i\, v_i^2 = 1/2\, k_BT\, (n_{dof}) = 1/2\, k_BT\, (3n_{atoms}) \tag{1.1.27}$$

where $m_i$ and $v_i$ are the mass and velocity of the $i$th atom.

Consider now the statistical mechanics definition of internal pressure:

$$P = 2/(3\,V)\,[E(\text{kin}) - \Xi] \tag{1.1.28}$$

$$\Xi = -1/2\, \Sigma_k\, \Sigma_l\, R_{k,l}\, F_{k,l} \tag{1.1.29}$$

$V$ is the volume of the system, and the quantity $\Xi$ is called the virial, a sum of products of the distance between any two atoms $k$ and $l$ in the system and of the force acting between them. Distances are always positive numbers so the virial is positive when forces are negative (that is attractive). For an ideal gas, $E(\text{kin}) = 3/2 RT$ per mole and since forces are all zero $\Xi = 0$ and $P = RT/V$; for real systems, including condensed phases, the attractive forces between molecules dictate the virial to subtract momentum from molecular action towards the surroundings. Equations (1.1.23)–(1.1.29) define the total energy of a molecular ensemble as a function of time, temperature and pressure: molecular dynamics samples the phase space and allows a calculation of complete $P/V/T$ phase diagrams.

### 1.1.2.5 Semi-classical Approaches: the SCDS-Pixel Method

The quantum mechanical approach cannot be used for the calculation of complete lattice energies of organic crystals, because of intrinsic limitations in the treatment of correlation energies. The classical approach is widely applicable, but is entirely parametric and does not adequately represent the implied physics. An intermediate approach, which allows a breakdown of the total intermolecular cohesion energy into recognizable coulombic, polarization, dispersion and repulsion contributions, and is based on numerical integrations over molecular electron densities, is called semi-classical density sums (SCDS) or more briefly 'Pixel' method. [12–14]

In the Pixel formulation, the basic concept is the electron density unit, or e-pixel. Let $\varrho_k$ be the electron density, calculated using a MO wavefunction for molecule A, in a volume $V_k$ centered at point $(k) = [x_k, y_k, z_k]$. Each of the e-pixels is assigned a charge $q_k = \varrho_k V_k$. Super-pixels are then formed by condensing $n \times n \times n$ original pixels, $n$ being called the condensation level. In addition, molecule A has nuclei of charge $Z_j$ at points $(j) = [x_j, y_j, z_j]$. A second molecule B has its e-pixels with $q_i = \varrho_i V_i$, at positions $(i) = [x_i, y_i, z_i]$, and nuclei of charge $Z_m$ at points $(m) = [x_m, y_m, z_m]$. Each molecule in the computational box is thus represented by a set of e-pixel charges, typical numbers being 10 000 for benzene to 25 000 for anthracene, plus a set of nuclear charges.

For the calculation of Coulombic energies, the electrostatic potential $\Phi_i$ generated by molecule A at point $(i)$ of the charge density of molecule B, and the potential $\Phi_m$ generated by molecule A at nucleus $m$ of molecule B, are calculated, and the total electrostatic potential energy between the two molecules is the sum of the electrostatic energies at points $(i)$, $E_i = q_i \Phi_i$, and at points $(m)$, $E_m = Z_m \Phi_m$.

Calling $\varepsilon_i$ the total electric field exerted by surrounding molecules at pixel $i$, $\alpha_i$ the polarizability at pixel $i$, and $\mu_i$ the dipole induced at pixel $i$ by that field, the linear polarization energy is:

$$E_{\text{POL},i} = -1/2 \ \mu_i \varepsilon_i = -1/2 \ \alpha_i \ \varepsilon_i^2 \tag{1.1.30}$$

In order to assign pixel polarizabilities, each pixel has to be first assigned to a particular atom in the molecule. Let $p$ be the number of atoms for which the nucleus–pixel distance is smaller than the atomic radius. If $p=1$, the charge pixel is within one atomic sphere only, and it is assigned to that atom. If $p > 1$, the pixel is assigned to the atom from which the distance is the smallest fraction of the atomic radius. If $p = 0$, the pixel is assigned to the atom whose atomic surface is nearest.

Atomic polarizabilities are taken from standard repertories, and pixel polarizabilities are calculated from them as $\alpha i = (q_i / Z_{atom}) \, \alpha_{atom}$.

Polarization contributions at very short distance between polarizer and polarized pose divergence problems, because of the $R^{-3}$ dependence of the polarization energy, and must be somehow damped. Electric fields in molecular crystals are mostly of the order of $10^{10}$ V m$^{-1}$, plus a small number in the $10^{10}$–$10^{13}$ V m$^{-1}$ range, plus a very small number of even higher ones; these high-field contributions are physically unrealistic, resulting from fortuitous short distances in the overlapping density meshes. The damped polarization energy at pixel $i$ is

$$E_{POL,i} = -1/2 \; \alpha_i [\varepsilon_i \, d_i]^2 \text{ for } \varepsilon < \varepsilon_{max}, \; d_i = \exp{-(\varepsilon_i / (\varepsilon_{max} - \varepsilon_i))} \qquad (1.1.31)$$

and $E_{POL,i} = 0$ for $\varepsilon > \varepsilon_{max}$. $\varepsilon_{max}$ is an adjustable empirical parameter in the formulation, set at $150 \times 10^{10}$ V m$^{-1}$. The total polarization energy at a molecule is the sum of polarization energies at each of its electron density pixels, $E_{POL,TOT} = \Sigma \, E_{POL,i}$, while the total polarization energy in an ensemble of molecules is the sum of all A...B and B...A polarization energies. Polarization energies are not pairwise additive over atomic contributions.

Intermolecular dispersion energies are calculated as a sum of pixel–pixel terms in a London-type expression, involving the above defined distributed polarizabilities and an overall 'oscillator strength', $E_{OS}$. To avoid singularities (as before described) due to very short pixel–pixel distances in an inverse sixth-power formula, each term in the sum is damped, as it is shown here for the molecule A...molecule B interaction:

$$E_{DISP,AB} = E_{OS} \; (-3/4) \Sigma\Sigma \, f(R) \; \alpha_i \alpha_j \, / [(4\pi c^{\circ})^2 \, (R_{ij})^6 ] \qquad (1.1.32)$$

where $f(R)$ is the damping function, $f(R) = \exp[-(D/R_{ij} - 1)^2]$ for $R_{ij} < D$ and $f(R) = 1$ for $R_{ij} > D$. $D$ is an adjustable empirical parameter, set at 3.50 Å. $E_{OS}$ was originally approximated by the molecular ionization potential, for very small molecules. In complex polyatomic molecules, $E_{OS}$ can be taken as the energy of the highest occupied molecular orbital, the interacting electrons being usually the peripheral ones and hence roughly at the HOMO energy level. This assumption works well with small molecules containing C, H, N or O atoms, but fails for example for heavily fluorinated aromatic compounds, when the interacting electrons belong with the fluorine atoms, whose ionization potential is much lower than the energy of the HOMO, which is usually a π-type molecular orbital. The London oscillator strength $E_{OS}$ for the interaction between any two molecules can then be calculated taking into account the different nature of the interacting electrons, as a weighted sum of atomic ionization potentials, weights being the atom–atom overlap integrals. This procedure amounts in fact to taking different dispersion energy coefficients according to the different kinds of approaching atomic basins.

The repulsion energy, $E_{REP}$, is modeled as proportional to intermolecular overlap. The total overlap integral between the charge densities of any two molecules, $S_{AB}$, is calculated by numerical integration over the original uncondensed densities, and is then subdivided into contributions from pairs of atomic species, $S_{mn}$, using the assignment of pixels to atomic basins. For each $m$–$n$ pair the repulsion energy is evaluated as

$$E_{REP,mn} = (K_1 - K_2 \, \Delta\chi_{mn}) \, S_{mn} \tag{1.1.33}$$

where $\Delta\chi_{mn}$ is the corresponding difference in Pauling electronegativity. The values of $K_1$ and $K_2$ were optimized at 4800 and 1200 respectively (for energies in kJ mol$^{-1}$ with electron densities in electrons Å$^{-3}$). The rationale behind this approach is that when atoms of different electronegativity meet, a larger reorganization of the electron density occurs, and the overlap repulsion must be smaller. The total repulsion energy is then the sum over all $m$–$n$ pairs.

The total intermolecular interaction energy is then:

$$E_{TOT} = E_{COUL} + E_{POL} + E_{DISP} + E_{REP} \tag{1.1.34}$$

For a crystal, a number of symmetry related molecules (nuclear positions plus charge density pixels) are generated around a reference molecule, and all the terms of Eq. (1.1.34) for all the molecular pairs are evaluated. Their sum is the lattice energy of the crystal.

### 1.1.2.6 Supramolecular Energies

The internal energy per mole of a chemical system is the sum of all energies, electronic, vibrational, and kinetic, possessed by one mole of molecules at a given temperature, pressure and volume. Its thermodynamic symbol is $U$, and the relationship to enthalpy, $H$, is given by $H = U + PV$. Chemists use enthalpies because for any process, $dH$ is equal to the heat exchanged, $dq$, without contributions from volume work.

Consider first the simple molecular recognition process in which molecule A meets molecule B and a molecular complex is formed. In the quantistic approach the energy variation upon complex formation can be written as:

$$\Delta E(\text{compl}) = \Delta E(\text{electronic}) + \Delta E(\text{pot,vibr}) + \Delta E(\text{kin}) \approx$$
$$\Delta E(\text{electronic}) + \Delta E(\text{zero point, vibr}) \tag{1.1.35}$$

$\Delta E(\text{electronic})$ must be calculated by quantum chemical methods by a difference between the total energy of the complex and the sum of the energies of the separated molecules. At the formally zero temperature of the quantum calculation, kinetic energies are zero, and vibrational terms are the zero-point contributions (i.e. those coming from vibrational levels of quantum number zero). These differences are very small and are often neglected, although they can be of the same order of magnitude as the difference in electronic energies.

Consider now an ensemble of molecules in the classical approach (force field + MD) to molecular energies, and the sum:

$$E(\text{pot,vibr}) + E(\text{nonbonded}) + E(\text{kin}) = U - U° \tag{1.1.36}$$

where the terms are taken as averages of the sampling of phase space during an MD simulation, and $U°$ is an unknown zero reference term, anyway unimportant since only energy differences are meaningful. Sum (1.1.36) is the total internal energy of the molecular system, and is conceptually equivalent to the statistical mechanics internal energy, or to the classical thermodynamics internal energy:

$$U - U° = -(\partial \ln Q / \partial \beta) \quad \text{statistical mechanics, ideal gas} \tag{1.1.37}$$

$$\Delta U = q + w \qquad \text{classical thermodynamics} \tag{1.1.38}$$

where $Q$ is a molecular partition function, and $q$ and $w$ are heat and work exchanges in a thermodynamic process. In molecular dynamics, $U$ is an average over the time evolution of the value of energy function, Eq. (1.1.36); the classical approach has the advantage that the simulation also shows the actual dynamics of the process, and can be made temperature- and pressure-dependent.

Other interesting properties, related to molecular cohesion in condensed phases, are the enthalpies of vaporization and sublimation. In the classical MD approach, the vaporization enthalpy of a liquid or of a solid is:

$$\Delta H(\text{vap}) = \Delta U(\text{vap}) + P\Delta V(\text{vap}) = \Delta U(\text{vap}) + PV(g) = \Delta E(\text{intra}) + \Delta E(\text{inter}) + RT \tag{1.1.39}$$

where the volume of the condensed phase is negligible with respect to the volume of the gas, considered ideal. Kinetic energies do not appear because $E(\text{kin})$ is the same for any phase at the same temperature. For rigid molecules, $\Delta E(\text{intra})$, the intramolecular potential energy difference between gas and condensed phase, is $\approx$ 0; the intermolecular potential in the gas is assumed equal to zero, so that the intermolecular potential energy difference, $\Delta E(\text{inter})$, is simply equal to the negative of the intermolecular energy in the condensed phase, so that for a crystal $\Delta H(\text{vap})$ is equal to $-E(\text{lattice}) + RT$. When the calculation is static, i.e. the lattice energy is derived by a single calculation on the equilibrium crystal structure, a somewhat different line of reasoning leads to $\Delta H(\text{subl}) = -E(\text{lattice}) - 2RT$. In practice these corrections are often absorbed into the parametrization and, conveniently, $\Delta H(\text{subl}) = -E(\text{lattice})$. Typical sublimation enthalpies of organic compounds are in the range 50–200 kJ mol$^{-1}$.

## 1.1.3
## Entropy

### 1.1.3.1 Statistical and Classical Entropy

Contrary to what happens for mechanical systems, chemical systems do not reach equilibrium by minimizing their internal energy. The essential difference is that in chemical systems molecular motion incessantly redistributes the available energies among the constituting molecules in a completely random fashion, and part of the driving force towards equilibrium comes from reaching the statistically most probable distribution of these energies. Entropy cannot be defined in terms of order and disorder, subjective concepts which cannot be defined in a proper way; entropy has to do with the distribution of available energies. This is evident from the statistical thermodynamic definition of entropy. Let $M$ be the total number of molecules in a system, and $m_i^*$ the number of molecules in a given quantum level $\varepsilon_i$ at equilibrium. The statistical weight of the equilibrium distribution, $W^*$, and the statistical entropy, $S$, are given by

$$W^* = M!/(\Pi m_i^*) \tag{1.1.40}$$

$$S = [k_B/N] \ln(W^*) \tag{1.1.41}$$

$N$ here is the number of particles in the system, or Avogadro's number for one mole. $W^*$ is also the number of different, indistinguishable ways in which the equilibrium distribution of energies can be obtained, or, in a quite similar way, in which the equilibrium distribution of spatial orientations can be obtained. Thus in a perfectly symmetrical crystal at zero temperature all molecules are in the ground energy level, and all molecules are oriented according to symmetry, so there is only one way of distributing energies and there is only one way of distributing molecular orientations; accordingly, the logarithm dependence makes entropy zero, independently of what the symmetry is (a P1 crystal with just translational symmetry has the same zero entropy as an Fdd2 crystal with 16 equivalent positions in the unit cell, as long as symmetry is obeyed in both). This is sometimes called the third principle of thermodynamics.

Further manipulations give the Boltzmann equilibrium distribution and the total entropy of an ideal gas:

$$m_i^* = M\exp(-\beta\ E_i)/Q;\ P_i^* = m_i^*/M \tag{1.1.42}$$

$$S = -\ k_B\Sigma\ [(m_i^*/M)\ln\ (m_i^*/M)] = -\ k_B\Sigma\ P_i^*\ \ln P_i^* \tag{1.1.43}$$

$$S = (U - U°)/T + k_B\ln(Q) = -1/T\ [(1/Q)(\partial Q/\partial\beta)_V] + k_B\ln(Q) \tag{1.1.44}$$

The partition function $Q$ has a translational, rotational and vibrational part, and so does entropy. The first two contributions can be calculated in a straightforward way, but the vibrational contribution to entropy requires a knowledge of the normal vibrational frequencies of the molecule or of the crystal.

The classical definition of entropy goes through the Clausius inequality:

$$dS > dq/T \quad \text{irreversible processes} \tag{1.1.45}$$

$$dS = dq/T \quad \text{reversible processes or equilibrium}$$

which, after rearranging, yield $(\partial S/\partial P)_T = -(\partial V/\partial T)_P$ and $(dS)_P = (dH/T)_P = (Cp/T)dT$, with access to entropy variations through measurable quantities like the expansion coefficient and the heat capacity.

### 1.1.3.2 Lattice Dynamics [15] and Lattice Vibration Frequencies

In the harmonic approximation for a molecular crystal, all atoms are oscillating about their equilibrium positions under the restraining action of a vibrational potential, $V$, which can be conveniently taken as just the intermolecular non-bonded potential of Eq. (1.1.25), and which obeys the translational invariance condition of Eq. (1.1.19). Let $x_i$ be the displacement of atom $i$ from its equilibrium position in the crystal, and consider all the atoms in the reference cell and all the atoms in the surrounding cells denoted by a real space vector $\mathbf{R}$. The mass-weighted force constants can be written as:

$$D_{ij,R} = (m_i\, m_j)^{-1/2}[\partial^2 V / \partial x_{\text{Ref},i}\, \partial x_{R,j}]^{\circ} \qquad (1.1.46)$$

where the $m_i$s are atomic masses, and the zero superscript recalls that the derivatives are calculated at the equilibrium position. These force constants have the same lattice periodicity as the potential, so they can be treated just as the electronic orbitals were treated, i.e. they are periodicized through a Bloch expansion:

$$D_{ij}(\mathbf{k}) = \Sigma_R \exp[i\mathbf{k}\mathbf{R}]\, D_{ij,R} \qquad (1.1.47)$$

The dynamic problem of vibrational spectroscopy must be solved to find the normal coordinates as linear combinations of the basis Bloch functions, together with the amplitudes and frequencies of these normal vibrations. These depend on $\mathbf{k}$, and therefore the problem must be solved for a number of $\mathbf{k}$-points to ensure an adequate sampling of the Brillouin zone. Vibrational frequencies spread in $\mathbf{k}$-space, just as the Bloch treatment of electronic energy gave a dispersion of electronic energies in $\mathbf{k}$-space. The number of vibrational levels whose energy lies between $E$ and $E+dE$ is called the vibrational density of states. Vibrational contributions to the heat capacity and to the crystal entropy can be calculated by appropriate integrations over the vibrational density of states, just like molecular heat capacities and entropies are obtained by summation over molecular vibration frequencies.

Just as for crystal orbitals, $\mathbf{k}$ is a reciprocal space vector that identifies the vibrational normal modes and counts the number of nodes in the corresponding combination of atomic vibrations. How the energy of a given vibration evolves in $\mathbf{k}$-space depends on the intermolecular potential restraints along that particular direction. In organic crystals, the forces are anyway moderate and no major anisotropies may arise – compare, for instance, with the case of an ionic crystal in which one finds alternations of positive and negative charges. In organic crystals lattice vibrational frequencies fall in a rather restricted range, something like 30–150 cm$^{-1}$, and even $\mathbf{k}$-spreading cannot bring about major differences. The vibrational density of states of organic crystals are all similar, and therefore the corresponding vibrational entropies cannot be too different. Enthalpy differences between polymorphs must be small because the potentials are anyway weak and almost isotropic, and entropy differences follow suit, because the force constants (the second derivatives of these potentials) also cannot be too different. Typical calculations show lattice-dynamical entropy differences between computational polymorphs in a range of $10–15\ \mathrm{J\ K^{-1}\ mol^{-1}}$, implying a $T\Delta S$ range at room temperature of less than $5\ \mathrm{kJ\ mol^{-1}}$, just the same range as that of the enthalpy differences.

### 1.1.3.3 Entropy and Dynamic Simulation

The internal energy (or enthalpy) of a liquid or crystalline molecular ensemble can be easily evaluated by a MD simulation, because even a moderate sampling of phase space is sufficient to provide an adequate representation and a meaningful average. Usually 100 ns are sufficient for liquids at not too high temperatures, and even 10 ns may be enough for a crystal, whose chances to roam over phase space are much more restricted. In fact, for a crystal even a single calculation over a static model gives a rather accurate evaluation of the lattice energy. Entropy, on the other hand, is connected with the size of phase space actually available to the system and cannot be obtained as an average of some ensemble property over a limited simulation; in principle, all phase space should be swept and checked for availability, an obviously impossible task. The estimation of entropies by molecular dynamics simulations uses statistical approaches to the evaluation of work and heat exchanges, but the involved calculations are neither simple nor affordable in a routine way to the casual investigator. [16, 17]

### 1.1.4
### Free Energy

#### 1.1.4.1 Complexation and Evaporation/Sublimation

Gibbs free energy, $G = H - TS$, or Helmholtz free energy, $F = U - TS$, are the key quantities in the description of chemical equilibrium. For a given chemical process, consider the following equations:

$$\Delta G^* = \Delta H - T\Delta S^* = -RT\ln(K_{eq}) \tag{1.1.48}$$

$$\ln(K_{eq}) = -(\Delta H/R)\ T^{-1} + \text{constant} \tag{1.1.49}$$

where $\Delta G^*$, $\Delta H$ and $\Delta S^*$ are the standard free energy, enthalpy and entropy changes, and $K_{eq}$ is the equilibrium constant, or the mass-dependent part of the equilibrium conditions. For A–B molecular complexation, $K_{eq} = [AB]/[A][B]$, a measurable quantity. For evaporation and sublimation, the mass-dependent part is the saturated vapour pressure, so $K_{eq} = (SVP)$, again a measurable quantity. Equation (1.1.49), the van't Hoff equation, allows an experimental determination of equilibrium enthalpies from equilibrium constants, for example, of sublimation enthalpies of crystals from measurements of their SVP. The latter measurement is however a difficult one, because these pressures are of the order of $10^{-5}$ to $10^{-10}$ bar. Uncertainties on measured sublimation enthalpies are never smaller than 5–10 kJ mol$^{-1}$, and this should be kept in mind when using these numbers to calibrate intermolecular potentials.

Free energies can be simulated by the so called free energy perturbation or thermodynamic integration methods. While, in principle, more revealing than enthalpy-based simulations, free energy simulations have a number of shortcomings. First, the sampling of phase space must be incomparably larger than in enthalpy simulations, so that the required calculations are computationally very demanding, and applications are limited to simple systems. Second, the choice of the computational parameters and of phase space sampling are critical for the significance and the numerical stability of the calculations. Finally, the theoretical underpinnings of free energy methods rely on subtle interpretations of statistical thermodynamics, so that great care must be applied when planning such calculations. Anyway, for reasons explained above, free energy differences between polymorphs cannot be too large, as they result from a sum of small $\Delta H$s and small T$\Delta$S contributions.

### 1.1.4.2 Melting and Polymorphism

Phase equilibria between condensed phases, like melting and crystal polymorphic transitions, have no mass-dependent terms (no equilibrium constants) since the activity of pure condensed phases is unity, and hence the equilibrium thermodynamics is represented by the simple relationships:

$$\Delta G = 0; \ \Delta S = \Delta H / T \tag{1.1.50}$$

Typical melting enthalpies of organic crystals are 10–50 kJ mol$^{-1}$, while the slope of a plot of experimental molar enthalpies of melting against melting temperature shows that the melting entropy of organic crystals is rather constant and is on average around 60 J K$^{-1}$ mol$^{-1}$. $\Delta H$(melt) can easily be simulated by MD by running separate simulations for the liquid and the crystal phase at the same temperature and taking the difference in total energies.

Enthalpy differences between polymorphic crystal forms of the same substance [18, 19] can be directly measured by calorimetry or by the difference in enthalpies of dissolution. A critical survey of these experiments shows that these enthalpy differences are very small, of the order of 0–10 kJ mol$^{-1}$, as expected since they must be a small fraction of the enthalpy of melting. In many cases the measured value is undistinguishable from experimental noise. These enthalpy differences can be estimated by just taking the differences in lattice energy between pairs of crystal phases whose complete structure is known, but the intrinsic uncertainty of such a calculation is also of the same order of magnitude of the property it tries to simulate.

### 1.1.5
### Tutorial Examples

#### 1.1.5.1  Dimerization Energies, a Scale of Intermolecular Interactions

Table 1.1.3 shows dimerization energies (Eq. (1.1.35), uncorrected for zero-point energy differences) of some sample bimolecular complexes, as calculated by Pixel, Eq. (1.1.34) [14]. Quite similar results are obtained by high-level *ab initio* calculations including electron correlation. These numbers constitute a relative scale of the strength of intermolecular interactions in organic compounds: the strongest is the hydrogen bond, with an energy of the order of 25–30 kJ mol⁻¹; next comes the arene–perfluoroarene stacking interactions, with about 20 kJ mol⁻¹ per ring, followed by aliphatic chain alignment, worth about 2 kJ mol⁻¹ per methylene group; next come the arene stacking interactions, whose strength per aromatic ring (around 10 kJ mol⁻¹) is about equal to some of the interactions between systems containing acidic C–H groups (e.g. the aromatics) and π-electron clouds of benzenoid rings. Of even lesser importance are most C–H···O or C–H···N type interactions, worth just a few kJ mol⁻¹, while other interactions (like Cl···Cl contacts, often invoked as important in crystal structures) are uninfluential. Even more relevant to the discussion of recognition modes is the amplitude of the corresponding potential energy wells: Table 1.1.3 shows that only the hydrogen bond has a relatively narrow width, corresponding to a really competitive binding power. All other interactions merge into a highly stretchable continuum of energies and forces, so that their use for prediction and control of molecular recognition is very problematic and must be carefully justified in each instance [20].

**Table 1.1.3** Equilibrium distances, binding energies and width of the potential energy well for hydrogen bonds.

| Dimer | $R$ °[a] | $\Delta E$[b] | Width[c] |
|---|---|---|---|
| benzoic acid, cyclic H-bond | 1.75 | 67 | 1.65–1.95 |
| formic acid, single H-bond | 1.75 | 38 | 1.55–2.05 |
| formamide, cyclic H-bond | 2.0 | 52 | 1.85–2.2 |
| formamide, single H-bond | 2.05 | 25 | 1.8–2.45 |
| methanol O–H··O–H | 1.95 | 23 | 1.7–2.3 |
| formic acid-acetone O–H···O=C | 1.8 | 31 | 1.55–2.1 |
| Water | 2.0 | 23 | 1.85–2.2 |
| parallel $C_6H_6$–$C_6F_6$ | 3.5 | 23 | 3.25–4.0 |
| parallel $C_6F_6$–$C_6F_6$ | 3.55 | 17 | 3.28–4.0 |
| parallel $C_6H_6$–$C_6H_6$ | 3.70 | 10 | 3.4–4.4 |
| parallel naphthalene | 3.6 | 26 | 3.35–3.9 |
| C–H···π acetylene–acetylene | 2.6 | 8 | 2.15–3.5 |
| C–H···πbenzene–benzene[i] | 2.65 | 11 | 2.25–3.5 |
| $CH_3$···π ethane-benzene[j] contacts | 3.2 | 7 | 2.75–4.3 |
| Acetone–acetylene C–H···O | 2.2 | 11 | 1.85–3.0 |

Table 1.1.3 (continued)

| Dimer | $R°^{[a]}$ | $\Delta E^{[b]}$ | Width[c] |
|---|---|---|---|
| Acetone–benzene C–H$\cdots$O | 2.45 | 6 | 2.05–4.0 |
| Acetone–methane C–H$\cdots$O | 2.6 | 2.5 | – |
| Acetonitrile–acetylene C–H$\cdots$N | 2.25 | 13 | 1.9–2.9 |
| Acetonitrile–benzene C–H$\cdots$N | 2.5 | 6.5 | 2.15–4.2 |
| Acetonitrile–methane C–H$\cdots$N | 2.6 | 2.5 | – |
| Benzoquinone, two C=O$\cdots$H–C | 2.35 | 15 | 2.1–3.0 |
| parallel chain hexane | 2.6 | 14 | – |
| $CH_3$–Cl$\cdots$Cl–$CH_3$ | – | 0 | – |

[a] $R°$ is the O$\cdots$H or N$\cdots$H distance in hydrogen bonds, or the inter-ring distance in parallel stacks, or the (C–)H$\cdots\pi$ or (C–)H$\cdots$O distance, Å units. [b] Binding energy, kJ mol$^{-1}$. [c] Range of the distance coordinate for an energy rise of 5 kJ mol$^{-1}$ from the minimum.

### 1.1.5.2 Calculation of Lattice Energies, Force Field Methods versus Pixel

The simplest crystal calculation consists of taking the atomic coordinates, cell data and space group for a crystal structure from an X-ray determination, building a static cluster of molecules to represent the crystal, and calculating all the interaction energies by some potential. This can be done by force field methods, Eq. (1.1.25) (time for one calculation a few milliseconds) or by Pixel, Eq. (1.1.34) (time for one calculation a few minutes). Figure 1.1.1 shows that the results are of comparable accuracy (remember that experimental values themselves are no more accurate than 5–10 kJ mol$^{-1}$). The UNI empirical force field [9] uses however about 100 parameters while the Pixel method uses only a few.

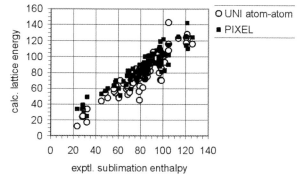

Figure 1.1.1 Experimental sublimation enthalpies and calculated lattice energies (kJ mol$^{-1}$) for 90 organic crystal structures.

### 1.1.5.3 **Energy Partitioning by Pixel**

Table 1.1.4 shows some typical examples of partitioned Pixel energies in molecular crystals. One sees there a clear predominance of dispersion in hydrocarbon crystals, an increased relevance of coulombic and polarization effects in heteroatom-containing molecules, and the clear predominance of coulombic-polarization energies in hydrogen-bonded crystals. In fact, a plot of the partitioned energies in hydrogen-bonded dimers shows that the total energy up to an $O\cdots H$ distance equal to the equilibrium value coincides almost exactly with the coulombic contribution. While of course the partitioning is to a large extent arbitrary and depends on the values of the parameters in the theory, the picture of intermolecular bonding is, within the present parametrization, quite convincing and helps explaining some of the main packing factors in crystals.

**Table 1.1.4** Typical values for Pixel partitioned energies in some organic crystals, units kJ mol⁻¹.

| Crystal name | Ecoul | Epol | Edisp | Erep |
|---|---|---|---|---|
| hydrocarbons | | | | |
| hexane | − 10.7 | − 3.8 | − 81.2 | 43.2 |
| octane | − 11.1 | − 3.7 | − 95.5 | 45.9 |
| benzene | − 13.3 | − 5.1 | − 59.5 | 30.1 |
| anthracene | − 26.1 | − 12.3 | − 115.9 | 68.4 |
| 1,4-benzene derivatives | | | | |
| dichloro | − 18.9 | − 7.5 | − 86.2 | 45.6 |
| dicyano | − 42.3 | − 10.1 | − 92.6 | 51.2 |
| dinitro | − 39.3 | − 10.7 | − 86.4 | 37.6 |
| benzoquinone | − 36.0 | − 8.6 | − 63.2 | 35.0 |
| polar compounds | | | | |
| succinic anhydride | − 47.0 | − 12.3 | − 56.0 | 30.0 |
| naphthoquinone | − 36.7 | − 11.4 | − 105.9 | 51.8 |
| triazine | − 24.4 | − 6.7 | − 68.1 | 34.9 |
| acids | | | | |
| acetic | − 78.2 | − 35.8 | − 37.7 | 92.9 |
| benzoic | − 94.3 | − 46.2 | − 69.0 | 120.9 |
| oxalic | − 115.5 | − 47.7 | − 55.6 | 115.8 |
| amides | | | | |
| acetamide | − 80.8 | − 27.7 | − 43.7 | 96.3 |
| benzamide | − 73.8 | − 25.9 | − 77.8 | 91.0 |
| urea | − 97.6 | − 32.3 | − 41.5 | 75.0 |

### 1.1.5.4  Analysis of Crystal Structures

The analysis of experimental crystal structures is at the same time the key to understanding the forces that determine one particular crystal structure, and a possible way to a better understanding of intermolecular forces in general. Methods based on geometrical features alone, like the identification of geometrical patterns, down to an analysis of atom–atom distances, are usually subjective and sometimes misleading. A comprehension of fine detail of the constitution of organic crystals requires objective, quantitative assessments. We propose the following systematic procedure for the quantitative classification of intermolecular interaction energies in crystals, consisting of the following sequential steps:

1. Using known atomic positions, cell dimensions and space group symmetry, find the closest molecular pairs in the crystal, for which the distance between centers of mass is smaller than the largest molecular dimension. The number of such neighbours is usually 10–12 in organic crystals.

2. Each molecular pair is characterized by the distance between centers of mass, $R_i$, a symmetry operator connecting the two molecules, $O_i$, and a molecule–molecule energy, $E_i$, calculated by some intermolecular potential method: quantum-chemical methods can be accurate, where affordable; the Pixel method has the advantage of good accuracy and energy partitioning; but even atom–atom force field energies can often be useful.

3. Rank these pairs in order of descending energetic relevance, and discuss the crystal structure in terms of these molecule–molecule determinants.

4. On the basis of these main interactions, proceed to the identification of extended motifs, like ribbons or layers within the crystal structure, where applicable.

We present here the analysis along these lines of some typical crystal structures [21]. Consider first naphthalene, a key example in aromatic interactions. Table 1.1.5 collects the interaction energies of closest neighbours, as calculated by the Pixel method, in comparison with the energies of some idealized, gas-phase dimers. In spite of many approximate and qualitative theories of aromatic packing, naphthalene in its crystal does not form close dimers, taking full advantage of the 50 kJ mol$^{-1}$ dispersive power of parallel stacking at short interplanar distance (about 4 Å). Naphthalene molecules cannot form extended parallel stacks in the crystal because then lateral inter-stack C–H$\cdots$H–C contacts would not be stabilizing enough; the molecules are thus compelled to approach at high interplanar angles, using a moderate coulombic stabilization between electron-poor hydrogen regions and electron-rich $\pi$-clouds. In this perspective, the herring-bone structure of crystals of benzene and other aromatics is dictated by avoided repulsion, rather than by an often invoked C–H$\cdots\pi$ attraction. Clearly, the observed crystal structure is a complex compromise between several demands, and does not lend itself to a simplification in terms of molecular geometry alone.

**Table 1.1.5** Pixel energies (kJ mol$^{-1}$) between neighbor pairs in the naphthalene crystal and between optimized gas-phase dimers[a] (PS and T).

| Symmetry operator | Distance, Å. | $E_{coul}$ | $E_{pol}$ | $E_{disp}$ | $E_{rep}$ | $E_{tot}$ |
|---|---|---|---|---|---|---|
| 4 screw (A) | 5.08 | − 5.1 | − 2.9 | − 23.5 | 13.2 | − 18.3 |
| 2 translation (B) | 5.97 | − 4.5 | − 1.1 | − 14.0 | 6.4 | − 13.2 |
| 4 screw (C) | 7.84 | − 1.9 | − 0.9 | − 7.7 | 5.0 | − 5.6 |
| PS | 3.50 | − 0.4 | − 5.0 | − 51.1 | 25.0 | − 31.6 |
| T | 4.88 | − 13.0 | − 7.9 | − 30.7 | 32.0 | − 19.6 |

[a] Idealized dimers: PS, parallel stacked, overlapping molecular planes; T, perpendicular molecular planes.

On the other hand, the most relevant molecular pairs in the crystal structure of naphthoquinone (Table 1.1.6) show a parallel stacked recognition mode. Molecules are related by a center of inversion, but contrary to naive electrostatic reasoning, the stacking arrangement is not stabilized by interaction between antiparallel dipoles, because the corresponding coulombic energies are very small. In contrast to naphthalene, naphthoquinone can use the parallel stacking mode because lateral inter-stack contacts are then stabilized by favorable coulombic adhesion between positively charged C–H moieties and oxygen basins (see determinants C, E, F in Table 1.1.6): note, however, that H$\cdots$O short contacts do not always imply a sizeable coulombic interaction energy (see determinant D).

**Table 1.1.6** Energies (kJ mol$^{-1}$) between neighbor pairs in the crystal of naphthoquinone.

| Label | Distance | Short O$\cdots$H distances, Å | $E_{coul}$ | $E_{pol}$ | $E_{disp}$ | $E_{rep}$ | $E_{tot}$ |
|---|---|---|---|---|---|---|---|
| 1 A | 4.07 | none | − 2.4 | − 2.5 | − 45.7 | 13.9 | − 36.7 |
| 1 B | 4.38 | none | − 1.5 | − 1.7 | − 37.1 | 10.0 | − 30.3 |
| 1 C | 8.13 | 2.52 | − 14.8 | − 3.3 | − 8.6 | 11.6 | − 15.0 |
| 2 D | 6.65 | 2.87, 2.71 | − 2.2 | − 1.6 | − 12.8 | 5.8 | − 10.9 |
| 1 E, 1 F | 7.40 | 2.47, 2.49 | − 13.2 | − 3.6 | − 10.2 | 14.9 | − 12.0 |

In the naphthoic acid crystal (Table 1.1.7), the by far most dominant structure determinant is, as expected, the hydrogen bonded cyclic dimer, A, described by the Pixel approach as a strongly coulombic interaction, while the polarization component supposedly represents the stabilization deriving from the dynamic rearrangement of the electron density – the partial covalent bond character. Next, at a considerable energetic distance, comes a determinant in which the dispersive stacking contribution dominates. The hydrogen bonded dimer is complemented by two in-plane molecular interactions (E–F), using the favorable C–H$\cdots$O coulombic pairing as in naphthoquinone. Whether one would like to call these interactions C–H$\cdots$O bonds, is a matter of taste; energy-wise, this may not be advisable, because even the sum of the REF-E and REF-F energies, $-14.4 \, kJ \, mol^{-1}$, is still smaller than the smaller of the energies in the stacked dimers.

**Table 1.1.7** Energies ($kJ \, mol^{-1}$) between neighbour pairs in the crystal of 2-naphthoic acid.

| Label | Distance, Å | $E_{coul}$ | $E_{pol}$ | $E_{disp}$ | $E_{rep}$ | $E_{tot}$ |
|-------|-------------|-----------|-----------|------------|-----------|-----------|
| 1 A | 9.68 | − 146.0 | − 78.2 | − 13.5 | 183.8 | − 53.8 |
| 2 B | 5.00 | −   6.0 | −  3.1 | − 37.6 | 19.6 | − 27.1 |
| 1 E | 8.29 | −   7.9 | −  3.0 | −  5.4 | 5.8 | − 10.4 |
| 2 F | 7.53 | −   1.9 | −  1.7 | −  7.5 | 7.1 | −  4.0 |

## References

1 Hehre, W. J., Radom, L., Schleyer, P. v. R., Pople, J. A., *Ab Initio Molecular Orbital Theory*, Wiley, New York 1986.

2 Ziegler, T. Approximate density functional theory as a practical tool in molecular energetics and dynamics, *Chem. Rev.* **1991**, *91*, 651.

3 Bickelhaupt, F. M., Baerends, E. J., Kohn-Sham density functional theory: predicting and understanding chemistry, *Rev. Comput. Chem.*, **2000**, *15*, 14.

4 Grimme, S., Accurate description of van der Waals complexes by density functional theory including empirical corrections, *J. Comput. Chem.* **2004**, *25*, 1463.

5 Dovesi, R., Civalleri, B., Orlando, R., Roetti, C., Saunders, V. R., Ab initio quantum simulations in solid state chemistry, *Rev. Comput. Chem.* **2005**, *21*, 1.

6 Williams, D. E., Starr, T. L., Calculation of the crystal structures of hydrocarbons by molecular packing analysis, *Comput. Chem.* **1977**, *1*, 173.

7 Kitaigorodski, A. I., in *Advances in Structure Research by Diffraction Methods*, Brill, R., Mason, R. (Eds.), Pergamon Press, Oxford 1970, Vol. 3, p.173.

8 G. Filippini, G., Gavezzotti, A., Empirical intermolecular potentials for organic crystals: the 6-exp approximation revisited, *Acta Crystallogr. Sect. B* **1993**, *49*, 868.

9 Gavezzotti, A., Filippini, G., Geometry of the intermolecular X–H$\cdots$Y (X,Y = N,O) hydrogen bond and the calibration of empirical hydrogen bond potentials, *J. Phys. Chem.* **1994**, *98*, 4831.

10 Williams, D. E., Net atomic charge and multipole models for the ab initio molecular electric potential, in *Rev. Comput. Chem.* **1991**, *2*, 219.

11 van Gunsteren, W. F., Berendsen, H. J. C., Computer simulation of molecular dynamics: methodology, applications, and perspectives in chemistry, *Angew. Chem. Int. Ed. Engl.* **1990**, *29*, 992.

12 Gavezzotti, A., Calculation of intermolecular interaction energies by direct numerical integration over electron densities. I. Electrostatic and polarization energies in molecular crystals, *J. Phys. Chem.* **2002**, *106*, 4145.

13 Gavezzotti, A., Calculation of intermolecular interaction energies by direct numerical integration over electron

densities. 2. An improved polarization model and the evaluation of dispersion and repulsion energies, *J. Phys. Chem. B*, **2003**, *107*, 2344.

**14** Gavezzotti, A., Quantitative ranking of crystal packing modes by systematic calculations on potential energies and vibrational amplitudes of molecular dimers, *J. Chem. Theor. Comput.* **2005**, *1*, 834.

**15** Pertsin, A. J., Kitaigorodski, A. I., *The Atom–Atom Potential Method, Applications to Organic Molecular Solids*, Springer-Verlag, Berlin 1987, Ch. 5 and 6.

**16** Kollmann, P., Free energy calculations: applications to chemical and biochemical phenomena, *Chem. Rev.* **1993**, *93*, 2395.

**17** Reinhardt, W. P., Miller, M. A., Amon, L. M., Why is it so difficult to simulate entropies, free energies, and their differences? *Acc. Chem. Res.* **2001**, *34*, 607.

**18** Gavezzotti, A., Filippini, G., Polymorphic forms of organic crystals at room conditions: thermodynamic and structural implications, *J. Am. Chem. Soc.* **1995**, *117*, 12299.

**19** Bernstein, J. *Polymorphism in Organic Crystals*, Oxford University Press, Oxford 2002.

**20** Dunitz, J. D., Gavezzotti, A., Molecular recognition in organic crystals: directed intermolecular bonds or nonlocalized bonding? *Angew. Chem. Int. Ed.* **2005**, *44*, 1766.

**21** Gavezzotti, A., Hierarchies of intermolecular potentials and forces: progress towards a quantitative evaluation, *Struct. Chem.* **2005**, *16*, 177.

**1.2**

**Understanding the Nature of the Intermolecular Interactions in Molecular Crystals.
A Theoretical Perspective**

*Juan J. Novoa, Emiliana D'Oria, and Maria A. Carvajal*

1.2.1
**Introduction**

All molecules undergo some sort of intermolecular interaction [1]. These interactions must be attractive, at least along some specific intermolecular orientation,
and consequently all known molecules aggregate to form solids and liquids. The
preference for a specific condensed state depends on the relative effects of the
interaction potential energy and of the thermal kinetic energy of nuclear vibrations,
at the temperature considered. It is worth pointing out here that molecules, as we
will see later, can also take relative orientations that give rise to repulsive intermolecular interactions.

Molecular crystals are ordered solids whose stability and structure is dictated by
the intermolecular interactions of their constituent molecules [2]. Each experimentally or theoretically obtained crystalline polymorphic form [3] is a minimum
in the free energy surface of that crystal. The enthalpic part of these minima results
from a compromise among all possible interactions the molecules may be engaged
in, not all of them attractive. At the same time, molecules are prevented from
collapsing into covalent linkages by the repulsive wall they encounter when they
get too close; this wall can also be instrumental in defining the preferred orientation between molecules, as originally stated by Kitaigorodski [2], and later by others
[4]. Thus, a proper approach to the analysis and design of crystals (the final aim of
Crystal Engineering [5]) is only possible if an adequate understanding of the nature
and properties of the intermolecular interactions present in the crystals of interest
has been obtained [2, 3, 5]. The aim of crystal engineering is the rational design of
crystals with specific properties, which is only possible if one is able to properly
tailor the crystal structure, on which depend most of the technological properties.
Although the large number of intermolecular interactions involved requires the
use of appropriate computer programs for the accurate and systematic prediction
of possible polymorphic structures [6], one still has to understand the logic behind
the relative stability of crystal forms. This is only possible by looking at the
predominant intermolecular interactions found in these crystals, with the aim of
obtaining a qualitative estimate of their nature and strength, and then understanding their interplay. We can thus obtain a qualitative view of the stability and
structure of existing crystals, and also of the forces that could lead to new crystal
structures.

Our aim in this work is to present a qualitative description of the properties of
the most relevant intermolecular interactions from the point of view of the packing
of molecular crystals. Therefore, we will not deal with the properties of metal–
ligand interactions, at the root of much of supramolecular chemistry, as they do not
play a leading role in the crystal packing of organometallic compounds. For the

same reasons, we will neglect those very weak interactions which do not play a relevant role in determining the packing, for instance, magnetic interactions. Information about intermolecular interactions will be mainly obtained from theoretical considerations, complemented, when appropriate, by statistical analysis of crystal packings. As we will see, the current knowledge provides a qualitative, but solid, understanding of the main interactions present in molecular crystals. Quantitative analysis requires the use of some theoretical method and of high-speed computers [6]; many of these quantitative studies are still approximate and sometimes lack adequate accuracy [8]. A fully rigorous quantitative study requires solving the Schrödinger equation for the crystal, which is impossible, but reliable results may be obtained using a powerful enough method that derives from the Schrödinger equation. This could be any of the methods that include electronic correlation; for example, even using density functional methods the calculation of the lattice energy of a molecular crystal is still very expensive and has been carried out for only a restricted number of crystalline systems [9]. Notice that the results obtained from these quantitative calculations are just the numerical value of the interaction energy and, when geometry optimizations are performed, the optimum geometry of the interacting system. But these numbers do not explain by themselves the preference for that optimum geometry, or why that interaction energy is stronger than that found in another system. In order to rationalize both types of numerical values one requires a separate qualitative analysis of the intermolecular interactions using the tools presented in this chapter. For instance, the preference of A–H $\cdots$ B hydrogen bonds to be linear can only be rationalized by assigning a dominant electrostatic nature to that interaction, and then looking at the most stable orientations of the dominant electrostatic components (in a hydrogen bonded dimer where the two monomers are neutral, the dipole–dipole component).

Hereafter, we present in concise and pedagogic terms the basic principles of the nature, strength and directionality of the intermolecular interactions found in molecular crystals. We mostly rely upon basic theoretical considerations, which will be connected, when appropriate, with the corresponding experimental observations. We will first analyze the nature of single types of intermolecular interactions, using sample systems like gas-phase dimers, where there is no competition among different interactions; an analysis will then be attempted of the changes that take place when more than one intermolecular interaction is present in the system under consideration.

## 1.2.2
### Intermolecular Interactions

Before proceeding further, let us mention that the current understanding of intermolecular interactions is not so good as that achieved in intramolecular chemical bonding – the attractive interactions responsible for the stability and structure of molecules and which also determine their reactivity. Decades of research have allowed us to understand why a carbon atom is tetravalent, that is, usually forms four bonds. However, penta-coordinated carbon atoms have been reported in the literature (for instance, $CH_5^+$), and even C atoms with higher coordination have been reported. Such a body of knowledge constitutes what is known as the "Theory of Valence". Coulson stated long ago [10] the essential conditions that these theories should satisfy to be reliable. They should (i) explain the principles that control the formation of the molecule, and (ii) explain the stability of the molecule, its maximum number of bonds, and the relative orientation of these bonds. These principles are well understood for most molecules of interest, although not always easily satisfied by following simple qualitative considerations. Since the basic principles of intermolecular interactions and bonds are similar to those at work in intramolecular interactions and bonds, there are no reasons for not extrapolating these conditions to intermolecular interactions, thus obtaining what could be called the "Theory of Intermolecular Valence". But when one considers how well intermolecular bonds fulfill the two essential requirements of the theory of valence, one finds that at present we are still far from predicting, for example, the maximum number of $A–H \cdots B$ hydrogen bonds that a given acceptor group can form, or the relative orientation of these bonds. Sometimes, even defining the attractive or repulsive nature of an $A–H \cdots B$ intermolecular interaction is still a matter of debate. These limitations show that the "Theory of Intermolecular Valence" is still under construction. However, it is already solid enough to allow, in many cases, a sound analysis of the strongest interactions present in crystals, that is, to allow one to "read the crystal" in its energetic terms.

As mentioned above, intermolecular interactions are the interactions that a stable molecule experiences in the presence of other molecules (not necessarily of the same type) [1]. We consider as stable molecules any aggregate of atoms having a long enough stability and lifetime, on a chemical scale. This definition also considers as stable molecules, open-shell radicals whose chemical structure makes them long-living, like those found in molecular magnets (for instance the family of the nitronyl nitroxide radicals).

Exemplary manifestations of the existence of intermolecular interactions [1] are the capillary rise of liquids in glass tubes and surface tension phenomena, already investigated in the 19th century, and explained in terms of mutual attraction between molecules in the liquid state. However, the first consistent identification of their attractive nature was achieved by van der Waals, in his study of real gases, but the origin and properties of the intermolecular interactions were only properly understood when the development of Quantum Mechanics allowed a full understanding of the electronic structure of atoms and molecules, and of their interaction.

Quantum mechanics indicates that when two molecules are close to each other, they experience a mutual interaction that depends on their relative orientation. This is illustrated in Fig. 1.2.1, where the interaction energy of two coplanar water molecules is computed as a function of $O \cdots O$ distance and orientation, defined in terms of the angle between the $C_{2v}$ axis of one molecule and the O atom of the other. It is clear from Fig. 1.2.1 that the interaction is attractive when the shortest contact is of the $O–H \cdots O$ type. But the interaction energy is not always attractive:

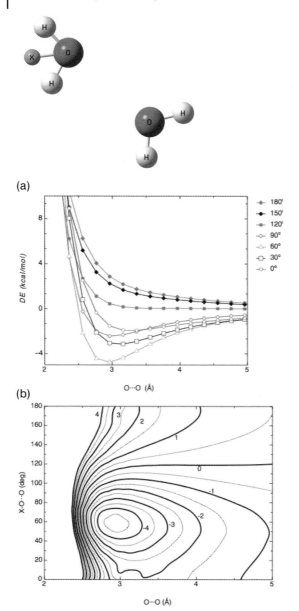

**Figure 1.2.1** Interaction energy (in kcal mol$^{-1}$, computed at the MP2/aug-cc-pVTZ level) of two coplanar water molecules oriented as shown in the upper diagram (the "dummy" atom X used to mark the $C_{2v}$ axis of the water molecule taken as reference in the calculations). The lower diagrams present two views of the interaction energy surface for changes in the O···O distance (in Å) and X–O···O angles (in degrees, where the X–O line marks the position of the $C_{2v}$ axis in one of the molecules). (a) A set of curves, one for each of the X–O···O angles; (b) a surface function of the O···O distance and the X–O···O angle (the minimum, located in the minimum energy region, corresponds to the hydrogen bonded dimer).

it becomes repulsive when the shortest contact is of the $O \cdots O$ type. It is also important to note the Morse-like shape of the interaction energy in the attractive region: the same shape found in intramolecular chemical bonds. This suggests that intermolecular and intramolecular interactions have a common fundamental background, namely, the interactions among all nuclei and all electrons. Their difference is found in: (i) the electronic structure of the interacting species (open shell atoms or radicals in chemical bonds, and closed shell molecules or radicals whose radical center is well protected, in intermolecular interactions), and (ii) in the values of the well depth and equilibrium distances in these Morse-like curves. Thus, interaction energy curves found in a chemical bond typically show a minimum around 1 Å with energies above the 100 kcal mol$^{-1}$ range, while attractive intermolecular interaction curves typically have a minimum above 2 Å with energies below 30 kcal mol$^{-1}$.

### 1.2.2.1 Interactions and Bonds: When Do Intermolecular Interactions Become Bonds?

Let us start this section by saying that "intermolecular interactions" and "bonds" are not equivalent concepts: *not all intermolecular interactions are bonds* (only attractive interactions can become bonds). Furthermore, as we will detail below, *not all attractive intermolecular interactions are bonds*.

The most fundamental definition of a bond is that introduced by Pauling in 1939 [11]: *"There is a chemical bond between two atoms or group of atoms in case that the forces acting between them are such as to lead to the formation of an aggregate with sufficient stability to make it convenient for the chemist to consider it as an independent molecular species"*. Therefore, according to this definition, the key issue for the existence of a bond is the energetic stability of the new aggregate formed. Consequently, *only energetically attractive interactions can be called bonds*. Pauling also listed a few examples of bonds [11]: the covalent bond, the ionic bond, and the coordination bond. Besides them, he also stated the following: *"In general we do not consider the weak van der Waals forces between molecules as leading to chemical-bond formation; but in exceptional cases, such as the weak bond that holds together the two $O_2$ molecules in $O_4$, it may happen that these forces are strong enough to make it convenient to describe the corresponding intermolecular interaction as a bond formation"*. Thus, he also made a clear distinction between chemical bonds and intermolecular bonds in terms of their strength and the entities involved (in intermolecular bonds, stable molecules).

Pauling's definition is exhaustive only for interactions between two atoms: for instance, the van der Waals bond in the $Ar \cdots Ar$ interaction, or the bond in the $H_2$ molecule. However, for more complex systems, it only recognizes the existence of a bond, but not the atoms involved in such a bond, or how many bonds are present. For instance, Pauling's definition recognizes that benzene is an entity with bonds, but does not determine their number or the atoms involved. The same is true for intermolecular bonds, where it cannot determine how many bonds can be found in the most stable geometry of the water dimer, for instance. Where are the bonds in complex structures?

A rigorous identification of bonds in complex aggregates is provided by the Atoms-in-Molecules (AIM) theory [12], which has proper quantum mechanical roots. According to AIM theory, the presence of a bond between two atoms requires: (i) the presence of a (3,–1) bond critical point [13] (or, in short, a bond critical point) in the electronic density, and (ii) a line of maximum curvature which links the two bonded atoms. Figure 1.2.2 shows a three-dimensional representation of the electronic density of an isolated benzene molecule, and a slice of that density in the plane that contains all atoms of the molecule. There one can see the (3,–1) bond critical points, which are the minima located along the side of the hexagon between the cusps of density associated with the C atoms, and between the C and their closest H atoms. We identify six C–C and six C–H bond critical points, thus giving support to the usual graphic representation for the benzene molecule, also plotted in Fig. 1.2.2.

Figure 1.2.3 shows similar graphs for the water dimer, where it is clearly seen that only one *intermolecular* bond critical point exists, corresponding to a H···O bond. In this figure one can also see two O–H *intramolecular* bond critical points, whose features are not fully shown due to the contours selected (those for the water molecule on the right side are not seen due to the out-of plane position of its two O–H bonds). As shown in Table 1.2.1, the intramolecular and intermolecular bond critical points differ in their properties (only one intramolecular bond is shown in Table 1.2.1, as all four bonds of this type have similar features). These data allow a

(a)

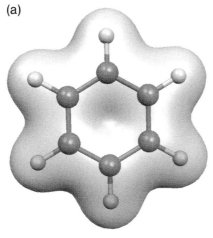

(b)

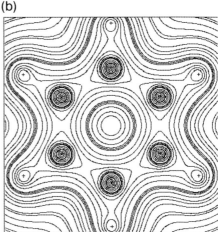

**Figure 1.2.2** Form of the electronic density of the benzene molecule. (a) Three-dimensional view of the plane of 0.002 atomic units, which encloses the region where the electron is located 90 % of the time (a graphical representation of the structure is also plotted);

(b) Cut of the electron density along the plane of the molecule (the bond critical points are the regions of minimum density between the C atoms along the sides of the hexagon, or between the C and H atoms).

(a)

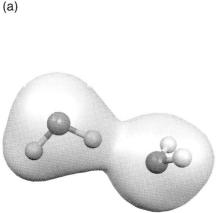

(b)

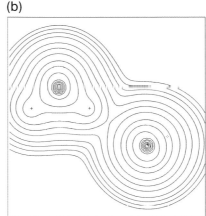

**Figure 1.2.3** Form of the electronic density of the water dimer complex. (a) Three-dimensional view of the plane of 0.002 atomic units, which encloses the region where the electron is located 90 % of the time (a graphical representation of the structure is also plotted); (b) cut of the electron density along the plane of the leftmost water molecule (the bond critical points are the regions of minimum density between the H and O atoms making the shortest H···O contacts).

distinction between these two types of bonds: (i) bond critical points of intermolecular bonds have a much smaller density (consistent with a larger bond distance), and (ii) intermolecular bond critical points have a positive Laplacian (the sum of the diagonal elements of the second derivative of the density), while intramolecular points present a negative Laplacian. Following the usual procedure, in the water dimer one can draw lines that connect the bonded atoms. The usual graphical representation of the dominant interactions for this dimer are thus obtained (see Fig. 1.2.3(a)), showing four intramolecular O–H bonds and one intermolecular H···O bond, the latter more commonly described as an O–H···O intermolecular bond. Given the energetic stability of the water dimer and the existence of only one critical point (the H···O bond critical point), the bond nature of the O–H···O intermolecular bond is well-established using rigorous quantum mechanical concepts.

**Table 1.2.1** Characteristic parameters of the intermolecular (O···H) and intramolecular (O–H) bond critical points found in the water dimer of Fig. 1.2.3. All values are given in atomic units.

| Critical point | Density at the critical point | Laplacian at the critical point |
| --- | --- | --- |
| O···H | 0.020 | 0.075 |
| O–H | 0.364 | − 0.262 |

Although the AIM analysis of the water dimer shows only one O–H···O intermolecular bond, even simple electrostatic arguments indicate that the H···O interaction is not the only attractive interaction. For instance, the interaction between the second O–H group of the left water molecule and the O atom of the right molecule (see Fig. 1.2.3) is also attractive. Similar situations are found in other systems. Thus, in benzene, the C–C interactions between atoms on opposite sides of the hexagon (origin of the non-Kekulé resonant structures) are also attractive, although they are not considered as bonds in the AIM analysis (and in the common graph of the benzene molecule). This is conventionally justified by saying that the energetically dominant resonant structures are the Kekulé structures. Consequently, *the bonds found in the molecular and supramolecular graphs do not show all the energetically stabilizing interactions, but only the dominant attractive interactions.* Why are bonds such a powerful concept? Because the graph that shows all bonds of a given system (the set of lines that we draw connecting the bonded atoms of this system) provides a rapid but accurate glimpse of the dominant forces responsible for the appearance of a given structure and for its stability. Consequently, bonds speak about dominant interaction energies, and energy is the driving force behind the existence of chemical structures, according to the Laws of Thermodynamics. (The change in free energy $\Delta G$ drives all chemical processes, including bond formation and rearrangement. $\Delta G = \Delta H + T\Delta S$, where $H$ is the enthalpy, $T$ the absolute temperature and $S$ the entropy; of these two terms, $\Delta H$ is usually the most important). The change in enthalpy of the system can be written as $\Delta U + P\Delta V$, where $\Delta U$, the change in internal energy of the system, is the dominant term, whose value is associated to the interaction energy of the parts that constitute the system. The dominant components of this interaction energy are just the energy of the bonds, either intra- or intermolecular. The same conclusions can be obtained by carrying out a quantum mechanics study of the microscopic systems and then applying statistical mechanics to the system.(It was demonstrated long ago that the laws of thermodynamics can be strictly rigorously derived from quantum mechanics by carrying out a statistical treatment of the system to describe its nonuniformity). This explains the cornerstone position that the bond concept has in chemistry.

The use of the bond concept in crystal engineering is still under discussion. Some authors [14] have suggested that it could be substituted by purely geometrical concepts (for instance, by using *hydrogen bridges* instead of *hydrogen bonds*). However, in our opinion, the use of purely geometrical terms causes a loss of information in the analysis of the interactions of interest, being a less powerful concept for understanding the structure and transformations of a crystal. Other authors [15], based on the fact that the interactions between molecules are weak and that many minima are possible, propose to abandon the analysis of intermolecular bonds when analyzing crystals, in favor of the analysis of molecule–molecule interactions as a whole. In our opinion, this is a useful proposal for an efficient analysis of crystal structures, but when, for instance, one wants to understand the reasons behind the relative strengths of the molecule–molecule interactions, the rationalization can only be properly done by looking at the intermolecular bonds and their properties (geometry, energy, cooperativity, ...).

It is also common in crystal engineering to talk about *contacts* and their geometry, because some intermolecular bonds show a preferential geometrical placement of their participating atoms. Contact is a topological term that results from the geometry of the intermolecular interactions. Usually one associates a preferential energetic behavior with a given contact, assumed to be similar when found between similar molecules. This has been the basis for the statistical analysis of crystals in terms of contacts [16]. Such preference results from the dependence of the interaction energy on the geometrical parameters that define the contact. For instance neutral $R–A–H \cdots O–R'$ interactions will not change too much when the chemical nature of the R and R' groups changes. However, when neutral R and R' groups are substituted by charged groups, the dependence of the interaction energy on the geometrical parameters can change dramatically. No correlations should exist in systems whose contacts are driven by interactions of very different chemical nature.

The existence of a (3,–1) bond critical point was originally considered as a necessary and sufficient condition for the existence of a bond [12]. However, recently there have been reports suggesting that it is a necessary but not always sufficient condition [17]. We have seen in our group that this is particularly true in many ionic crystals, where critical points are found for anion $\cdots$ anion interactions despite their repulsive nature (this situation is thought to arise primarily from the much larger size of the anion compared to the cation, although the matter is more complex, as recently discussed by Bader and Fang and others [18]). Therefore, in the rest of this chapter we will assume that the existence of a bond critical point is one of the two necessary conditions for the existence of a bond, but it is not a sufficient condition.

It is worth pointing out here that the validity of the AIM theory was recently questioned [19] on the basis that it provides the same results as the promolecule model (an independent atom model in which the density is obtained by adding the spherical densities of all atoms that constitute the molecule). This is not surprising for some weakly interacting complexes, where the polarization of the density is generally small, but this observation would have raised serious doubts on the accuracy of the AIM theory if found to be always valid. However, recent systematic studies [20] on the NANQUO02 crystal (the crystal of 3,4-bis(dimethylamino)-3-cyclobutene-1,2-dione) have shown that the critical points obtained using the promolecule and real electron densities are different, and that there are fundamental reasons why the two densities are different (Fig. 1.2.4 shows all the short C–H$\cdots$O contacts present in the NANQUO02 crystal for the in-the-plane interactions; a star was placed on those that are not hydrogen bonds). It was found that the electron density computed using the promolecule model and the real electron density present different bond critical points in four of the 19 bond critical points. This example also illustrates the power of an AIM analysis in the search for bonds in complex aggregates.

The quest for (3,–1) bond critical points in a complex or crystal is nowadays carried out not only using computed electron densities, but also using experimental ones [21]. The bond critical points obtained with the two densities are similar in

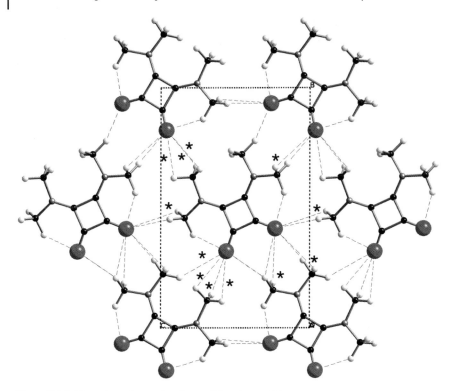

**Figure 1.2.4** View along the a axis of one of the planes of the NANQUO02 crystal, the crystal of the 3,4-bis(dimethylamino)-3-cyclobutene-1,2-dione. The shortest C–H···O contacts are shown (the dotted lines, all in the 2.0–3.5 Å range of distances). We have marked with a * those contacts that are not bonds, as they do not present (3,−1) bond critical points.

all the cases where such comparison has been done [21]. Correlations have been proposed between the density at the bond critical point and various bond parameters, among them the strength of the bond [22]. Bond critical points are also very helpful in complex geometrical arrangements to establish if a short A–H···B contact is a hydrogen bond or a van der Waals bond, a controversial choice in some cases. The AIM analysis solves this problem in a natural form: it is a hydrogen bond when a H···B bond critical point is located, but a van der Waals bond when an A···B bond critical point is found [23].

### 1.2.2.2 The Nature of the Intermolecular Interactions

The most appropriate approach to establishing the nature of any intermolecular interaction is to look at the dominant components of its interaction energy. As we will show in this section, such a procedure provides a fingerprint identification of any intermolecular interaction.

Using perturbation theory [24] it is possible to derive expressions for the inter-action energy of two molecules and write such energy as the sum of terms that have a useful physical meaning. One of these expressions has been provided by the IMPT method [25] (an acronym for Inter Molecular Perturbation Theory). Similar expressions can be obtained by employing other perturbation formalisms [26]. Within the IMPT theory [25], the interaction energy between two closed-shell molecules is the sum of the following five components:

$$E_{int} = E_{er} + E_{el} + E_{p} + E_{ct} + E_{disp} \qquad (1.2.1)$$

The first term $(E_{er})$ is the *exchange–repulsion component*, a combination of the Pauli repulsion that two electrons experience when forced into the same region of space, and the attractive exchange component. The sum of the two components is repulsive, and is responsible for the existence of the so-called repulsive wall in the interaction curves. The second term $(E_{el})$ is the *electrostatic component*, associ-ated with the electrostatic interaction of two unpolarized molecules (having the electronic distribution of the isolated molecule). The third term is the *polarization component* $(E_{p})$, which describes the change in the electrostatic interaction compo-nent due to the mutual polarization of electronic distributions caused by proximity. In classical terms, such polarization is proportional to the polarizability of the two molecules and to the square of the electric field. The fourth term is the *charge-transfer component* $(E_{ct})$, associated with transfer of electronic charge from one interacting molecule to the other. The last term is the *dispersion component* $(E_{disp})$, a nonclassical term whose existence is due to the correlated motions of the electrons. This term originates from quantum effects that have no exact classical translation, but its closest classical analogy is with an interaction between the instantaneous dipole moments induced in one molecule by electron motions of the other molecule [1]). When the interacting molecules have an open-shell structure, another term is needed, which we can call the *bond component* $(E_{bond})$, whose origin is the tendency for spin pairing of the unpaired electrons from each molecule when their orbitals are allowed to overlap. This term is found whenever two stable (that is, long-lived) radicals interact, for instance, two TCNE·· anion-radicals. The bond-ing component can be so important that very long C–C bonds are produced between these anion-radicals [27]. These long bonds show all the properties of the conventional covalent C–C bonds, but at much larger separations, around 2.9 Å [27]. Notice that long-distance C–C bonds can exist only in ionic crystals where the cation···anion interactions outweigh the sum of the anion···anion and cati-on···cation repulsions (for this reason, they have also been called charge-mediated bonds [27], as the cations "mediate" the existence of these bonds).

The values of all the above energy components can be accurately computed, and from there one can obtain useful qualitative trends. Figure 1.2.5 illustrates, in the water dimer, how their values change when the shortest intermolecular H···O distance is increased (no $E_{bond}$ term is present in this dimer, as the two molecules are closed-shell). Some points are worth mentioning about the trends in these values: (i) the exchange–repulsion component is repulsive at all values of the H···O distance,

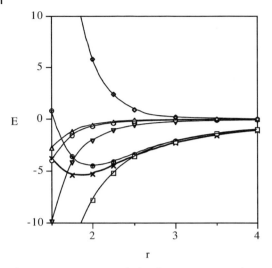

**Figure 1.2.5** Variation with the shortest H···O distance (*r*, in Å) of the IMPT components of the interaction energy for the optimum structure of water dimer (see Fig. 1.2.3). The value of *r* was increased while preserving the relative orientation of the two fragments. The components are identified as follows: electrostatic (□), exchange–repulsion (♦), polarization (○), charge transfer (▲), and dispersion (▼). We have also plotted the sum of the previous components ($^\times$), and the MP2 interaction energy (●). All energies are in kcal mol$^{-1}$. The calculations have been done using the aug-cc-pVDZ basis set in all cases.

(ii) all other terms are attractive, (iii) the dominant attractive term is the electrostatic component, followed by a much smaller but non-negligible dispersion term, (iv) the total energy curve is similar to the MP2 interaction energy curve, since the minima are in the same range of distances and energies. It is always important to check this similarity for the geometry of interest, as the perturbative nature of the IMPT method could induce a lack of convergence in the perturbative series in some cases. It is worth mentioning here that the IMPT analysis has been done on many dimer complexes, with good agreement between the MP2 energy value and the sum of the IMPT components (some of these cases are shown below).

Perturbative analysis of the interaction energy can be performed on any dimer found within a crystal. However, such an analysis requires some specialized background that is not generally available for most crystal engineers. An alternative is to perform a qualitative analysis of the interaction energy, looking for main trends in the interaction energy, that is, looking for the energetic interactions that dominate the crystal packing. We should then examine their nature and see how geometrical changes in the intermolecular geometry affect these interactions. As already mentioned, the experimental structure is a minimum of the total interaction energy of the crystal, that is, the optimum structure is driven by the interaction energy. But understanding how such energy is affected by geometrical changes is useful to rationalize how the crystal selects its optimum energy structure as result of a compromise of the energies of all interactions (within the crystal, an intermolecular interaction is not necessarily found in its minimum energy structure when isolated, as many other interactions are present and the crystal packs to optimize the total interaction energy).

To carry out such qualitative analysis one needs reliable qualitative expressions for the energetic components found in Eq. (1.2.1) in order to reproduce the main trends obtained in accurate calculations. Here we describe those presently thought to be most reliable.

The exchange-repulsion term ($E_{er}$), which is always repulsive, can be taken as proportional to the overlap between the densities of the interacting molecules [28]:

$$E_{er} \approx KS_{AB}, \text{ where } S_{AB} = <\varrho_A|\varrho_B> \tag{1.2.2}$$

where $K$ is a proportionality constant that depends on the specific interacting molecules. Due to the properties of the $S_{AB}$ overlap integral, the $E_{er}$ term follows exponential behavior as a function of the shortest intermolecular atom–atom distances, that is,

$$E_{er} \approx \exp(-r_{ij}) \tag{1.2.3}$$

Its exponential behavior makes this term the dominant one when short atom–atom distances between the interacting molecules are produced. Consequently, this term prevents molecules from getting closer than some limiting distance; this is the physical principle behind the hard-sphere model and Kitaigorodsky's close-packing principles [2]. Moreover, the shortest atom–atom distances that one can find in different intermolecular interactions for the same pair of atoms always fall in a restricted range, a fact that allows one to define atomic radii. They differ in ionic and neutral crystals due to the different electronic structure of ionic and neutral species, as easily shown when comparing the contours at 90% probability in electron density maps for isolated atoms and their ions.

The electrostatic term ($E_{el}$) describes the electrostatic energy between molecules A and B with nondeformed electronic structure. Using classical electrostatics, the electron density of a molecule can be expanded in a series of multipoles centered on one point, usually the center of mass of the molecule [1]. For quantitative studies, where more accuracy is required, the multipole expansion is done on all atoms of the interacting molecules (the so-called distributed multipole expansion [29]). However, for qualitative analysis, the central multipole expansion provides sufficient accuracy. One then uses the multipole values found in the literature for isolated molecules [1]. The electrostatic energy in the central multipole expansion can be written as a series, whose leading terms up to the dipole level are:

$$\begin{aligned} E_{el} = &\frac{1}{4\pi\varepsilon_o} \frac{q_A q_B}{r} \\ &- \frac{1}{4\pi\varepsilon_o} \frac{q_A \mu_B}{r^2} \cos\theta_1 - \frac{1}{4\pi\varepsilon_o} \frac{q_B \mu_A}{r^2} \cos\theta_2 \\ &- \frac{1}{4\pi\varepsilon_o} \frac{\mu_A \mu_B}{r^3} (2\cos\theta_1 \cos\theta_2 - \sin\theta_1 \sin\theta_2 \cos\phi) - \dots \end{aligned} \tag{1.2.4}$$

Here $r$ is the distance between the two centers of charges of the interacting molecules, $\theta_1$ and $\theta_2$ are the angles between the dipole of each molecule and the axis that links their centers of charges, and $\phi$ the dihedral angle of the dipoles around that axis. The first row component is the *charge–charge* ($E(q,q)$) term, in the second row the two components are (charge$_A$–dipole$_B$ plus charge$_B$–dipole$_A$) of the *charge–dipole* ($E(q,\mu)$) term, and the third row component is the dipole–dipole ($E(\mu,\mu)$) term. The expression was truncated at the dipole level; usually this is adequate for a qualitative analysis, although higher multipoles must be considered, at least at quadrupole level, for instance in $CO_2$ dimers, where the charge and dipole are zero. Equation (1.2.4) can be written in a more compact and physically meaningful form as:

$$E_{el} = E(q,q) + E(q,\mu) + E(\mu,\mu) + \dots \tag{1.2.5}$$

At the typical equilibrium distances for intermolecular interactions (1.2–3 Å, where the shortest limit is taken as the shortest intermolecular H$\cdots$O distance found in neutron diffraction crystals), the first term dominates over the remaining two (due to its $1/r$ dependence). Such a situation is found when the two fragments are charged. When one fragment is charged and the other is neutral but has a non-negligible dipole, the first term is zero and the second term dominates (due to its $1/r^2$ dependence). When two neutral fragments which present a non-negligible dipole moment interact the only remaining term is the third one (with a $1/r^4$ dependence). Examples of each type of behavior are the interaction between ions in ionic salts, ion–solvent interactions, and solvent–solvent interactions.

The polarization term ($E_p$) takes into account the electrostatic effect of the mutual polarization of the electronic density of the interacting molecules. Notice that $E_{el} + E_p$ is the true electrostatic energy between two molecules, and that $E_p$ is usually smaller than $E_{el}$. Within the central multipole expansion Eq. (1.2.6) provides an analytical expression (up to the dipole moment) [1]:

$$E_p = -\frac{1}{2\left(4\pi\varepsilon_o\right)^2} \frac{\left(q_A^2 \alpha_B + q_B^2 \alpha_A\right)}{r^4}$$
$$-\frac{1}{2\left(4\pi\varepsilon_o\right)^2} \frac{\left(\mu_A^2 \alpha_B\left(1 + 3\cos^2\theta_1\right) + \mu_B^2 \alpha_A\left(1 + 3\cos^2\theta_2\right)\right)}{r^6} - \dots \tag{1.2.6}$$

where $\alpha_I$ is the polarizability of molecule $I$, and all other symbols have the same meaning as in Eq. (1.2.4). Physically, the first term corresponds to the polarization induced by a charge, while the second term is the polarization induced by a dipole. Obviously, the first term disappears when the molecules have no charge and the second disappears when they have no dipole.

Finally, the dispersion term between molecules A and B can be approximated by the following expression due to Drude [29]:

$$E_{disp} = -\frac{3}{2(4\pi\varepsilon_o)^2} \frac{\left[I(A)\,I(B)/(I(A) + I(B))\right]\alpha_A\alpha_B}{r^6} \tag{1.2.7}$$

Here $\alpha_J$ is the polarizability of molecule $J$, $I(J)$ is the ionization energy of molecule $J$ from its ground state, and all other symbols have the same meaning as in Eq. (1.2.4). No general analytic expressions are available for the bond component ($E_{bond}$). When such a component exists one has to evaluate it by doing quantum chemical calculations [24, 31] on the appropriate models.

The previous equations allow a qualitative and semiquantitative analysis of the relative importance of the interaction between any pair of molecules, once the geometry, dipole moment, and polarizability are known. At the usual ranges of intermolecular distances (1.2 to 3.5 Å), the electrostatic terms are expected to dominate over the polarization and dispersion terms due to their power in $r$. However, dispersion will be seen to be more relevant than usually expected. Obviously, it is the dominant term when the interacting molecules have no net charge and no large dipoles, as in the $\pi$–$\pi$ interaction between two benzene molecules. Below, we will describe in detail various examples of how the previous equations can be useful in the analysis of the intermolecular interactions found in crystals.

One should note that Eqs. (1.2.4)–(1.2.7) are valid for molecules in vacuum. When one works in a condensed phase system one has to substitute $\varepsilon_o$ by $\varepsilon_r$ in the above expressions, where $\varepsilon_r$ is the relative dielectric constant of the condensed medium. We also remind the reader that these expressions describe the interaction of molecules fixed in space (as is the case in crystals). When the molecules are freely moving, as in liquids, one can obtain more simplified expressions [1] that show no angular dependences.

### 1.2.2.3 Types of Intermolecular Interactions and Intermolecular Bonds Found in Molecular Crystals

As already mentioned above, intermolecular interactions can be attractive or repulsive. Usually, in chemistry we are interested in analyzing structures of energetically stable aggregates, where, by definition, the sum of all intermolecular interaction energies is attractive. Within these aggregates, one can rationalize their energetic stability and structure by looking for their main attractive intermolecular interactions and how they vary between different conformations, in other words, at the properties of the intermolecular bonds that the aggregate presents. As mentioned before, the optimum geometry of the aggregate results from a compromise among all attractive interactions. The existence and shape of the repulsive wall, which prevents atoms from collapsing into each other at the usual range of energies of chemical processes, is also relevant (only considering the attractive components our aggregate would collapse).

There are many classifications of intermolecular interactions in the literature [1, 2, 10–12, 21, 32] (long- and short-range interactions, covalent, steric, electrostatic, polarization, van der Waals, ...) that can be applied to molecular crystals. Here, we will classify the intermolecular interactions in molecular crystals by looking at the dominant term of their interaction energy. According to such a criterion, intermolecular interactions can be classified as: (i) *exchange–repulsion* (sometimes known as steric interactions); (ii) *electrostatic interactions* (also called Coulombic or ionic interactions); (iii) *polarization interactions*; (iv) *dispersion interactions*, also called van der Waals interactions; and (v) *bonding interactions* (sometimes called covalent interactions). Except for the last class, their nature is described by the analytical expressions (1.2.4)–(1.2.7) shown above. These categories are not rigid, since, as one component dominates, all others may also be non-negligible. Very often one also finds complex situations where the relative importance of the components is similar, thus making possible more than one classification.

Based on the dominant component of the interaction energy, intermolecular bonds found in molecular crystals can be classified within one of the following three main classes: (i) *ionic bonds*, (ii) *hydrogen bonds*, and (iii) *van der Waals bonds*. Each class presents a clearly differentiated specific properties and electronic structure (see Fig. 1.2.6)

- *Ionic bonds* are found whenever at least one of the species that form the molecular crystal has a net charge (that is, we are dealing with ions). The simplest case involves two atoms of opposite sign, as in the $X^+ \cdots Y^-$ bond (the prototypical example is the $Na^+ \cdots Cl^-$ interaction), but there are many examples of molecular ionic crystals, for instance in conductors, superconductors or magnetic molecule-based magnets. Ion–solvent interactions are also a type of ionic bond, found for instance in solvated ionic salts (notice, however, that when these ion–solvent interactions present $A–H \cdots X$ hydrogen bonds, they should be better considered as *charge-induced hydrogen bonds* [33]). The interaction energy in ionic bonds (Table 1.2.2) is dominated by the electrostatic term: in ion–ion interactions by the charge–charge term, and in ion–neutral interactions by the charge–dipole term (or when no dipole is present by the charge–quadrupole term). When the two ions involved have the same charge the interaction is generally repulsive. In this case, we cannot talk about the existence of a bond, as it fails to satisfy the criteria of energetic stability required for an interaction to become a bond. Their electronic structure is shown in Fig. 1.2.6 for the cation–anion and anion–anion interactions, in a molecular orbital (MO) and valence-bond (VB) representation.

- *Hydrogen bonds* are intermolecular bonds that present an $A–H \cdots X$ topology. The A–H group is called the proton donor, and the X atom (or group of atoms) is called the proton acceptor. The A–H group has a local dipole moment and points towards the region of accumulation of electronic density on X (for instance, the lone-pair electrons in $O–H \cdots O$ bonds, or the bonding $\pi$-electrons in $O–H \cdots \pi$ bonds). These bonds also require the existence of an $H \cdots X$ bond critical point between the H and X atoms. Complex situations are possible, like in bifurcated hydrogen bonds, where one A–H groups makes two bond critical points to two different X acceptor atoms, or when the same X acceptor atom is hydrogen bonded to more than one A–H group. The prototypical example is the $O–H \cdots O$ bond found in water dimers, which is a hydrogen bond of moderate strength. The interaction energy in the water dimer is dominated by the electrostatic term (Table 1.2.2), in particular by the dipole–dipole term of Eq. (1.2.4). This fact has

(a)

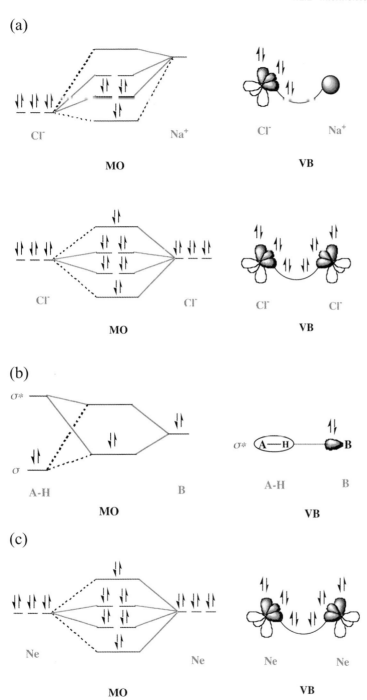

Figure 1.2.6 Diagram representing the electronic structure of the ionic, hydrogen bond and van der Waals interactions.

suggested to some authors that hydrogen bonds are always electrostatic in nature. However, not all hydrogen bonds show the same relative weight of the interaction energy components (see Table 1.2.2). Very strong hydrogen bonds (as those found in F–H$\cdots$F⁻, for instance) have an important bonding component, which induces the formation of two covalent F–H bonds. At the other extreme, in very weak hydrogen bonds the relative weight of the dispersion term becomes more important. Sometimes, it is even the dominant term (as in C–H$\cdots$O or F–H$\cdots$Ar hydrogen bonds). It is also worth pointing out here that the existence of a short A–H$\cdots$X contact does not always mean that it is a hydrogen bond. This is the case in aggregates where the two molecules have net charges of the same sign (where the short A–H$\cdots$X contact results from the anion$\cdots$cation interactions and not from the stability of the short A–H$\cdots$X contact) [27, 34, 35]. When the two molecules have net charges of opposite sign, or when one is neutral and the other is charged, the aggregate is more stable than when the two molecules are neutral, a fact that explains why these bonds are sometimes called charge-assisted hydrogen bonds. The electronic structure of the hydrogen bond, represented in Fig. 1.2.6(b), is characterized as the interaction between the empty antibonding orbital of A–H and the doubly occupied nonbonding orbital of B that hosts the unpaired electrons.

- *van der Waals bonds* are intermolecular bonds that result from the direct overlap of regions of strong electronic accumulation (lone pairs, π-orbitals) in atoms or molecules, when no electrostatic energetic component is possible because their charge, dipole, quadrupoles, ... are negligible. These bonds are usually indicated as A$\cdots$B (for instance, Ar$\cdots$Ar, or π$\cdots$π), where A and B are the atoms (or group of atoms) with strong electronic localization. The A$\cdots$B interaction must be energetically attractive and the A and B atoms (or group of atoms) must present a bond critical point that connects them (more than one bond critical point is found connecting π$\cdots$π dimers). Van der Waals bonds originate from the dispersion terms (Table 1.2.2) and are usually weaker than hydrogen bonds and ionic bonds (notice, however, that some strong van der Waals bonds can be stronger than weak hydrogen bonds). In complex geometrical arrangements AIM analysis helps to define the atoms that participate in the bond, and to check whether one is dealing with a hydrogen bond or a van der Waals bond (for instance, in short A–H$\cdots$X contacts with an A–H$\cdots$X angle deviating from collinearity) [23]. The electronic structure of the van der Waals bonds is characterized by the diagram of Fig. 1.2.6(c), where the only orbitals overlapping are doubly occupied orbitals (in π$\cdots$π interactions the overlapping orbitals would be those of the π system of each molecule).

**Table 1.2.2** Value of the energetic components of the interaction energy for the indicated systems, computed using the IMPT method. $E_{el}$, $E_{er}$, $E_p$, $E_{ct}$ and $E_{disp}$ are, respectively, the electrostatic, exchange–repulsion, polarization and dispersion components. $E_{tot}$ is the sum of these components and $E_{MP2}$ is the BSSE-corrected interaction energy, shown to serve as reference to calibrate the quality of the IMPT calculation. The MP2 intermolecular optimum distance is also shown ($r_{opt}$). Distances are given in Å and energies in kcal mol$^{-1}$.

| System | $r_{opt}$ | $E_{el}$ | $E_{er}$ | $E_p$ | $E_{ct}$ | $E_{disp}$ | $E_{tot}$ | $E_{MP2[a]}$ |
|---|---|---|---|---|---|---|---|---|
| | | | *ionic bonds* | | | | | |
| $Na^+\cdots Cl^-$ | 2.412 | −142.3 | 25.4 | −10.5 | −1.4 | −35.5 | −164.3 | −128.0 |
| $HC_2O_4^-\cdots HC_2O_4^-$ | 1.537 | 25.4 | 31.4 | −7.7 | −4.2 | −11.6 | 33.2 | 40.9[b] |
| | | | *hydrogen bonds* | | | | | |
| $NH_3CH_2COOH^+\cdots SO_4^-$ | 1.520 | −155.1 | 31.4 | −14.4 | −6.2 | −12.4 | −156.6 | 160.0 |
| $H_3O^+\cdots H_2O$ | 1.202 | −45.5 | 54.3 | −27.4 | −9.6 | −31.7 | −59.9 | −31.8 |
| $H_2O\cdots F^-$ | 1.415 | −38.3 | 30.6 | −10.1 | −4.3 | −8.5 | −30.6 | −25.3 |
| $CH4\cdots F^-$ | 1.873 | −8.3 | 13.7 | −6.7 | −1.4 | −4.0 | −6.8 | −5.8 |
| $FH\cdots H_2O$ | 1.716 | −16.3 | 12.2 | −1.0 | −0.6 | −3.2 | −8.9 | −7.9 |
| $H_2O\cdots H_2O$ | 1.990 | −6.7 | 5.0 | −0.7 | −0.5 | −2.0 | −3.9 | −4.2 |
| $C_2H_2\cdots H_2O$ | 2.174 | −4.1 | 2.6 | −0.4 | −0.2 | −1.4 | −3.5 | −2.7 |
| $CH_4\cdots H_2O$ | 2.553 | −0.9 | 1.5 | −0.2 | −0.1 | −0.9 | −0.6 | −0.4 |
| $FH\cdots Ar$ | 2.634 | −0.1 | 0.5 | −0.2 | −0.1 | −0.4 | −0.3 | −0.4 |
| | | | *van der Waals* | | | | | |
| $Ar\cdots Ar$ | 3.842 | −0.04 | 0.15 | 0.00 | 0.00 | −0.28 | −0.16 | −0.16 |
| $CO_2\cdots CO_2$ | 3.058 | −1.7 | 1.9 | −0.2 | −0.1 | −1.8 | −1.8 | −1.0 |
| $C_6H_6\cdots C_6H_6$ | 3.800[c] | 1.2 | 2.7 | −0.2 | −0.2 | −5.2 | −1.7 | −1.8 |

[a] BSSE corrected values. [b] The interaction is repulsive, and therefore cannot be a bond. It is only shown here for illustrative purposes. [c] Value obtained after a partial geometry optimization using frozen fragments.

Table 1.2.2 also collects the values obtained by IMPT computations for some representative complexes at their equilibrium MP2/aug-cc-pVDZ geometry (as shown in Fig. 1.2.7–1.2.9). These values can be qualitatively rationalized using Eqs. (1.2.4)–(1.2.8). The trends that are found in the data in Table 1.2.2 help to find qualitative trends for similar complexes: (i) ionic bonds are the strongest bonds, their interaction energies generally located in the –200 to +200 kcal mol$^{-1}$ range [36, 37], where positive values indicate repulsive interactions that cannot be called bonds, while negative values indicate attractive interactions (their values depend strongly on the net charge of the interacting fragments; the values in Table 1.2.2 lie within a narrower range); (ii) the interaction between ions of the same charge is repulsive [27, 33], as in the case of the $HC_2O_4^-\cdots HC_2O_4^-$ dimer in Table 1.2.2, and consequently they cannot be a hydrogen bond even when they have an A–H$\cdots$X short contact with the geometry usually associated with a hydrogen bond; (iii) in ionic interactions, the electrostatic component is the dominant one (accounting for more than 90% of the interaction energy), but the dispersion component is far

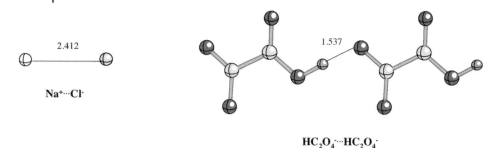

$HC_2O_4^-\cdots HC_2O_4^-$

**Figure 1.2.7** Optimum geometry of the complexes in the ionic bond class of Table 1.2.2. Notice that the $HC_2O_4^-\cdots HC_2O_4^-$ complex presents a short O–H$\cdots$O contact that is not energetically attractive, so cannot be considered as a bond. Otherwise, it should be better included within the hydrogen bond class.

from negligible; (iv) hydrogen bond strengths lie in the $-160$ to $-0.4$ kcal mol$^{-1}$ range, the lowest value being weaker than some van der Waals interactions (this range includes the charge-induced hydrogen bonds, also called ionic hydrogen bonds; very weak hydrogen bonds in the $-0.4$ kcal mol$^{-1}$ range could only be

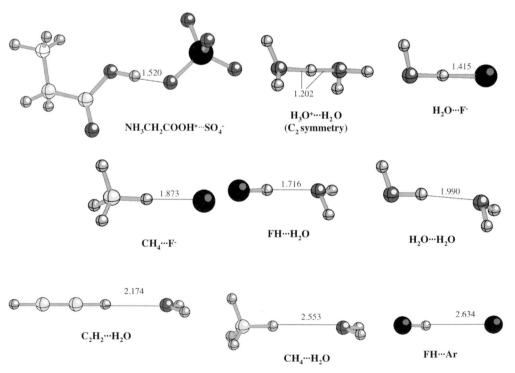

**Figure 1.2.8** Optimum geometry of the complexes in the hydrogen bond class of Table 1.2.2.

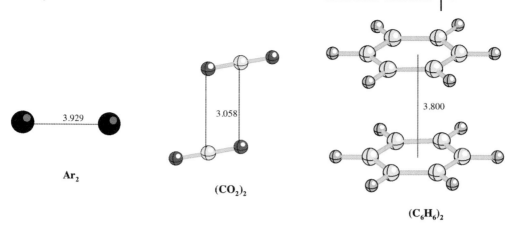

**Figure 1.2.9** Optimum geometry of the complexes in the van der Waals bond class of Table 1.2.2.

observed at very low temperatures and in the absence of stronger bonds); (v) the strength of the hydrogen bonds depends strongly on the net charge of the fragments (compare the interaction energy for $CH_4 \cdots F^-$ and $CH_4 \cdots H_2O$), and more weakly on the donor group acidity (compare $H_2O \cdots F^-$ and $CH_4 \cdots F^-$) and acceptor group basicity (compare $FH \cdots H_2O$ and $FH \cdots Ar$); (vi) the weight of the dispersion component in the interaction energy of some hydrogen bonds is of the same order as the electrostatic component (see, for instance, the $H_3O^+ \cdots H_2O$ complex) and becomes the dominant one in weak hydrogen bonds (for instance, $FH \cdots Ar$), thus, generally speaking, the attractive nature of hydrogen bonds results from the addition of the electrostatic and dispersion components; (vii) van der Waals interactions are dominated by the dispersion component, but the electrostatic component is also important (for instance, in the $CO_2$ dimer both components have the same weight). Studies on many other complexes have shown the general validity of these rules.

It is also worth pointing here that the data in Table 1.2.2 show that some hydrogen bonds are much stronger than sometimes previously considered [32, 38]. We have also found that some charged-assisted hydrogen bonds can be of similar strength to classical ionic bonds, as is clearly illustrated in Table 1.2.2 by looking at the $NH_3CH_2COOH^+ \cdots SO_4^-$ complex. Such a fact originates from the strong opposite molecular charges. Therefore, in this complex the charge–charge adds to the dipole–dipole term, the first one being by far the larger.

### 1.2.2.4 Hydrogen Bonds

The results described in the previous section give a general picture of the nature of hydrogen bonds. However, given the relevance of such a bond to crystal engineering, we will elaborate a bit further on various specific properties of hydrogen bonds.

Among the many ways of subdividing hydrogen bonds, a common one has been suggested by Jeffrey [32], where they are subdivided according to their strength. He proposed to define three subgroups: *strong, moderate* (also called *normal*), and *weak* hydrogen bonds. The subdivision is made by taking as reference the normal class, which is formed by all hydrogen bonds similar to water, whose interaction energy is taken to lie in the $-4$ to $-14$ kcal mol$^{-1}$ range. Strong hydrogen bonds are stronger than $-14$ kcal mol$^{-1}$, and weak hydrogen bonds are weaker than $-4$ kcal mol$^{-1}$. The subdivision is in some sense arbitrary, as is also the selection of energy ranges, and a matter of personal taste.

An analysis of the data of Table 1.2.2 shows that many strong hydrogen bonds are only found when one of the two interacting molecules is charged. Therefore, it seems to us that a more natural subdivision would be one that takes into account the nature of the interaction energy of the hydrogen bond. In this case, we can distinguish three cases: (i) *neutral* $\cdots$ *neutral* hydrogen bonds, where both interacting fragments have no net charge and the dominant electrostatic term is the dipole–dipole term (the last five hydrogen bond complexes in Table 1.2.2), (ii) *charge* $\cdots$ *neutral* hydrogen bonds (second to fourth hydrogen bonded complexes in Table 1.2.2), where the electrostatic component is dominated by the charge–dipole term of Eq. (1.2.4), stronger than the dipole–dipole term, and (iii) *charge* $\cdots$ *charge* hydrogen bonds (the first hydrogen bonded complex in Table 1.2.2), where the dominant electrostatic component is the charge–charge term (much stronger than the dipole–dipole term). The different energetic nature of these three classes of hydrogen bonds is reflected in their very different interaction energy ranges. Hereafter, when required we will identify precisely their subclass indicating the effective charge in the interacting molecules (that is, A–H$^-$ $\cdots$ X and A–H$^-$ $\cdots$ X$^+$ will identify two bonds of the charge $\cdots$ neutral and charge $\cdots$ charge subclasses). As the acidity and basicity of the A–H and X groups also plays an important role (compare $H_2O$ $\cdots$ F$^-$ and $CH_4$ $\cdots$ F$^-$ interactions), the interaction energies within each class can overlap in some cases, but this is not a problem when defining their nature of membership. Notice that A–H$^-$ $\cdots$ X$^-$ and A–H$^+$ $\cdots$ X$^+$ interactions are not likely to be hydrogen bonds, as the charge–charge term is repulsive, and this term is the dominant one in general. Only in cases where the net charge is concentrated in a small group of atoms of the interacting molecules (for instance, in the COO$^-$ group of a deprotonated molecule) is the charge–charge term found to be smaller than the dipole–dipole term.

Within the neutral subclass one can tune the strength of the hydrogen bond by varying the acidity and basicity of the A–H and X groups. In this class one can speak of weak hydrogen bonds for bonds where the dispersion term is important, although such relevance can, to some extent, be arbitrary.

### 1.2.2.5 Existence of Intermolecular Bonds in Crystals

One of the problems sometimes found when working with intermolecular interactions is establishing whether or not they are bonds (see, for instance, the arguments used in Refs. [27, 34, 35]). As we will show in this section, the problem arises when one uses indirect criteria to establish the presence of bonds, many of them based on empirical trends.

The definition of bond stated above provides two *direct criteria of existence* for intermolecular bonds: (i) the aggregate resulting from that intermolecular interaction must be energetically stable with respect to the separate fragments, and (ii) a (3,–1) bond critical point must exist, connecting at least one atom from each of the bonded molecules. Therefore, it is possible to obtain a direct proof of existence of any intermolecular bond by measuring [39] (or computing [24, 31]) the energetic stability of the aggregate, and finding a bond critical point in the experimental [21] (or computed [11]) electron density linking atoms of the interacting molecules.

Over the years, instead of these direct criteria, it has been very common to use some *indirect criteria* that have been taken as evidence of the presence of intermolecular bonds. Hereafter we summarize some of the most popular ones, to allow the reader to evaluate their strong and weak points:

### Geometry of the A–H···X short contacts

- Based on empirical observations (see Fig. 1.2.10 and 1.2.11), it has been found that A–H···X short contacts show a preference for <AH···X angles close to 180°. This fact has been justified by the electrostatic nature of the hydrogen bond (the A–H dipole prefers to point towards the local negative charge located on the X atom). Such behavior has suggested the use of cut-off values for the <AH···X angle [32, 38] (for instance, cut-off angles close to 90° have been proposed, where values smaller than this cut-off are indicative of van der Waals bonds). However, as shown in Table 1.2.2, the dispersion is important in many hydrogen bonds,

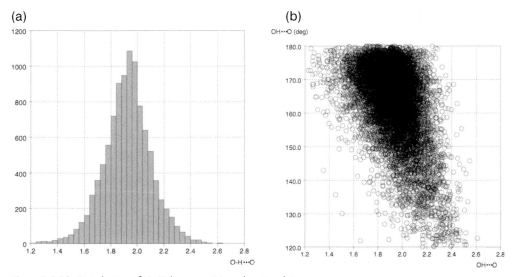

**Figure 1.2.10** Distribution of H···X distances (a), and scatterplot of the O–H···O angles versus the H···X distances for the crystal structures deposited in version 5.26 of the Cambridge Crystallographic database (November 2004). The van der Waals radii of hydrogen and oxygen are 1.2 and 1.52 Å.

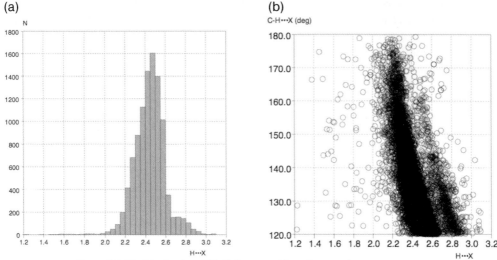

**Figure 1.2.11** Distribution of H···X distances (a), and scatterplot of the C–H···O angles versus the H···X distances (b) for the crystal structures deposited in version 5.26 of the Cambridge Crystallographic database (November 2004). The van der Waals radii of hydrogen and oxygen is 1.2 and 1.52 Å.

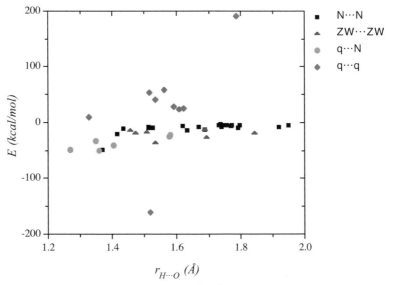

**Figure 1.2.12** Interaction energy computed at the MP2/6–31+G(d) level for all dimers that present O–H···O contacts shorter than the sum of the O and H van der Waals radii located. These dimers were located by systematically searching the whole subset of neutron diffraction crystals within the Cambridge Crystallographic Database. (N = neutral, ZW = zwitterionic, q = charged fragments.)

and it does not present an energetic preference for collinear geometries. In weak hydrogen bonds, such as the C–H$\cdots$O bonds, the dispersion component is the dominant one. Therefore, the reasons for a cut-off in the angle are not general and rather weak.

- It has been proposed that A–H$\cdots$X contacts having H$\cdots$X distances shorter than the sum of the H and X van der Waals radii are hydrogen bonds [32, 38]. Based on empirical observations, such cut-off was later relaxed, adding 0.5 Å or even more to the sum of the H and X van der Waals radii [32, 38]. However, recent studies [37] on the energy of short O–H$\cdots$O contacts found in neutron diffraction crystals deposited in the Cambridge Crystallographic Database [40] indicate that the rule "the shortest-the strongest" (sometimes called the bond strength–bond length relationship) is not fulfilled in general (see Fig. 1.2.12 and Table 1.2.3). The strength–length relationship can be valid in crystals of the same type, where the dominant forces are similar and therefore follow a similar $1/r^n$ law (see the trends for neutral$\cdots$neutral hydrogen bonds in Fig. 1.2.11). One typical example of such failure is found when A–H$^-\cdots$X$^-$ contacts between anions are possible. Those contacts have an H$\cdots$X distance shorter than the sum of the H and X van der Waals radii despite being repulsive (see Fig. 1.2.12). Their existence is due to cation$\cdots$anion attractions that are stronger than the anion$\cdots$anion and cation$\cdots$cation repulsion, as has been numerically demonstrated in small aggregates. Thus, they do not follow the same power law as neutral$\cdots$neutral hydrogen bonds, a fact that explains the different scatter-plots of Fig. 1.2.13.

(a)

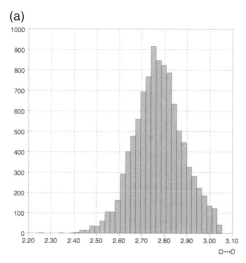

(b)

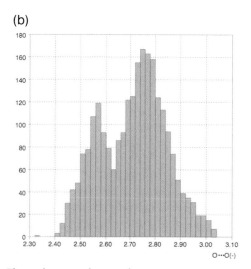

**Figure 1.2.13** Number of contacts as a function of the O$\cdots$O distance for the O–H$\cdots$O contacts. (a) neutral$\cdots$neutral contacts; (b) neutral$\cdots$anion and anion$\cdots$anion contacts (the left peak is that due to the anion$\cdots$anion contacts). The analysis was done on the crystal structures deposited in version 5.26 of the Cambridge Crystallographic database (November 2004). The van der Waals radii of oxygen is 1.52 Å.

**Table 1.2.3** Values of the H···O distance (in Å), IMPT energy components of the interaction energy (in kcal mol$^{-1}$; $E_{el}$, $E_{er}$, $E_p$, $E_{ct}$ and $E_{disp}$ are, respectively, the electrostatic, exchange-repulsion, polarization and dispersion components, while $E_{tot}$ is the sum of these components), type of hydrogen bond (four groups are present in this table: N···N where both fragments are neutral, ZW···ZW where both are zwitterions, $q$···N and $q$···$q$; in the last two groups, the charge in each fragment is indicated, and the BSSE-corrected MP2 interaction energy ($E_{MP2}$). Each complex is identified by the REFCODE of the crystal where the complex is found, sometimes followed by a number, different complexes are found within the same crystal.

| REFCODE | $r_{(H···O)}$ | $E_{el}$ | $E_{er}$ | $E_{pol}$ | $E_{ct}$ | $E_{disp}$ | $E_{tot}$ | Type | $E_{MP2}$ |
|---|---|---|---|---|---|---|---|---|---|
| ACYGLI11 | 1.529 | − 26.1 | 25.8 | − 4.9 | − 4.7 | − 4.4 | − 14.1 | N···N | − 10.1 |
| KECYBU02-dim1 | 1.435 | − 24.5 | 37.1 | − 7.1 | − 5.2 | − 16.2 | − 19.9 | N···N | − 11.0 |
| KECYBU02-dim2 | 1.514 | − 21.7 | 28.2 | − 4.7 | − 3.9 | − 11.7 | − 13.7 | N···N | − 7.6 |
| LYXOSE01 | 1.793 | − 11.5 | 10.6 | − 1.5 | − 1.8 | − 2.3 | − 6.4 | N···N | − 5.0 |
| NALCYS02 | 1.513 | − 25.1 | 27.1 | − 4.3 | − 5.6 | − 4.4 | − 12.4 | N···N | − 9.0 |
| OXALAC06 | 1.918 | − 9.5 | 8.0 | − 1.2 | − 0.7 | − 3.9 | − 7.2 | N···N | − 7.8 |
| PERYTO10 | 1.790 | − 25.2 | 19.7 | − 2.2 | − 2.7 | − 4.6 | − 15.0 | N···N | − 9.2 |
| RESORA | 1.756 | − 14.8 | 14.3 | − 1.7 | − 1.5 | − 7.0 | − 10.7 | N···N | − 5.0 |
| SUCACB02 | 1.688 | − 35.3 | 33.6 | − 5.5 | − 4.3 | − 14.5 | − 26.0 | N···N | − 13.4 |
| UREAOH12 | 1.669 | − 19.5 | 19.1 | − 2.9 | − 2.3 | − 6.9 | − 12.5 | N···N | − 8.5 |
| UREXPO11 | 1.619 | − 19.5 | 21.1 | − 3.2 | − 2.5 | − 6.7 | − 10.2 | N···N | − 6.4 |
| UROXAL01 | 1.416 | − 43.1 | 46.2 | − 9.5 | − 6.2 | − 22.0 | − 33.7 | N···N | − 20.5 |
| AEPHOS02 | 1.535 | − 73.8 | 59.1 | − 12.6 | − 8.5 | − 72.2 | − 111.1 | ZW···ZW | − 36.0 |
| DLASPA02 | 1.508 | − 34.0 | 27.4 | − 6.6 | − 5.8 | − 4.2 | − 23.3 | ZW···ZW | − 17.3 |
| DLSERN11 | 1.692 | − 46.9 | 30.2 | − 6.4 | − 4.2 | − 17.0 | − 44.4 | ZW···ZW | − 27.0 |
| HOPROL12 | 1.843 | − 26.1 | 10.5 | − 3.1 | − 2.5 | − 2.5 | − 23.7 | ZW···ZW | − 19.3 |
| KECROT01 | 1.348 | − 46.9 | 53.6 | − 17.1 | − 9.1 | − 30.1 | − 49.6 | (−1)···N | − 33.7 |
| KHDFRM12 | 1.270 | − 70.9 | 72.1 | − 22.6 | − 12.0 | − 33.1 | − 66.6 | (−1)···N | − 48.7 |
| KOXPHY12 | 1.579 | − 35.3 | 28.9 | − 8.9 | − 4.8 | − 10.4 | − 30.4 | (−2)···N | − 25.2 |
| NAOXAP11 | 1.582 | − 30.9 | 28.1 | − 8.7 | − 4.8 | − 10.4 | − 26.8 | (−2)···N | − 21.6 |
| TGLYSU11-dim2 | 1.406 | − 26.2 | 54.4 | 15.5 | − 13.5 | − 22.4 | − 23.1 | (+1)···N | − 41.6 |
| 18183(hco3-) | 1.562 | 46.0 | 31.3 | − 7.8 | − 4.5 | − 11.3 | 53.6 | (−1)···(−1) | 58.9 |
| EDATAR01 | 1.784 | 188.0 | 11.6 | − 8.6 | − 5.0 | − 2.7 | 183.4 | (−2)···(−2) | 190.0 |
| LGLUTA | 1.624 | 16.3 | 14.8 | − 3.9 | − 2.5 | − 2.6 | 22.0 | (+1)···(+1) | 25.3 |
| NHOXAL02 | 1.537 | 25.4 | 31.4 | − 7.7 | − 4.2 | − 11.6 | 33.2 | (−1)···(−1) | 40.9 |
| PUTRDP11 | 1.517 | 33.5 | 29.0 | − 7.9 | − 4.6 | − 11.2 | 38.8 | (−1)···(−1) | 52.1 |
| TGLYSU11-dim1 | 1.520 | − 155.1 | 31.4 | − 14.4 | − 6.2 | − 12.4 | − 156.6 | (−2)···(+1) | − 160.0 |

### Spectroscopy of the A–H···X short contacts

- The existence of a complex and its geometry has been determined for a few complexes using rotational spectroscopy [41]. In some cases, these studies can even give information on the strength of the hydrogen bond.

- In most cases the aggregate is in solution or in a pure condensed phase and no direct evidence can be obtained. Then, one usually looks at differences between the isolated species behavior and the situation where the aggregation exists, using IR [32, 42] or NMR techniques [39, 42, 43]. For instance, the hydrogen bond is manifested by a decrease in the value of the A–H stretching frequency ($\Delta v_{AH}$) with respect to the free A–H value, together with a broadening of this vibrational band. In some cases, the so-called *blue-shifted hydrogen bonds*, the shift of the A–H stretching frequency is towards higher wavenumbers. The reasons for this exceptional behavior are now clear [44]. However, it is worth pointing out here that this vibrational frequency shift also exists in short A–H$\cdots$X contacts that are repulsive, as in A–H$^-\cdots$X$^-$ contacts between anions. Therefore, the presence of these shifts is not always indicative of A–H$\cdots$X hydrogen bonds. Equivalent considerations also apply to NMR spectroscopic results.

### 1.2.2.6  Intermolecular Bonds in Crystals

Up to now we have studied the properties of intermolecular bonds as if they were independent entities, that is, without taking into account the effect that nearby bonds can induce in the bond of interest. Crystals are the result of a complex network of intermolecular bonds. Thus it is appropriate to discuss here some important specific issues that derive from the fact that intermolecular bonds form complex connected networks.

#### Identification of the existence of intermolecular bonds

According to our previous considerations, the existence of bonds can be established when two direct criteria are fulfilled: (i) energetic stability of the aggregate, and (ii) the existence of bond critical points. The existence of the crystalline solid is a direct indication of its overall interaction energy being stable (in fact, being larger than the thermal energy at the working temperature, without entering into the complexities of phase equilibria). Thus, looking at bonds just requires one to search for the existence of bond critical points in their electron density. Figure 1.2.4 shows the result of one of these analyses in one of the planes of the NANQUO02 crystal [21]. When this is not possible, one can use indirect techniques, complemented with qualitative analysis of the energy of the interaction energy, to be sure that they are attractive interactions.

#### Energy scale of the interaction energy in solids

When one thinks about intermolecular interactions in solids there is a tendency to associate them with values in the 0 to 40 kcal mol$^{-1}$ range of energies [32, 38], based on values found for gas phase complexes. However, the results collected in Table 1.2.3 show that such a scale is much wider than expected. This table collects the molecule–molecule interaction energies and their components for the shortest O–H$\cdots$O contacts found in neutron diffraction crystals deposited in the Cambridge Crystallographic Database (also plotted in Fig. 1.2.13). The complexes were separated in groups according to their charge (see above). The following ranges of energies are found within each group: (i) N$\cdots$N group: (−3.4, −48.5); (ii) N$\cdots$ZW group: just has one value −36.0; (iii) ZW$\cdots$ZW group: (−12.6, −27.0); (iv) $q\cdots$N group: (−21.6, −50.8); (v) $q\cdots q$ group: (190, −160). The first four groups always have energetically stable O–H$\cdots$O interactions, but the *$q\cdots q$ group contains interactions that have the metric expected for hydrogen bonds but are energetically unstable* (these interactions cannot be considered hydrogen bonds, according to our previous considerations). Notice that some interactions are energetically destabilizing. Obviously, they can be found in crystals due to the existence of other attractive interactions within the same crystal.

**Cooperativity**

It has been demonstrated that the properties of multiple, interconnected intermolecular bonds are different from those of isolated bonds. This effect is called cooperativity and its origin is the polarization that the formation of an intermolecular bond induces in the electron density of the interacting molecules. When a molecule makes more than one bond, the second one is made with the polarized molecule. The relevance of the effect is clearly shown in Table 1.2.4 for the O–H$\cdots$O bond in water aggregates. But such a polarization effect is expected to be present in all kinds of intermolecular bonds, and will be proportional to the electronic polarizability of the molecule. Experimentally, the existence of polarization effects can be demonstrated by comparing the formation energy of an isolated water dimer (−5.44 kcal mol$^{-1}$ [45], that is, 2.72 kcal mol$^{-1}$ per water molecule) with the formation energy per water molecule in ice at 0 K (−11.3 kcal mol$^{-1}$ [46]).

**Table 1.2.4** Formation energy *per water molecule* computed for various (H$_2$O)$_n$ clusters at the HF/aug-cc-pVDZ [46], MP2/aug-cc-pVDZ [46] and BLYP/TZVP levels [47]. The first two sets are BSSE-corrected results, while those for the BLYP data are non-corrected values (for a DFT calculation with a basis set of this quality, the BSSE is expected to be very small [48]). All values in kcal mol$^{-1}$.

| Complex | HF | MP2 | BLYP |
|---|---|---|---|
| (H$_2$O)$_2$ | − 1.83 | − 2.32 | − 2.63 |
| (H$_2$O)$_3$ | − 3.67 | − 4.62 | − 5.58 |
| (H$_2$O)$_4$ | − 4.88 | − 6.08 | − 7.56 |
| (H$_2$O)$_5$ | − 5.19 | | − 8.03 |
| (H$_2$O)$_6$ | − 5.40 | | − 8.13 |
| (H$_2$O)$_7$ | | | − 8.56 |
| (H$_2$O)$_8$ | | | − 9.64 |
| (H$_2$O)$_9$ | | | − 9.18 |
| (H$_2$O)$_{10}$ | | | − 9.43 |
| (H$_2$O)$_{12}$ | | | − 10.12 |
| (H$_2$O)$_{16}$ | | | − 10.40 |
| (H$_2$O)$_{20}$ | | | − 10.57 |

## Analysis of crystals: Graph sets

Crystals and aggregates make complex networks of intermolecular bonds. A useful tool to describe these complexities is graph sets analysis, looking for packing motifs [49]. According to this procedure, any complex network of hydrogen bonds can be reduced to combinations of the simple patterns shown in Fig. 1.2.14: *chain, rings, self* (for intramolecular bonds), and *finite* (for patterns not propagated over the crystal). These patterns are represented by the letters **C, R, S** and **D**, respectively. The specification of a pattern is done using the following convention: $L_d^a(n)$, where **L** is a letter designating the simple pattern (either **C, R, S** or **D**), the subscript $d$ indicates the number of hydrogen bond donor groups in the pattern, the superscript $a$ indicates the number of hydrogen bond acceptor groups in the pattern, and $n$ is the number of atoms in the pattern. Patterns containing only one type of hydrogen bond are called *motifs*. The list of all motifs found in a crystal constitutes the first-level or unitary graph set of that crystal, and is a sort of fingerprint of a crystal that can help to distinguish between different polymorphs of the same crystal (for instance Form 2 of the iminodiacetic acid crystal contains six motifs and its first-level graph set is $N_1 = DDC(5)C(5)C(8)C(8)$, while, Form 1 has only three motifs and its first-level graph set is $N_1 = C(5)C(8)R_2^2(10)$).

In the original proposal [49] hydrogen bonds were identified using distance cutoffs, but a more strict form is now possible using the AIM method [11]. The graph set analysis can be applied not only to crystals but also to any kind of hydrogen bonded aggregate. In principle, it is also possible to use it in aggregates presenting other types of intermolecular bonds.

One can also use the motifs as conceptual tools to understand the structure of aggregates and crystals. This is done by substituting the concept of "bond" by the concept of "motif" as the central concept in our analysis. This is the idea behind the concept of *supramolecular synthons* [51], defined as energetically robust motifs that serve as the basis for the crystal structure by propagation of their structure. A

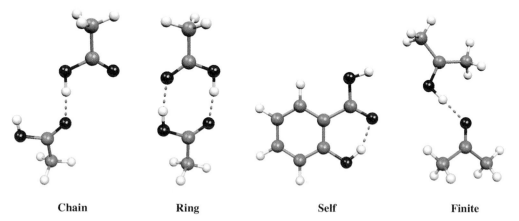

**Chain**  **Ring**  **Self**  **Finite**

**Figure 1.2.14** Examples of the four basic motifs in graph set analysis. From left to right we represent the **C(4)**, **R$_{22}$(8)**, **S** and **D** motifs.

proper analysis of crystals using motifs (or supramolecular synthons) also requires the computation of the strength of the motifs (obtained from the interaction energy of the molecules of the motif). Additionally, a full understanding of the properties of the synthons requires an identification of the bonds present in them, and also a qualitative or quantitative estimation of their energy (which can be done, in a first approach, by adding the energies of the constituent bonds, if possible corrected by the cooperative effects).

### 1.2.3
### Summary

Bonds are very powerful tools in the analysis of the structure of molecules. They can also be very helpful in analyzing supramolecular structures and, in particular, molecular crystals. These aggregates result from a complex interplay among attractive and repulsive interactions driven by a quest for a minimum in the free energy surface (entropy can also be important, though usually is not considered the dominant force in most cases, and thus in practice most analyses are done in terms of the enthalpy). When trying to find the reason for their structure and energetic stability one can look for the attractive interactions, trying to understand how these interactions compete among themselves and determine the relative disposition of the molecules in the crystal (repulsive components are always found at short distances between the interacting molecules, due to quantum effects that originate when the densities of the interacting molecules overlap, and avoid the collapse of the aggregate). Such analysis can be simplified by looking at the dominant attractive intermolecular interactions, in other words, at the intermolecular bonds.

Here we have reviewed the nature of the intermolecular interactions and analyzed when they become bonds. Bonds are the dominant attractive interactions; not all attractive interactions are bonds, but to be a bond they must be attractive. We have analyzed their types according to the dominant term in the intermolecular energy, and found qualitative expressions that allow us to make a qualitative estimate of their values.

Building on these considerations, we have paid special attention to the hydrogen bond and its classification, also analyzing the special issues associated with these bonds in solids, namely, range of energy, identification, and cooperativity.

### Acknowledgments

The authors acknowledge the continuous support of the "Ministerio de Ciencia y Tecnología" (projects BQU2002–04587-C02–02 and CTQ2005–02329/BQU) and CIRIT (2001SGR-0044 and 2005-SGR-00036) for funding. They also thank CEPBA-IBM Research Institute, CEPBA and CESCA for allocation of CPU time on their computers. E. D'O. also acknowledges the Spanish "Ministerio de Educación y Ciencia" for the award of a PhD grant.

## References

**1** G. C. Maitland, M. Rigby, E. B. Smith, W. A. Wakeham, *Intermolecular Forces. Their Origin and Determination*, Clarendon Press, Oxford, 1981; J. Israelachvili, *Intermolecular and surface forces*, 2nd edn., Academic Press, Amsterdam, 1992; A. J. Stone, *The Theory of Intermolecular Forces*, Clarendon Press, Oxford, 1996.

**2** A. I. Kitaigorodsky, *Molecular Crystals and Molecules*, Academic Press, New York, 1973; J. D. Wright, *Molecular Crystals*, 2nd edn., Cambridge University Press, Cambridge, 1995.

**3** J. Bernstein, *Polymorphism in Molecular Crystals*, Clarendon Press, Oxford, 2002.

**4** J. D. Dunitz, A. Gavezzotti, *Acc. Chem. Res.* 1999, *32*, 677.

**5** G. R. Desiraju, *Crystal Engineering. The Design of Organic Solids*, Elsevier, Amsterdam, 1989; D. Braga, F. Grepioni, A. G. Orpen (Eds.), *Crystal Engineering: From Molecules and Crystals to Materials*, Kluwer, Dordrecht, 1999; J. A. Howard, F. H. Allen, G. P. Shields (Eds.), *Implications of Molecular and Materials Structure for New Techonolgies*, Kluwer, Dordrecht, 1999.

**6** A. Gavezzotti, *Acc. Chem. Res.* 1994, *27*, 309; A. Gavezzotti (Ed.), *Theoretical Aspects of Computer Modeling of the Molecular Solid State*, Wiley, Chichester, 1997.

**7** J. M. Lehn, *Supramolecular Chemistry, Concepts and Perspectives*, VCH, Weinheim, 1995.

**8** Crystal packing prediction is nowadays mostly based on empirical atom–atom potentials. Although reasonable results are found in many crystals, they fail to predict the experimental polymorph in many of the cases. See for instance: W. D. S. Motherwell, H. L. Ammon, J. D. Dunitz, A. Dzyabchenko, P. Erk, A. Gavezzotti, D. W. M. Hofmann, F. J. J. Leusen, J. P. M. Lommerse, W. T. M. Mooij, S. L. Price, H. Scheraga, B. Schweizer, M. U. Schmidt, B. P. van Eijck, P. Verwer, D. E. Williams, *Acta Crystallogr. Sect. B*, 2002, *58*, 647.

**9** Such calculation can be done by codes such as CRYSTAL (R. Doversi R, C. Pisani, F. Ricca, C. Roetti, V. R. Saunders, *Phys. Rev. B*, 1984, *30*, 972) and others, but they are not done commonly on molecular crystals; C. Rovira, J. J. Novoa, *J. Chem. Phys.* 2000, *113*, 9208; C. Rovira, J. J. Novoa, *J. Phys. Chem. B*, 2001, *105* 1710.

**10** C. A. Coulson, *Valence*, Oxford University Press, Oxford, 1952.

**11** L. Pauling, *The Nature of the Chemical Bond*, 3rd edn., Cornell University Press, Ithaca, 1960. The first edition of this seminal book was published in 1939.

**12** R. F. Bader, *Atoms in Molecules. A Quantum Theory*, Clarendon Press, Oxford, 1990.

**13** A (3,–1) bond critical point in the electronic density is a point where the gradient of the density is zero, and the three eigenvalues of the Hessian of the density (the second derivative of the density with respect to the cartesian coordinates) at such a point are different from zero, two being negative and one positive.

**14** G. R. Desiraju, *Acc. Chem. Res.* 2002, *35*, 565. Herein it is proposed to go back to the term "hydrogen bridge" instead of "hydrogen bond", on the following argument: "if the term bond has other hallowed connotations in chemistry, it might be far preferable to refer to hydrogen bonds as hydrogen bridges, for so different are they from covalent bonds. ... The terminology of a hydrogen bridge does not carry with it the unnecessary and incorrect implication that a hydrogen bond is like a covalent bond only much weaker". In our opinion, the term bond is flexible enough to include all known classes of energetically stable interactions, and does not assume any consideration about its nature. Within its classes it includes the ionic bond and the covalent bond. Hydrogen bonds are just one more type, whose nature is partly ionic and partly covalent, in a degree that changes with the specific hydrogen bond. So, there is no sound quantum chemical reason to substitute bonds by bridges at this moment.

**15** J. Dunitz, A. Gavezzotti, *Angew. Chem. Int. Ed.* 2005, *44*, 1766.

**16** H. B. Bürgi, J. D. Dunitz (Eds.), *Structure Correlation*, vols. 1 and 2, VCH, Weinheim, 1994.

**17** A first report of such a situation was presented in: J. Cioslowski, S. T. Mixon, *J. Am. Chem. Soc.* **1992**, *114*, 4382.

**18** R. F. W. Bader, D.-C. Fang, *J. Chem. Theor. Comput.* **2005**, *1*, 403; Y. A. Abramov, *J. Phys. Chem. A*, **1997**, *101*, 5725.

**19** M. A. Spackman, *Chem. Phys. Lett.* **1999**, *301*, 425.

**20** C. Gatti, E. May, R. Destro, F. Cargnoni, *J. Phys. Chem. A*, **2002**, *106*, 2707.

**21** V. G. Tsirelson, R. P. Ozerov, *Electron Density and Bonding in Crystals*, Institute of Physics Publishing, Bristol, 1996.

**22** E. Espinosa, E. Molins, C. Lecomte, *Chem. Phys. Lett.* **1998**, *285*, 170; S. J. Grabowski, *J. Phys. Chem. A*, **2001**, *105*, 10739; I. Alkorta, J. Elguero, *J. Phys. Chem. A*, **1999**, *103*, 272.

**23** J. J. Novoa, P. Lafuente, F. Mota, *Chem. Phys. Lett.* **1998**, *290*, 519.

**24** A. Szabo, N. S. Ostlund, *Modern Quantum Chemistry: Introduction to Advanced Electronic Structure Theory*, Macmillan, New York, 1982.

**25** I. C. Hayes, A. J. Stone, *J. Mol. Phys.* **1984**, *53*, 83.

**26** One of the most successful alternatives is SAPT, described in: B. Jeziorski, R. Moszynski, K. Szalewicz, *Chem. Rev.* **1994**, *94*, 1887.

**27** J. J. Novoa, P. Lafuente, R. Del Sesto, J. S. Miller, *Angew. Chem. Int. Ed.* **2001**, *40*, 2540; J. J. Novoa, P. Lafuente, R. Del Sesto, J. S. Miller, *CrystEngComm*, **2002**, *4*, 373; R. Del Sesto, J. S. Miller, J. J. Novoa, P. Lafuente, *Chem. Eur. J.* **2002**, *8*, 4894.

**28** M. Born, J. E. Mayer, *Z. Phys.* **1932**, *75*, 1.

**29** A. J. Stone, *Chem. Phys. Lett.* **1981**, *83*, 233; A. J. Stone, M. Alderton, *Mol. Phys.* **1985**, *56*, 1047.

**30** P. K. L. Drude, *The Theory of Optics*, Longman, London, 1933.

**31** F. Jensen, *Introduction to Computational Chemistry*, John Wiley, New York, 1998.

**32** G. A. Jeffrey, *An Introduction to Hydrogen Bonding*, Oxford University Press, New York, 1997.

**33** P. Gilli, V. Bertolasi, V. Ferretti, G. Gilli, *J. Am. Chem. Soc.* **1994**, *116*, 909.

**34** D. Braga, F. Grepioni, E. Tagliavini, J. J. Novoa, F. Mota, *New. J. Chem.* **1998**, *22*, 755; D. Braga, F. Grepioni, J. J. Novoa, *Chem. Commun.* **1998**, 1959; D. Braga, C. Bazzi, F. Grepioni, J. J. Novoa, *New. J. Chem.* **1999**, *23*, 577; J. J. Novoa, I. Nobeli, F. Grepioni, D. Braga, *New. J. Chem.* **2000**, *24*, 5; D. Braga, L. Maini, F. Grepioni, F. Mota, C. Rovira, J. J. Novoa, *Chem. Eur. J.* **2000**, *6*, 4536; D. Braga, J. J. Novoa, F. Grepioni, *New. J. Chem.* **2001**, *25*, 226; D. Braga, E. D'Oria, F. Grepioni, F. Mota, J. J. Novoa, C. Rovira, *Chem. Eur. J.* **2002**, *8*, 1173.

**35** These claims were initially objected to by various authors, but their objections were satisfactorily solved in later publications (see Refs. [24] and [30]). For details, see: T. Steiner, *Chem. Commun.* **1999**, 2299; M. Mascal, C. E. Marajo, A. J. Blake, *Chem. Commun.* **2000**, 1591; P. Macchi, B. B. Iversen, A. Sironi, B. C. Chokoumakas, F. K. Larsen, *Angew. Chem. Int. Ed.* **2000**, *39*, 2719.

**36** These limit values have been estimated by looking at the interaction energy of dimers found in crystals that present short O–H$\cdots$O contacts (see Ref. [34]).

**37** E. D'Oria, J. J. Novoa, *CrystEngComm*, **2004**, *64*, 367.

**38** T. Steiner, *Angew. Chem. Int, Ed.* **2002**, *41*, 48; G. R. Desiraju, T. Steiner, *The Weak Hydrogen Bond in Structural Chemistry and Biology*, Oxford University Press, Oxford, 1999

**39** F. Hibbert, J. Emsley, *Adv. Phys. Org. Chem.* **1990**, *26*, 255; J. S. Chickos, W. E. Acree, Jr., *J. Phys. Chem. Ref. Data*, **2002**, *31*, 537.

**40** F. H. Allen, *Acta Crystallogr. Sect. B*, **2002**, *58*, 380.

**41** A. C. Legon, D. J. Millen, *Acc. Chem. Res.* **1987**, *20*, 39.

**42** G. C. Pimentel, A. L. McClellan, *The Hydrogen Bond*, W. H. Freeman, San Francisco, 1960; G. C. Pimentel, A. L. McClellan, *Annu. Rev. Phys. Chem.* **1971**, *22*, 347.

**43** E. Brunner, U. Sternberg, *J. Prog. NMR Spectrosc.* **1998**, *32*, 21.

**44** P. Hobza, Z. Havlas, *Chem. Rev.* **2000**, *100*, 4253.

**45** K. Kuchitsu, Y. Morino, *Bull. Chem. Soc. Jpn.* **1965**, *38*, 805.

**46** E. Whalley, in *The Hydrogen Bond*, P. Shuster, G. Zundel, C. Sandorfy Eds., North-Holland, Amsterdam, 1976, p. 1427.

**47** S. S. Xantheas, *J. Chem. Phys.* **1994**, *100*, 7523.

**48** C. Lee, H. Chen, G. Fitzgerald, *J. Chem. Phys.* **1995**, *102*, 1266.

**49** J. J. Novoa, C. Sosa, *J. Phys. Chem.* **1995**, *99*, 15837.

**50** M. C. Etter, *Acc. Chem. Res.* **1990**, *23*, 120; J. Bernstein, R. E. Davis, L. Shimoni, N.-L. Chang, *Angew. Chem. Int. Ed.* **1995**, *34*, 1555.

**51** G. R. Desiraju, *Angew. Chem. Int. Ed.* **1995**, *34*, 2311.

## 1.3
## Networks, Topologies, and Entanglements
*Lucia Carlucci, Gianfranco Ciani, and Davide M. Proserpio*

### 1.3.1
### Introduction

Making crystals by design requires both a detailed knowledge of the nature of the interactions involving the selected building blocks and also some general concepts concerning the types of the supramolecular extended arrays that could be generated by self-assembly, i.e. the possible target networks. The presence of suitable functional groups in an organic building block or free coordination sites in a metal-based center, all having well defined directions for interactions, can allow for some predictions about the resulting architecture, thus representing the fundamentals of the modern area of crystal engineering of networks. These points are treated in great detail in other parts of this book. However, in spite of this rather optimistic *incipit*, we are presently faced with many problems in the field and we are still far from being able to deliberately build networks with desired properties. A major difficulty arises from the range of possible different structures that can form in the self-assembly process, due to the occurrence of many, often unpredictable and subtle, concurrent factors.

Some advantage can derive from the analysis of related structures reported in the literature, searching for prototypical models that can fit the specific problem. The increasing number of networked species reported in the last years offers a rich variety of new structural types that continuously amplify our knowledge of these organic or metal–organic extended systems.

From the structural point of view much attention has been devoted to the rationalization and classification of the different topological types of single extended motifs [1] and to the categorization of the frequently encountered interpenetrating [2] and entangled [3] species. Indeed, the "network approach" or topological approach to crystal chemistry [4, 5] is an important achievement in this area, as a useful tool for the analysis of network structures, in that it simplifies complex species to schematised nets, in order to make comparisons and identify packing trends that may help in the rational design of functional materials.

In what follows we will describe some approaches to the rationalization of these supramolecular systems, proceeding in order of increasing complexity. We will deal first with the simplification of crystal structures, in order to carry out a topological analysis of the individual idealized networks. Problems concerning the unambiguous assignment of the topology are presented, and topological approaches towards making crystals are described. Various types of entanglements of single motifs, including interpenetrated, polycatenated, Borromean, polythreaded and other networked arrays, will be then considered and classified.

1.3.2
**Rationalization and Simplification of the Extended Structures**

A correct analysis of the crystal structures is fundamental in order to avoid mis-interpretations, that can easily occur, about the network topology and, when present, the type and extent of the entanglement of distinct motifs. The rationalization of these, often complicated, structures implies a sequence of steps leading to the basic nets of linked nodes.

The steps, in general, can be summarized as follows: (i) simplification process, (ii) identification and separation of the individual motifs, (iii) topological analysis of these motifs, and (iv) topological analysis of the whole entanglement.

The simplification phase, previously exemplified by some authors [6], consists in the rather obvious operation of removing all the unnecessary elements that have no topological relevance, thus leaving only the essentials, represented by nodes and links (vertices and edges, respectively, in graph theory [7]). For instance, polyatomic nodes (like metal clusters or polyfunctional ligands) can be replaced by their barycenters.

The simplified model must be effectively representative of the connectivity of the real network. In some cases, however, this step leads to structural descriptions that are rather subjective in the selection of both the nodes and the connections (links) joining the nodes. Alternative rationalizations can be accomplished, that result in motifs of different topology, depending upon the choice of the chemical or/and mechanical bonds and the nodes of the architecture (see Fig. 1.3.1) [8]. These "alternative" motifs, however, are not something unnatural; they reflect the possibility of consideration of a crystal from different points of view or at various levels of structure organization.

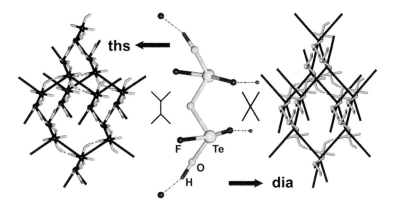

**Figure 1.3.1** An alternative selection of nodes gives two distinct topological descriptions, here illustrated for the 2-fold interpenetrated H-bonded structure of $H_2Te_2O_3F_4$ (J. C. Jumas, M. Maurin, E. Philippot, *J. Fluor. Chem.*, **1976**, *8*, 329): 3-connected ThSi$_2$ (**ths**) with Te, and diamond (**dia**) with the molecule centroid.

In coordination networks concurrent chemical forces are often operative, including coordinative bonds, hydrogen bond bridges, secondary bonds involving the anions, metal–metal interactions, π-interactions and others. Thus we are faced with the problem of deciding when a link is a topologically significant link. A current useful practice consists in assuming a level within this hierarchy of forces to be applied in the description of the net. For instance, we have previously observed [3] that in [Zn(4,4'-bipy)$_2$(H$_2$O)$_2$](SiF$_6$) [11] the Zn(4,4'-bipy)$_2$ coordination sublattice is a 3D polycatenated array of two sets of inclined 2D square layers, but including also H$_2$O–SiF$_6$–H$_2$O hydrogen bond bridges that link the two different sets results in a more complex 3D single net having 4- and 8-connected nodes, with the peculiar feature of self-catenation (see later). Thus the insertion in a model of additional weaker interactions can have drastic effects, like changing the topology of the individual motifs, the type of entanglement or also the dimensionality of the whole array.

Mechanical bonds, like those shown in Fig. 1.3.2, are responsible for the formation of the majority of the known extended periodic entanglements. A debatable point concerns the difficult distinction between rotaxanes and pseudo-rotaxanes. At the molecular level rotaxanes are characterized by the presence of bulky stoppers on the rods, that must inhibit dethreading, while pseudo-rotaxanes can be separated by chromatography or at high temperatures. Also some rotaxanes, however, can have a low energy barrier to dethreading, so that the difference with respect to pseudo-rotaxanes remains a delicate problem and the boundary is not well defined [12]. The extended polymeric analogs of these species present problems of similar nature; moreover, for them no experiment can be envisaged to differentiate between the two situations.

We could, therefore, conclude that establishing the topology of these species is not only a formal process but rather an interplay between chemistry and mathematical theory. The process of simplification or abstraction that represents the passage from real objects to their idealized models needs chemical considerations. After simplification the problem remains exclusively based on a mathematical ground, and the successive process consists in the identification and classification of the topology of the individual motifs contributing to the entanglement. Their separation can be accomplished in a computer-aided process, whose output is a set of distinct "colored" nets [13–15].

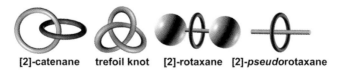

**[2]-catenane**   **trefoil knot**   **[2]-rotaxane**   **[2]-*pseudo*rotaxane**

**Figure 1.3.2** Models of links and knots for finite entities.

1.3.3
**Topological Classification of Networks**

### 1.3.3.1 Nomenclature for Single Nets: Schläfli and Vertex Symbols

The topological classification of an individual extended motif can be accomplished according to generally accepted criteria used in a variety of structural contexts that are here described. First, however, we want briefly to mention the pioneering work of the researchers who introduced the basic concepts for the rationalization, enumeration and classification of the extended crystal structures, using different approaches.

A fundamental contribution is represented by a famous series of articles and books on crystal chemistry published many years ago by A. F. Wells [4], who analysed and classified a great number of nets. He emphasized the importance of describing a crystal structure in terms of its basic topology: such a description not only provides a simple and elegant way of representing the structures but also evidences relations between structures that are not always apparent from the conventional descriptions. Wells introduced a method for the systematic generation of 3D arrays from 2D nets and also described many hypothetical motifs that were successively discovered within the realm of coordination polymers or of other extended systems. His results included a list of many simple nets described with only one kind of node (uninodal) or with two nodes of different connectivity (mainly binodal 3,4-connected).

The studies of J. V. Smith focused on the classification of zeolites and related materials with tetrahedral nodes of the $TO_4$ type [16]. He described a prolific number of frameworks that were discovered mostly by hand treatment, using systematic methods. Such methods include permutation of up–down linkages between identifiable two-dimensional layers in zeolites, as well as permutations of linkages between identifiable polyhedra.

An empirical search by O'Keeffe [17] for uninodal four-connected nets resulted in the recognition of 168 distinct types (many of which were unseen by Wells).

A thousand uninodal nets were described over the years by Fischer, Koch and Sowa in their enumeration of homogeneous sphere packings, exploring systematically all crystal systems [18].

Treacy and coworkers collected a hypothetical zeolite database, using a method that enumerates all possible 4-connected nets within each crystallographic space group given the number of unique tetrahedral vertices. The database contains $10^{10}$ graphs and is available at the website www.hypotheticalzeolites.net [19].

The different network topologies can be recognized following the graph theoretical approach. We can assign to the simplified 2D or 3D crystal structure a net, that is an infinite graph (or a periodic connected graph) described with translational symmetry in exactly two or three independent directions (2-periodic or 3-periodic nets) [20] and represented by nodes and links (vertices and edges, following the nomenclature of graph theory [7, 21]). Figure 1.3.3 illustrates some common net topologies.

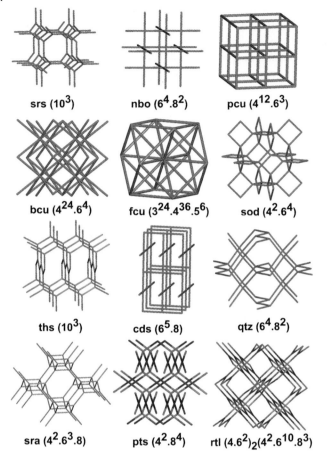

**Figure 1.3.3** Nets with common topologies and their nomenclature (see text). The symbols are related to prototypical structures or lattices as follows: SrSi$_2$ (**srs**), NbO (**nbo**), primitive cubic (**pcu**), body centered cubic (**bcu**), face centered cubic (**fcu**), sodalite (**sod**), ThSi$_2$ (**ths**), CdSO$_4$ (**cds**), quartz (**qtz**), SrAl$_2$ (**sra**), PtS (**pts**), rutile (**rtl**).

In the analysis of the topology of such nets we can look for a classification scheme that will allow one to uniquely assign equal nets (up to isomorphism) derived from totally different crystal structures. Remember that the topology is not influenced by the metrical properties of the structure (angles, distances), so that a 4-connected diamondoid net is such, even if highly distorted (the geometry around the nodes could be far from tetrahedral), and also, obviously, is not dependent on the chemical nature of each node/vertex.

To make it easy we suggest here to follow the lower-case three-letter code nomenclature for nets proposed by O'Keeffe and coworkers [22, 23] in an attempt to overcome a certain confusion in the field, due to the fact that some networks have been described with many names and symbols. For example the **srs** net observed for the Si atoms in the $SrSi_2$ structure (see Fig. 1.3.3) has been called (10,3)-a, Laves net, Y*, 3/10/c1, or labyrinth graph of the gyroid surface. The suggested nomenclature is designed to parallel the widely accepted upper-case three-letter codes used for zeolite frameworks [24]. Moreover, more than 1300 nets with these symbols have been collected in a very useful list described in detail at the RC3R (Reticular Chemistry Structure Resource) website (http://reticularchemistry.net/RCSR/ [25]).

For most of the nets of fundamental importance in crystal chemistry, there are only a few different kinds of nodes (by "kind of nodes" we mean a set of vertices related by symmetry operations, including translations) and the assignment of the local topology for 2D or 3D nets is based on the analysis of the "circuits" (or cycles) at the angles of a node. A circuit is a closed path beginning and ending at each node, characterized by a size equal to the number of edges in the path (3-circuit, 4-circuit and so on). Any two edges at a node define an angle, and with $n$-connected nodes [26] there are $[n(n-1)/2]$ such angles in a 3D net (but only $n$ in a 2D simple layer [27]), for each of which we can find a large number of different circuits of different sizes.

An $n$-connected node can be identified by a Schläfli Symbol (short symbol or point symbol) of the form $A^a.B^b.C^{c...}$, in which $A < B < C < ...$ and $a + b + c ... = n(n-1)/2$, that represents the sizes ($A$, $B$, $C$...) and numbers ($a$, $b$, $c$...) of the "shortest circuits" contained at each angle. Thus for the 4-connected diamond net (**dia**), where all the shortest circuits are 6-cycles, the symbol is $6^6$. For the 2D honeycomb and square-grid nets and for the 3D net of the 6-connected primitive cubic lattice (**pcu**), all uninodal (i.e. containing a single type of node), the Schläfli Symbols are $6^3$ [or (6,3)], $4^4$ [or (4,4)] and $4^{12}.6^3$, respectively.

For multinodal nets the Schläfli Symbols are grouped according to the multiplicity of the node in the unit cell; for example the net formed by the Si atoms in the silica form moganite (**mog**) is described as the 4-connected binodal net $(4.6^4.8)_2(4^2.6^2.8^2)$ because there are two crystallographic distinct nodes with multiplicity 2:1.

A special type of point symbol is assigned to uniform nets, i.e. nets in which the shortest circuits at each angle are all equal in size. It follows that the Schläfli Symbol for each node in an $n$-connected uniform net with circuits of size $A$ is $A^{n(n-1)/2}$. Examples include the series of 3-connected $10^3$ nets (**srs**, **ths**, **bto**, **utp** and others), the 4-connected $6^6$ nets (**dia**, **lon**, **gsi**, **lcs** and others) and also the binodal 3,4-connected $(8^3)_4(8^6)_3$ nets of $C_3N_4$ (**ctn**) and $Pt_3O_4$ (**pto**).

The "point symbol" was extensively used in a slightly different typographic form by Wells [4], who was aware that this classification was far from being sufficient to unambiguously identify a net. In fact he added letters to the point symbol in order to assign different topologies, as in the case of the well known 3-connected nets of the Si atoms in the $SrSi_2$ and $ThSi_2$ structures (**srs** and **ths** with the same Schläfli Symbol $10^3$), called (10,3)-a and (10,3)-b, respectively. Another case of ambiguity is given by the NbO net (**nbo**) and the arrangement of Si atoms in quartz (**qtz**), that have the same Schläfli Symbol $6^4.8^2$.

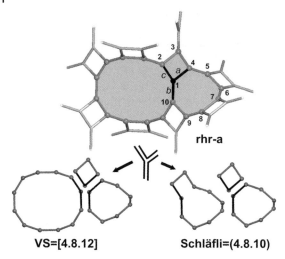

**rhr-a**

**VS=[4.8.12]**                    **Schläfli=(4.8.10)**

**Figure 1.3.4** In the 3-connected uninodal net **rhr-a** the vertex **1** with the three edges *a,b,c* has Vertex Symbol 4.8.12 (angles *ac*: 4-ring, *ab*: 8-ring , *bc*: 12-ring, with single multiplicity for all three rings: $4_1.8_1.12_1$). If we consider the shortest circuits we get the Schläfli Symbol 4.8.10: in fact, the angle *bc* is included in the circuit with 10 edges [1,2,3,4,5,6,7,8,9,10] that is not a ring because it is the sum of two shortest rings (there is a short-cut represented by edge *a*).

Given this inadequacy other descriptors are needed. To differentiate nets with the same Schläfli Symbol a modified version of it, the Vertex Symbol (VS, or long symbol introduced by O'Keeffe [1]) appears more useful. In a VS the size of the shortest *rings* at each angle is given with a subscript to denote the number of such rings. Solid state chemists define a *ring* as a circuit that has the property that there is no shorter path ("short cut") between any two vertices on the circuit than the shortest one that is part of the circuit (see Fig. 1.3.4) [28].

Thus in the 3-connected **srs** 3D net there are five 10-rings at each angle and the VS is $10_5.10_5.10_5$, and for the **ths** net, with two, four and four 10-rings at each angle, the VS is $10_2.10_4.10_4$ (remember that for both nets the Schläfli Symbol is $10^3$).

In the case of 4-connected nets there are six angles at a node. The diamond net (**dia**) with two 6-rings at each angle has VS = $6_2.6_2.6_2.6_2.6_2.6_2$. In 4-connected nets only, the angles are grouped into three pairs of opposite angles, and subject to that constraint, the smallest numbers are given first. The feldspar structure (**fel**) contains a 4-connected net with two kinds of vertices both having the same Schläfli Symbol $4^2 6^3 8$, but different Vertex Symbols, $4.6.4.6.8_2.10_{10}$ and $4.6_2.4.8.6.6_2$. Here some 6- and 8-circuits have short-cuts and become 8- and 10-rings in the VS. Analogously we can now distinguish NbO, with VS(**nbo**) = $6_2.6_2.6_2.6_2.8_2.8_2$, from quartz, with VS(**qtz**) = $6.6.6_2.6_2.8_7.8_7$. The Vertex Symbol of PtS (cooperite **pts**), that contains "square-planar" Pt and "tetrahedral" S centers with the same Schläfli Symbol $4^2.8^4$, also allows one to differentiate between the two nodes: Pt $(4.4.8_2.8_2.8_8.8_8)$ and S $(4.4.8_7.8_7.8_7.8_7)$. In this case the multiplicity of the 8-rings makes the difference. More details on these symbols are given in Refs. [1, 21, 29]. They are used also in the *Atlas of Zeolite Framework Types* [24 a,b].

Not all the angles of a 4-connected net necessarily possess rings. When there is not a ring an asterisk is inserted in the VS. Thus for the 4-connected $CdSO_4$ net (the net of the Cd and S atoms, with the –O– links considered as edges, **cds**, Schläfli Symbol $6^5 8$) the Vertex Symbol is $6.6.6.6.6.6_2.*$ .

Vertex Symbols can be computed for higher connectivities, but since the number of angles increases as the square of the coordination number, they soon become cumbersome; the Vertex Symbol for the 6-coordinated net of the primitive cubic lattice (**pcu**) is $4.4.4.4.4.4.4.4.4.4.4.4.*.*.*$, and a 12-coordinated net has $(12\times11)/2 =$ 66 angles.

The Vertex Symbols, however, are also unable to fully characterize nets. For example diamond and lonsdaleite (hexagonal diamond, **lon**) have the same $VS = 6_2.6_2.6_2.6_2.6_2.6_2$ and, therefore, other descriptors must be considered.

It is easy to introduce the "Coordination Sequence" $CS(k)$ [1] of a node, defined as a sequence of numbers $(n_1, n_2, n_3, ..., n_k, ...)$ in which the $k$th term is the number of nodes in "shell" $k$ that are connected to nodes in "shell" $k - 1$. Shell 0 [CS(0)] consists of a single node and the number of nodes in the first shell [CS(1)] is the connectivity of the node (or, more rigorously, the coordination number of the vertex [26]). Another equivalent definition states that $CS(k)$ gives the number of "topological neighbors" as the number of nodes that are $k$ edges/links away from a given node. It follows that a $k$th neighbor of a node is one for which the shortest path to that node consists of $k$ edges. Though there is no limit for $k$ the CS is usually computed up to $k = 10$. With this new descriptor we can now distinguish **dia** (CS= 4,12,24,42..) from **lon** (CS= 4,12,25,44...). The Coordination Sequences differ already at the third neighbor, see Fig. 1.3.5. We must say that CS alone is also not sufficient, because there are nets with the same CS but different VS (for example the zeolites Linde A **lta** and Rho **rho**).

For many years the structural analysis for assigning VS and CS was carried out by hand for simple nets but nowadays it can be performed computationally with the use of computer programs such as TOPOS [14, 15, 30]. This program gives VS

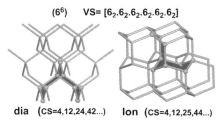

$(6^6)$    VS= $[6_2.6_2.6_2.6_2.6_2.6_2]$

**dia**  (CS=4,12,24,42...)    **lon**  (CS=4,12,25,44...)

**Figure 1.3.5** The structures of diamond **dia** (left) and lonsdaleite (hexagonal diamond) **lon** (right). The two nets have the same Schläfli and Vertex Symbol VS (top), but different Coordination Sequence (CS). The rings multi- plicity for one selected angle is evidenced in gray; in both nets for each angle there are two 6-rings, in **dia** both have the same *chair* conformation, in **lon** the conformation is different, one *chair* and one *boat*.

and CS for each (crystallographically independent) node. If all nodes result equal the net is classified as uninodal, otherwise binodal, trinodal, ... *n*-nodal. A uninodal net is one in which all the nodes have the same VS and CS, i.e. are congruent. In its maximum symmetry form there is only one node in the asymmetric unit and all nodes will be related by symmetry operations (i.e. are equivalent). In real crystal structures, on the other hand, it is common to find low symmetry embeddings of uninodal nets (an embedding of a graph is a realization of the graph in Euclidean space with a given geometrical description, atomic coordinates, distances and angles) where there are crystallographic distinct nodes but all have the same VS and CS, and therefore we can describe the topology of the derived net as that of an uninodal net (see Fig. 1.3.6).

Then the results can be compared with a list of known nets (e.g. more than 1300 nets described in RCSR and the zeolite databases are stored in TOPOS) to find (if any) the topological type, with its given name and ideally most symmetrical structure.

This pair of descriptors (Vertex Symbol and Coordination Sequence for all nodes) appears to be unique for a particular net topology, i.e. they can be effectively employed to distinguish different framework types unambiguously. This statement has not been proved rigorously but long experience with databases has shown that, at present, only a few theoretical different nets with the same VS and CS descriptors are known and these can still be differentiated if all the rings (not only the shortest) are computed. All these considerations belong to the context of graph theory: we can always identify two nets as isomorphic ignoring edge crossings. Two different embeddings of a net that can be deformed into each other only with edge crossings are called non-"ambient isotopic" [21, 31] and therefore cannot be distinguished in graph theory. In recent years, indeed, some examples of non-"ambient isotopic" nets have been identified (see Fig. 1.3.7) [3, 32–34] for which new topological descriptors are needed, taken outside the graph theory e.g. in the context of knot theory [31].

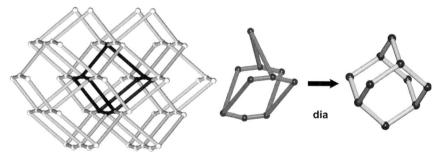

**Figure 1.3.6** The highly distorted diamondoid net observed in the 2-fold interpenetrated structure of SbCl$_3$($p$-diacetylbenzene) with a diamondoid cage in black (W. A. Baker, D. E. Williams, *Acta Crystallogr., Sect. B* **1978**, *34*, 1111) (left). The observed diamondoid cage (center) is compared with the undistorted adamantane cage as in diamond (right).

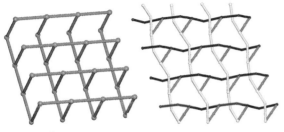

$(8^2.10)$    VS= $[8_2.8_2.*]$    CS= 3,6,12,20,28,36...

**Figure 1.3.7** Two different layered nets (2-periodic) characterized by the same VS and CS that are not ambient isotopic. We cannot transform one into the other without bond/edge crossing. The layer on the left is rather common while that on the right is observed only in $[Ag_2L_2(\mu\text{-}PO_2F_2)](PF_6)$ [L = 1-(isocyanidomethyl)-1H-benzotriazole]. (I. Ino, J. C. Zhong, M. Munakata, T. Kuroda-Sowa, M. Maekawa, Y. Suenaga, Y. Kitamori, *Inorg. Chem.* **2000**, *39*, 4273.)

Isomorphic nets can be rigorously recognized with the SYSTRE program by Delgado-Friedrichs [35], that can establish the "equilibrium" configuration of nets and assign topologically unique coordinates to the vertices. By this we mean that each vertex has coordinates that are the average of the coordinates of its neighbors. Once an origin is chosen, barycentric coordinates are unique for a given choice of basis vectors. Such a placement is valuable in determining properties such as combinatorial symmetry. "Collisions" occur when two vertices have the same barycentric coordinates, but such nets are still very rare in crystal chemistry (for further discussion of "collisions" see Ref. [35a]). Thus, for nets without collisions, SYSTRE solves the graph isomorphism problem: it unambiguously determines whether two nets have the same or different topology.

It has been suggested [5, 25, 36] that a limited number of "common" topologies should be sufficient to represent the large majority of organic, inorganic and metal–organic (MOFs) networks; thus furnishing a useful "network approach" to the crystal engineering of these complex systems. This important feature is true not only for single networks but also for interpenetrating 3D architectures. The distribution of the topologies found for the 445 interpenetrated structures collected in CSD and ICSD databases shows that the most common nets by far are the **dia** nets (ca. 42%), followed by **pcu** nets (ca. 18%) and by **srs** nets (ca. 7%) [14, 15].

### 1.3.3.2 Tiling Theory and Topological Approaches to Making Crystals

The current growing interest in network topologies has led to investigations in novel directions, especially for enumeration purposes, but also for crystal engineering. In the search for a rational design of coordination polymers/MOFs the importance of the concept of Secondary Building units (SBU), that had been used sometimes to describe conceptual fragments of zeolites, appeared evident. When polytopic units are copolymerized with metal ions, it is useful to recognize linked cluster entities in the assembled solid, each such cluster being considered a

SBU. The successful design of rigid frameworks based on SBUs is the rationale for "Scale Chemistry" [37] and "Reticular Chemistry" [36b].

The theoretical background for reticular synthesis arises from looking at nets with a different mathematical approach, i.e. moving from the periodic graph description to the tile description of nets (based on combinatorial tiling theory) [38]. The tiles are generalized polyhedra (cages) which generate the entire structure when packed together. For each periodic net there are many possible tilings but a unique type called "natural" tiling [23 a, 23 b, 39]. The cube with eight vertices and six faces formed by four-membered rings is the natural tile of the primitive cubic net (**pcu**) and is symbolized as $[4^6]$. The adamantane unit with ten vertices is the natural tile of the diamond (**dia**) net. Its four faces are six-membered rings and hence the symbol is $[6^4]$. The natural tiling of the lonsdaleite net (**lon**) has equal numbers of two tiles that are $[6^3]$ and $[6^5]$, as shown in Fig. 1.3.8. This approach classifies the nets according to the "transitivity" of the structure: if there are $p$ kinds of vertices, $q$ kinds of edges (links), $r$ kinds of rings (faces of tiles) and $s$ kinds of tiles, the transitivity is [pqrs]. The five structures **srs, nbo, dia, pcu, bcu** whose natural tilings have transitivity [1111] are called "regular". Their importance in reticular synthesis is comparable to that of the five platonic solids in other areas of chemistry. There is only one structure with transitivity [1112] called "quasiregular": the face-centered cubic net (**fcu**). The next class includes the 14 "semiregular" nets with transitivity [11rs] [23b]. All together these twenty nets are the most likely to form with one kind of SBU joined by one kind of link [40].

The field of infinite periodic minimal surfaces (IPMS), that was introduced a few decades ago for the analysis of the topology of crystal structures [41], is a different approach to the analysis of nets: many common nets are related to the known intersection-free IPMS [42]. The IPMS studies have also produced a systematic enumeration of nets that has been recently proposed by the EPINET project (Euclidean Patterns in Non-Euclidean Tilings, see http://epinet.anu.edu.au): instead of working directly in three dimensions, the intrinsic hyperbolic geometry of IPMS is used to map 2D hyperbolic patterns into 3D Euclidean space [43].

The interested reader can find more information on the topology of networks in a recent review, with more details on what we now know about nets [33].

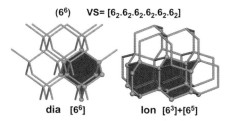

$(6^6)$    VS= $[6_2.6_2.6_2.6_2.6_2.6_2]$

**dia**  $[6^6]$            **lon**  $[6^3]+[6^5]$

**Figure 1.3.8** The structures of diamond (**dia**) (left) and lonsdaleite (hexagonal diamond, **lon**) (right) showing in grey the natural tile with 10 vertices for **dia** (adamantane cage) with transitivity [1111] and the two natural tiles with 8 and 12 vertices for **lon** with transitivity [1222].

#### 1.3.3.3 Self-catenated Networks

In the topological analysis of a network one can encounter the unusual phenomenon of self-catenation or self-penetration. Self-catenated nets are single nets that exhibit the peculiar feature of containing rings through which pass other components of the same network. In more detail, we must refer to the topological classification of nets, represented by the Vertex Symbol, assuming that if one of the "shortest rings" is catenated by other "shortest rings" of the same net we can speak of a true case of self-catenation.

Different terms have been employed for this type of self-entangled network. Batten and Robson [44] have underlined the relationship of these species with molecular knots (see Fig. 1.3.2) but they have preferred to classify them as self-catenating or self-penetrating nets. Champness, Hubberstey, Schröder and coworkers have used the term "polyknotted coordination polymers" [45]. The term "*intra*penetrating networks" has also been employed [46]. Here we prefer to adopt the term self-catenation since the catenation happens within the net itself and there are no different nets involved. This unusual topological feature was previously also observed in simple mineral/inorganic frameworks, as in coesite (**coe**), a high-pressure polymorph of silica [described as a 4-connected binodal net, with Schläfli Symbol $(4^2.6^3.8)(4^2.6.8^2.9)$], recognized by O'Keeffe to be "unique among nets found in nature in that it contains 8-circuits that are linked as in a chain" (see Fig. 1.3.9) [47]. Later it was also recognized that two structures of ice, considering hydrogen bonding, are self-catenated: Ice IV (**icf**) and Ice XII (**itv**) [48].

Probably the first examples of 3D self-catenated frameworks within coordination polymers were described by Robson and coworkers [49]. In the species $[Cd(CN)_2L]$ [L = pyrazine or 1,4-bis(4-pyridyl)butadiene] the $Cd(CN)_2$ sublattices form $4^4$ undulated layers that are joined in a criss-cross fashion by the L spacers. Catenated hexagonal chair-like circuits are observed in the interlayer connections (see Fig. 1.3.9). The Schläfli Symbols of these 6-connected nets are $4^8.6^6.8$ (**rob**) for L = pyrazine, and $4^4.6^{10}.8$ for L = 1,4-bis(4-pyridyl)butadiene, respectively, so that the requirements for self-catenation are fulfilled, since the catenated 6-rings are the shortest rings [50]. Perhaps one of the most fascinating examples is represented by $[Ni(tpt)(NO_3)_2]$ (tpt = tri-4-pyridyl-1,3,5-triazine) [51]. This contains a remarkable 3-connected chiral uniform net of $12^3$ topology (**twt**, VS=$12_4.12_7.12_7$) (see Figure 1.3.9) that was enumerated by Wells [4b] but never found previously in real species.

**Figure 1.3.9** Three examples of self-catenated nets with two catenated rings (Hopf links) in black: 6-rings (**rob**), 8-rings (**coe**), 12-rings (**twt**).

Within uninodal 3-connected nets this is special in that it contains relatively "large" shortest rings (12-gons), the largest *n*-gons considered by Wells.

An intriguing feature of coordination network chemistry is its ability to produce rare and even only hypothetical structural motifs, given the correct building blocks. Moreover the models are reproduced on a larger scale. This is, inter alia, the case of [Ag(2-ethpyz)$_2$](SbF$_6$) (2-ethpyz = 2-ethylpyrazine) that shows the **coe** topology [52].

Some coordination polymers exhibit notable structures containing independent polycatenated motifs (vide infra) that are joined by bridging counterions or via supramolecular weak interactions, thus resulting in an unique self-catenated net. A beautiful example of this class has been observed in [Cd$_2$(4,4'-pytz)$_3$(μ-NO$_3$)(NO$_3$)$_3$](MeOH) [4,4'-pytz = 3,6-bis(pyridin-4-yl)-1,2,4,5-tetrazine [45]. The structure consists of molecular ladders that give inclined catenation forming a 3D array. The ladders of the two sets are cross-linked by μ-NO$_3$_ anions bridging the Cd(II) centers, resulting in the formation of a single 3D polymer, that the authors describe as a "polyknot". It should be now obvious that bridging of catenated layers (vide infra) results in a single self-catenated net [53].

## 1.3.4
## Entangled Systems

### 1.3.4.1 Types of Entanglements

Many of the intricate organic and metal–organic networks reported in the literature are particularly intriguing because of the presence of independent motifs entangled together in different ways. After the topological classification of the individual nets we must pass to a higher level of complexity, i.e. to the analysis of the "topology of entanglements".

Entangled systems are extended arrays, more complex than their constituents, that are comprised of individual motifs forming, via interlocking or interweaving, periodic architectures infinite in at least one dimension. Simple interdigitation is not considered here. As previously stated, most of the entangled arrays can be considered regularly repeated infinite versions of finite molecular motifs like catenanes, rotaxanes and pseudo-rotaxanes.

Such molecular species are the subject of many topological investigations [9, 10], especially concerning the classification of isomers within complex organic molecules [54]. Thus, in treating the complexity of entangled arrays, we will try to use concepts derived both from crystal chemistry, that has classified inorganic nets, and from the mathematical theory of knots and links. Before investigating the different structural classes we will examine some general preliminary aspects concerning the topology of entanglements.

The most common type of entanglement is represented by the numerous family of "interpenetrating networks" [2]. Interpenetration can be ascribed to the presence of large free voids in a single network, though it has been shown that this phenomenon does not necessarily prevent the formation of open porous materials [55]. Wells introduced the theme of interpenetrating nets (identical or of two or more kinds) by stating that they "cannot be separated without breaking links" [4b].

Only a few examples were known at that time, including neptunite (2-fold inter-penetrated **ths**), cuprite $Cu_2O$ (one of the first crystal structures determined, 2-fold **dia**)), β-quinol (H-bonded 2-fold hexagonal decorated primitive cubic **pcu-h** $6.10^2$) , [Cu(adiponitrile)$_2$](NO$_3$), (6-fold **dia**) and some others. Since then a great number of structural reports have appeared on interpenetrating nets, sustained by both coordinative or hydrogen bonds. The most important and comprehensive contri-butions in the area can be found in a review by Batten and Robson [2] and in the successive highlights by Batten [48 a, 56]. According to these authors, interpene-trating structures, that "can be disentangled only by breaking internal connec-tions", are characterized by the presence of infinite structurally regular motifs that must contain rings "through which independent components are inextricably entangled" [2].

However, within structures that are consistent with all the conditions described above, a distinct subclass can be recognized, namely polycatenated nets. "Poly-catenanes" have the peculiar features that all the constituent motifs have lower dimensionality than that of the resulting entangled architectures and that each individual motif is catenated only with the surrounding ones but not with all the others, like a single ring of a chain. In principle, also 0D (finite) motifs could be included within the possible component motifs. A comparison of the main aspects of polycatenation vs. interpenetration is given in Fig. 1.3.10. In between these two classes additional intricate intermediate situations are also possible, like some recently found examples of interpenetrating 3D *plus* 2D [57] and 3D *plus* 1D frameworks [58].

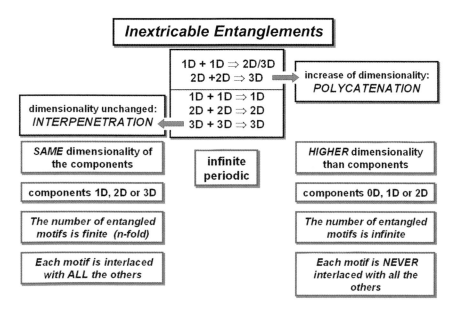

**Figure 1.3.10** The differences between interpenetration and polycatenation.

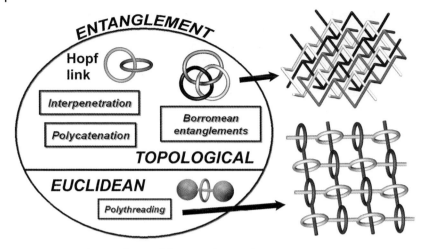

**Figure 1.3.11** The overall view of the entanglement phenomena in extended compounds (coordination and H-bonded).

   In both interpenetrated and polycatenated species the individual motifs cannot be separated "without breaking *rings*" and, according to the concepts of chemical topology, both classes give nontrivial entanglements, in the sense that the whole arrays can be considered "topological isomers" [9, 10] of their component motifs (like an *n*-catenane vs. the separated *n*-rings).

   In contrast to the above species, other entangled systems, like polyrotaxanes, poly-pseudo-rotaxanes, interweaved chains and infinite multiple helixes, are all trivial (separable) entanglements in a strict topological sense. Polyrotaxanes deserve some comments. They are currently described as interpenetrating systems, i.e. ones that cannot be disentangled [2], but this relies on metrical (dimensions of the components, geometrical rigidity) or energy (barriers to dethreading) considerations rather than on topological ones. By analogy with the topological nomenclature for stereoisomers [10] we have suggested [3] to call interpenetrated and polycatenated nets "Topological" entanglements, while polyrotaxane arrays are called "Euclidean" entanglements (see Fig. 1.3.11).

### 1.3.4.2   Interpenetrating Networks

The "topology of interpenetration" is, at present, a still poorly explored field of chemical topology [59]. Batten and Robson [2] have treated this phenomenon through the analysis of a number of different real cases, and have introduced a currently accepted nomenclature on this subject.

   The classification is based on recognizing the dimensionality (1D, 2D, 3D), the connectivity of the nodes and the topology of the individual interpenetrating motifs, the degree of interpenetration and the possible modes of interpenetration. The last point, the less known, is concerned with the relative disposition of the individual motifs in the three-dimensional space and the consequent mutual interlacing modes. It is presently difficult to answer questions like: how many topologically distinct interpenetration modes are allowed for an $n$-fold diamondoid network? or how can we recognize whether two arrays with the same degree of interpenetration and the same topology of the individual motifs are "topologically identical" or not? We can only state that an acceptable definition could be the following: two arrays comprised of the same number of interpenetrating nets with the same topology can be consider "topologically identical" when distorsions can be performed that bring all the nodes to coincidence without breaking/crossing links (i.e. we could say that, by analogy with what was stated for single nets, they are "ambient isotopic" – see above).

Distinct modes of interpenetration for 2D layers [2] and **dia** nets [60] have been discussed.

According to the concepts introduced in Fig. 1.3.10 we can have three types of interpenetrating nets based on the dimensionality (1D, 2D and 3D).

Only one case of 1D interpenetration based on hydrogen-bonded ladder motifs has as yet been reported [61]. On the other hand, a large variety of 2D interpenetrating networks is presently known (with a maximum of 6-fold interpenetrated $6^3$ hexagonal layers [62]). 3D interpenetrating networks are quite numerous, with interpenetration degree ranging from 2 to 18. For some time the world records for the degree of interpenetration belonged to two diamondoid networks: within coordination polymers to the 10-fold interpenetrated $[Ag(ddn)_2](NO_3)$ (ddn = 1,12-dodecanedinitrile) [60b], while within hydrogen bonded organic supermolecules to an 11-fold interpenetrated framework containing molecules of a tetraphenol as tetrahedral centers and benzoquinone units as rods [63]. Only very recently Zaworotko and coworkers have described the exceptional structure of the H-bonded net formed by trimesic acid and 1,2-bis(4-pyridyl)ethane in the ratio 2:3, containing a 18-fold **srs** net [64].

The analysis and comparison of interpenetrated networks is often a difficult task and time expensive work because of the structural complexity of these systems and the enormous growth in their number, that requires the use of some computer-aided procedure. A recent program suited for this purpose is TOPOS. The automatic investigation with this package [30] has many advantages: (i) the possibility to process a large amount of structural data; (ii) the automatic determination of the interpenetration degree and network relationships; (iii) the automatic simplification and topological classification. This analysis of the crystal structures has evidenced that distinct classes can be recognized, corresponding to the different modes in which individual identical 3D motifs can interpenetrate, that are represented by the operations generating the overall array from a single net. The approach is strictly related to the actual crystallographic structures rather than to the "idealized" simplified networks in their highest symmetry. Three classes are defined, independently from the network topology (see Fig. 1.3.12) [for the full details on these classes see Refs. [14, 15].

Class I (Translational): The individual nets are exclusively related by translations. The degree of interpenetration $Z$ corresponds to the translational degree of interpenetration $Z_t$. There are two distinct subclasses (Ia and Ib) depending on the presence or not of a full interpenetration vector. In class Ia all the independent nets are related by a single vector and the whole interpenetrated array is generated by translating a single net $(Z_t - 1)$ times this vector. Note that many different full interpenetration vectors can exist in species belonging to this class; of them the shortest one is selected. On the other hand, in class Ib a full interpenetration vector does not exist and the whole array is generated by application of more than one translational operation. While class Ia is numerous the examples of class Ib are much rarer (less than 3% of class I).

Class II (Non-translational): The individual nets are related by means of space group symmetry elements, mainly inversion centers, but also proper rotational axes, screw axes and glide planes. The degree of interpenetration $Z$ corresponds to the non-translational degree $Z_n$, i.e. the order of the symmetry element that generates the interpenetrated array from the single net. In almost all cases $Z_n$ is 2, but a few examples with $Z_n$ up to 4 are known.

The existence of compounds containing more than one interpenetration symmetry element cannot be ruled out. This requires the introduction of an additional sub-classification, similar to the division of class I, i.e. class IIa with a unique full interpenetration symmetry element and class IIb with different partial interpenetration symmetry elements (only one real example is known [15]).

Class III (Translational and non-translational): The overall entanglement is generated both by pure translations and by space group symmetry elements. The value of $Z$ is given by the product $Z_t \times Z_n$.

The complete search for 3D equivalent interpenetrating networks with TOPOS in the CSD and ICSD structural databases has produced lists of 301 and 144 reference codes, respectively.

It was observed that 57% of the structures fall in class I of translational interpenetration (that favors high degrees of interpenetration), while 40% belong to class II (non-translational) with degree of interpenetration equal to 2 for almost all (there are only 5 known structures with degree 3 or 4). Class III is rather unusual

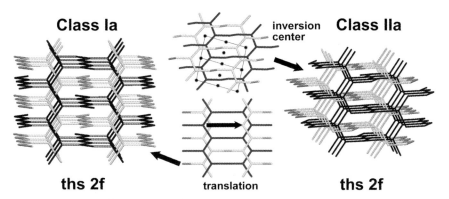

**Figure 1.3.12** The different classes of interpenetration (translational and non-translational) illustrated with an example of 3-connected 2-fold **ths**.

(3%) and presents always a degree of interpenetration of 4 or more; the highest known case of 18-fold **srs** interpenetration quoted above belongs to this class.

It was recently proposed [65] that interpenetrating networks (called "catenated" therein) could be differentiated on the basis of the relative displacement of the subnets: so we have "interpenetration", when the frameworks are maximally displaced from each other, or "interweaving" when they are minimally displaced and exhibit many close contacts. However, a quantitative criterion to establish when the separation is maximal or minimal is still lacking, and difficult to find, especially with a degree of interpenetration exceeding 2.

### 1.3.4.3 Polycatenated Networks

The main feature of polycatenation is that the whole catenated array has a higher dimensionality then that of each of the component motifs. These motifs can, in principle, be 0D, 1D or 2D species that must contain closed loops and that are interlocked, as for interpenetrating nets, via topological Hopf links [31] (see Fig. 1.3.11). Each motif can be catenated with a finite or also with an infinite number of other independent motifs *but not* with all (see Fig. 1.3.10).

The classification of these entanglements can be established by assigning in each case: (i) the dimensionality and topology of the individual motifs; (ii) the mode of catenation, i.e. the mutual orientation and interrelation of the component motifs, and (iii) the "degree of entanglement".

The last point requires some comment. While in $n$-fold interpenetrated networks we can easily assign the degree of interpenetration since the whole array contains a finite number of independent nets, in a polycatenated species we are forced to introduce different concepts since each individual motif can give finite or infinite interlocking with the other motifs (Fig. 1.3.10). In the former case we must specify only the number $n$ of motifs effectively entangled with each motif, while in the second case, the number of rings of other motifs interlocked with a single ring of each component should be specified (and in general different values are possible for the different components).

The different types of polycatenated species can be enumerated and classified on the basis of the increasing dimensionality of the component motifs (0D, 1D or 2D) [3]. Finite (0D) motifs, containing closed circuits, can give, in principle, catenation into infinite periodic arrays (from 1D up to 3D), though no real example has as yet been characterized. Examples of polycatenated species containing many molecular rings have been described within complex organic species [54], proteins [66] and synthetic DNA assemblies [67].

Polycatenation of 1D and 2D nets can occur essentially in two modes that are described as "parallel" and "inclined", according to the nomenclature introduced by Batten and Robson [2].

Different types of 1D motifs can give polycatenation, including chains of alternating rings and rods, ribbons of rings, molecular ladders and more complex species. Some examples are illustrated in Fig. 1.3.13. Almost all the known real cases are, at present, based on infinite molecular ladders interlocked to give 2D or

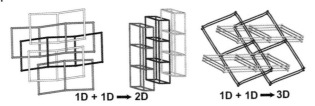

**1D + 1D ➡ 2D**　　　**1D + 1D ➡ 3D**

**Figure 1.3.13** Polycatenation of 1D motifs: 2D entangled layers from parallel catenation of undulating ladders (laterally catenated by two other ladders by each side) (left), and of nanotube motifs (center); 3D superstructure from catenation of ladders running in different directions (right).

3D entanglements [68]. An example of catenated tubular motifs giving 2D layers has also been recently reported [69] (see Fig. 1.3.13).

Interlocked 2D layers can exhibit different relative orientations of the planes of the independent motifs. When all the entangled sheets are parallel and have a common average plane the resulting array is an *n*-fold interpenetrated 2D structure, with no increase in dimensionality and a finite number of interweaved motifs (see an example in Fig. 1.3.14). If, otherwise, the independent layers show average planes that are parallel but displaced in a perpendicular direction, they generate a polycatenated 3D architecture. This "parallel" catenation can occur either because the sheets are undulated simple layers or because they are multiple layers, in both cases exhibiting some thickness (see Fig. 1.3.14, left, and 1.3.15). Many cases are known and have been reviewed previously [3].

Polycatenation can also involve different sets of 2D sheets (usually two) that cross at a certain angle, in the mode that has been called "inclined" [2] (see Fig. 1.3.14, right). The examples of this class are numerous, the majority consisting of two identical sets of 2D parallel layers, of $6^3$ or $4^4$ topology, spanning two different stacking directions. In these species there is an increase in dimensionality (2D → 3D) and each individual layer is catenated with an infinite number of other inclined layers but, obviously, not with all the frames contained in the 3D array. Moreover,

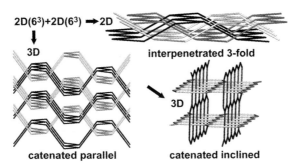

$2D(6^3)+2D(6^3)$ ➡ 2D

3D

interpenetrated 3-fold

3D

catenated parallel　　　catenated inclined

**Figure 1.3.14** Three possible topological entanglements of hexagonal $6^3$ layers showing interpenetration versus catenation (parallel and inclined).

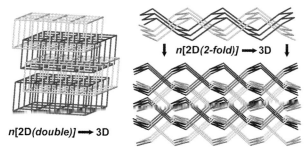

$n[2D(double)] \longrightarrow 3D$

$n[2D(2\text{-}fold)] \longrightarrow 3D$

**Figure 1.3.15** The 3D entanglement of parallelly catenated double layers (left) and an intricate case of parallel catenation of undulated 2-fold $6^3$ layers (right). (S. Banfi, L. Carlucci, E. Caruso, G. Ciani, D. M. Proserpio, *Cryst. Growth Des.* **2004**, *4*, 29.)

in the analysis of these species different modes of interlocking of the two independent sets have been envisaged for both topologies [2]. For $4^4$ layers three possible arrangements have been suggested by Zaworotko and coworkers [70], called parallel–parallel (p–p), parallel–diagonal (p–d) and diagonal–diagonal (d–d), depending on how the networks orient and penetrate through each other (see some examples in Fig. 1.3.16). Systematic enumerations of these entangled species have already appeared in some reviews [2, 44, 48 a, 56, 70].

Only in a few cases were major variations of the above inclined polycatenation scheme observed. For instance, catenation can involve two sets of 2D layers of different topology, as in [Ni(azpy)$_2$(NO$_3$)$_2$]$_2$[Ni$_2$(azpy)$_3$(NO$_3$)$_4$].4CH$_2$Cl$_2$ [azpy = *trans*-4,4'-azobis(pyridine)], containing both $6^3$ and $4^4$ layers [71]. Another variation is represented by the presence of more than two distinct sets of layers. No real case was known in 1998 when Batten and Robson [2] suggested the possibility of finding three different mutually perpendicular stacks; since then a few examples have been discovered and one is illustrated in Fig. 1.3.16, left, namely [Ni$_6$(bpe)$_{10}$ (H$_2$O)$_{16}$](SO$_4$)$_6 \cdot x$H$_2$O [bpe = 1,2-bis(4-pyridyl)ethane] [72], that contains three distinct sets of $4^4$ layers spanning three different spatial orientations and giving inclined mutual catenation.

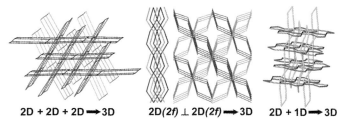

2D + 2D + 2D $\longrightarrow$ 3D      2D(2f) $\perp$ 2D(2f) $\longrightarrow$ 3D      2D + 1D $\longrightarrow$ 3D

**Figure 1.3.16** The 3D entanglements of different inclined catenations: three sets of 2D $4^4$ layers (left), perpendicular catenation of 2-fold $4^4$ layers (center), (K. A. Hirsch, S. R. Wilson, J. S. Moore, *Chem. Commun.* **1998**, 13), $4^4$ layers catenated by perpendicular 1D ribbons of rings (right).

Polycatenation involving motifs of different dimensionality is also known, though quite rare, as in $[Cu_5(bpp)_8(SO_4)_4(EtOH)(H_2O)_5](SO_4) \cdot EtOH \cdot 25.5H_2O$ [bpp = 1,3-bis(4-pyridyl)propane] [73] containing two different polymeric motifs entangled to give a unique 3D array: ribbons of rings and two-dimensional tessellated sheets of $4^4$ topology (Fig. 1.3.16, right).

### 1.3.4.4   Borromean Networks

Molecular motifs containing different rings could give inextricable entanglements via other than Hopf links. An alternative way involving at least three closed circuits at a time is represented by the Borromean link [3, 32] (see Fig. 1.3.11). This is a nontrivial link in which three rings are entangled in such a way that any two component rings form a trivial link, i.e. if any one ring is cut the other two are free to separate. Chemists for a long time have considered the realization of a molecular Borromean link to be a synthetic goal of great interest. An example has been very recently reported by Stoddart and coworkers, comprised of three macrocyclic Borromean rings obtained by metal templated synthesis [74]. Previously, only one species was known, constructed from single-stranded DNA by Seeman and coworkers in 1997 [75]. Using Borromean links, however, infinite 1D, 2D or 3D arrays could be imagined. Links like these could appear to be only mathematical curiosities; indeed they have been identified in some real examples of infinite interlocked nets. Borromean links are present in two types of arrays: (2D $\rightarrow$ 2D) entanglements of three layers and (2D $\rightarrow$ 3D) entanglements of infinite layers [32].

Three cases of Borromean 3-fold entangled 2D layers have been recognized, all exhibiting a $6^3$ layer topology [32,76] (see Figure 1.3.11, upper right; one can easily realize that the black net is completely located above the gray one, the gray net above the white one, but the gray net lies above the white one).

A Borromean (2D $\rightarrow$ 3D) entanglement can be envisaged in the structures of two isomorphous polymeric silver(I) complexes reported by Chen and coworkers, $[Ag_2(H_2L)_3](NO_3)_2$ and $[Ag_2(H_2L)_3](ClO_4)_2$ [$H_2L$ = *N,N*'-bis(salicylidene)-1,4-di-ami-nobutane] [77] and in another silver species with a similar formula [78]. They are comprised of highly undulated $6^3$ layers of 3-connected silver centers (Fig. 1.3.17). The silver atoms exhibit interlayer unsupported Ag···Ag interactions and, taking into account these contacts, a single 4-connected 3D array results, that can be described as self-catenated embedding of **dia** (regular **dia** – see e.g. Fig. 1.3.5, 1.3.7 and 1.3.8 – and self-**dia** are non-"ambient isotopic" and have obviously the same Schläfli Symbol, Vertex Symbol, and Coordination Sequence, but **dia** cannot be deformed into self-**dia** without edge crossings) [32, 34]. On the other hand, and much more interestingly, if the Ag···Ag interactions are neglected, the layers are not (2D $\rightarrow$ 3D) polycatenated but are interlinked via Borromean links involving three layers at a time. The whole array represents the first infinite case of *n*-Borromean links (Fig. 1.3.17).

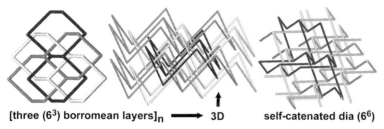

[three (6³) borromean layers]ₙ ➡ 3D     self-catenated dia (6⁶)

**Figure 1.3.17** A Borromean (2D → 3D) entanglement of undulated $6^3$ layers (left and center, see Refs. [32, 77, 78]). On the right is illustrated the related self-**dia** net discussed in the text (see also Refs. [32, 34]).

### 1.3.4.5 Other Entanglements

Other known types of infinite entanglements are comprised of finite or infinite components that can be "ideally" separated without breaking effective topological links. We are speaking of polythreaded arrays, i.e. extended periodic analogs of molecular rotaxanes and pseudo-rotaxanes. The term "Euclidean entanglements" was suggested for polyrotaxanes, meaning that only the presence of geometrical or energetical (non-topological) constraints can prevent the separation of the individual motifs [79]. On the other hand, poly-pseudo-rotaxanes contain motifs that are threaded by chains or strings which, in principle, can be slipped off; they display a type of entanglement that is neither topological nor Euclidean.

Polythreaded systems contain motifs that are interweaved via rotaxane-like mechanical links and this implies the presence of closed loops as well as of elements that can thread the loops (Fig. 1.3.2). These two types of moieties may belong to the same unit or may be separately supplied by motifs having different structures. The constituent motifs could be, in principle, 0D species, 1D polymers or arrays of higher dimensionality. The resulting array can show the same or an increased dimensionality with respect to that of the polythreaded units.

Known examples involving 0D motifs, i.e. molecular "beads" (cucurbituril molecules), threaded by coordination polymers of different topology have been described by Kim and coworkers [80].

Many examples of extended systems that "cannot be disentangled" have been reported very recently. The 1D chain motif comprised of alternating rings and rods

## POLYTHREADING

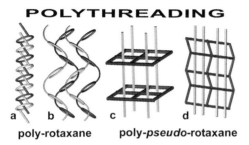

a       b       c       d

**poly-rotaxane**      **poly-*pseudo*-rotaxane**

**Figure 1.3.18** Examples of polyrotaxanes and poly-pseudo-rotaxanes.

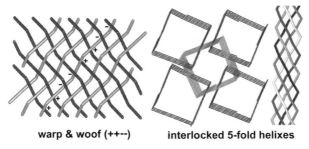

**warp & woof (++--)**          **interlocked 5-fold helixes**

**Figure 1.3.19** A warp-and-weft like 2D layer of 2-over/2-under chains (left) (Y.-H. Li, C.-Y. Su, A. M. Goforth, K. D. Shimizu, K. D. Gray, M. D. Smith, H.-C. zur Loye, *Chem. Commun.* **2003**, 1630) and interlaced 5-fold helixes (right) (Y. Cui, S. J. Lee, W. Lin, *J. Am. Chem. Soc.* **2003**, *125*, 6014).

is particularly suitable for polythreading since it contains both the elements needed in a rotaxane-like mechanical linkage. Chains of this type have been found to give (1D → 1D) [81], (1D → 2D) parallel [82] (see Fig. 1.3.18(a) and (b)) and inclined [82, 83] (see Fig. 1.3.11, bottom right) polythreaded arrays.

Numerous entangled arrays containing "separable" motifs (poly-pseudo-rotaxanes), like those shown in Fig. 1.3.18(c) and (d) have also been reported [see Ref. [3]].

Architectures of interwoven 1D motifs display different types of entanglements that are neither topological nor Euclidean. Polymeric chains, both rigid or flexible, represent the most simple and common structural extended motif. The packing of 1D polymers usually occurs with parallel orientation of all the chains; less commonly they can span two different directions on alternate layers. Independent chains can sometimes be connected in pairs or in extended 2D or 3D arrays by weak supramolecular interactions. Quite rare are, on the other hand, other types of associations of the chains resulting in entangled systems of unusual topologies, including chains woven like warp-and-weft threads in a cloth to give 2D sheets (Fig. 1.3.19, left) and infinite multiple helices (Fig. 1.3.19, right).

## 1.3.5
## Conclusions

We have attempted to rationalize here the structures of extended frames at different levels of complexity in terms of the network topology. With the aim of establishing useful relationships between structures and properties [84] a careful analysis of the topology is unavoidable, particularly for the complicated species that are increasingly being discovered. Indeed, phenomena like interpenetration are not only structural curiosities, with some esthetical appeal, but can play an effective role in the control of the properties of materials, as shown, inter alia, by their influence on network porosity [55 a, 65a] and by their capability to produce anomalous magnetic properties [85].

# References

1 M. O'Keeffe, B. G. Hyde, *Crystal Structures I: Patterns and Symmetry*, Mineralogical Society of America, Washington 1996.

2 S. R. Batten, R. Robson, *Angew. Chem. Int. Ed. Engl.* **1998**, *37*, 1460.

3 L. Carlucci, G. Ciani, D. M. Proserpio, *Coord. Chem. Rev.* **2003**, *246*, 247.

4 (a) A. F. Wells, *Structural Inorganic Chemistry*, 5th edn., Oxford University Press, Oxford 1984; (b) A. F. Wells, *Three-dimensional Nets and Polyhedra*, Wiley, New York 1977; (c) A. F. Wells, *Further Studies of Three-dimensional Nets*, ACA monograph 8, 1979.

5 M. O'Keeffe, M. Eddaoudi, H. Li, T. Reineke, O. M. Yaghi, *J. Solid State Chem.* **2000**, *152*, 3.

6 B. F. Hoskins, R. Robson, *J. Am. Chem. Soc.* **1990**, *112*, 1546.

7 J. W. Essam, M. E. Fisher, *Rev. Mod. Phys.* **1970**, *42*, 272.

8 A similar problem is encountered in the construction of the graph of a molecule for the study of its topology. While the identification of the vertex set is straightforward, the relationship of the edges in the graph to the bonds in the molecule is less well defined, and we should establish which bonds in the molecule can be regarded as "topologically significant". According to Mislow, a considerable arbitrariness is inevitably implicit in any graph, and "whether or not a given geometrically chiral molecular model is considered to be also topologically chiral depends on which subset of bonds in the molecule is considered to be topologically significant" [9]. Following the common usage in organic chemistry, Walba suggested to consider topologically significant only covalent bonds [10], but there is no obvious reason to limit the "edge set" to these bonds, thus ignoring the whole area of supramolecular chemistry.

9 (a) K. Mislow, *Croat. Chem. Acta* **1996**, *69*, 485; (b) K. Mislow, *Bull. Soc. Chim. Belg.* **1977**, *86*, 595.

10 D. M. Walba, *Tetrahedron*, **1985**, *41*, 3161.

11 R. W. Gable, B. F. Hoskins, R. Robson, *J. Chem. Soc., Chem. Commun.* **1990**, 1677.

12 (a) D. B. Amabilino, J. F. Stoddart, *Chem. Rev.* **1995**, *95*, 2725; (b) P. R. Ashton, I. Baxter, M. C. T. Fyfo, F. M. Raymo, N. Spencer, J. F. Stoddart, A. J. P. White, D. J. Williams, *J. Am. Chem. Soc.* **1998**, *120*, 2297.

13 V. A. Blatov, A. P. Shevchenko, V. N. Serezhkin, *J. Appl. Crystallogr.* **2000**, *33*, 1193.

14 V. A. Blatov, L. Carlucci, G. Ciani, D. M. Proserpio, *CrystEngComm.* **2004**, *6*, 377.

15 I. A. Baburin, V. A. Blatov, L. Carlucci, G. Ciani, D. M. Proserpio, *J. Solid State Chem.* **2005**, *178*, 2452.

16 (a) J. V. Smith, *Chem. Rev.* **1988**, *88*, 149; (b) J. V. Smith, *Tetrahedral Frameworks of Zeolites, Clathrates and Related Materials*, Landolt-Börnstein New Series IV/14 Subvolume A, Springer, Berlin 2000.

17 (a) M. O'Keeffe, N. E. Brese, *Acta Crystallogr. Sect. A* **1992**, *48*, 663; (b) M. O'Keeffe, *Acta Crystallogr. Sect. A* **1992**, *48*, 670; (c) M. O'Keeffe, *Acta Crystallogr., Sect. A* **1995**, *51*, 916.

18 For the most recent results see: (a) H. Sowa, E. Koch, *Acta Crystallogr., Sect. A* **2005**, *61*, 331; (b) W. Fischer, *Acta Crystallogr., Sect. A* **2005**, *61*, 435, and the references therein.

19 (a) M. M. J. Treacy, K. H. Randall, S. Rao, J. A. Perry, D. A. Chadi, *Z. Kristallogr.* **1997**, *212*, 768; (b) M. M. J. Treacy, I. Rivin, E. Balkovsky, K. H. Randall, M. D. Foster, *Microporous Mesoporous Mater.* **2004**, *74*, 121.

20 For enumeration and computational purposes, the infinite graph (i.e. our net) is reduced to a finite "labelled quotient graphs", closely related to a crystallographic primitive unit cell, where the labelled edges in the quotient graph describe the connectivity of the nodes. For a given number of nodes per unit cell all labelled quotient graphs which have a predefined connectivity can be found in: S. J. Chung, Th. Hahn, W. E. Klee, *Acta Crystallogr. Sect. A* **1984**, *40*, 42; W. E. Klee, *Cryst. Res. Technol.* **2004**, *39*, 959; J.-G. Eon, *Acta Crystallogr.* Sect. A **2005**, *61*,

501; ref. 21. This approach has been applied, for instance, for the search of novel carbon polymorphs, see: G. Thimm, *Z. Kristallogr.* **2004**, *219*, 528; R. T. Strong, C. J Pickard, V. Milman, G. Thimm, B. Winkler, *Phys. Rev. B* **2004**, *70*, 45101.

**21** O. Delgado-Friedrichs, M. O'Keeffe, *J. Solid State Chem.* **2005**, *178*, 2480.

**22** This nomenclature includes some simply related nets such as the augmented net and the edge net called **xxx-a** and **xxx-e**, respectively. An **augmented net** is one in which the vertices of the original net are replaced by a polygon or polyhedron corresponding to the original coordination figure. The **edge net** (also expanded net) is obtained from the original net by placing vertices in the middle of the edges, discarding the original vertices and edges, and joining the new vertices to enclose the coordination figure of the original vertices. Moreover **decorated nets** (**xxx-f,g** etc.) are obtained by replacing the vertices of the original net by groups of vertices. Augmented nets are special cases of decorated nets. See RCSR web page and Ref. [23].

**23** (a ) O. Delgado Friedrichs, M. O'Keeffe, O. M. Yaghi, *Acta Crystallogr., Sect. A* **2003**, *59*, 22; (b) O. Delgado Friedrichs, M. O'Keeffe, O. M. Yaghi, *Acta Crystallogr., Sect. A* **2003**, *59*, 515.

**24** (a) C. Baerlocher, W. M. Meier, D. H. Olsen, *Atlas of Zeolite Framework Types*, Elsevier, Amsterdam, 2001; (b) http://www.iza-structure.Org/databases/ ; (c) M. M. J. Treacy, *Microporous Mesoporous Mater.* **2003**, *58*, 1.

**25** N. W. Ockwig, O. Delgado-Friedrichs, M. O'Keeffe and O. M. Yaghi, *Acc. Chem. Res.* **2005**, *38*, 176.

**26** In a more rigorous graph-theoretical approach we should use the term *n*-coordinated instead of *n*-connected, but we keep here the word *connected* to indicate what is clear to chemists: for example, a metal could be 6-coordinated but give only a 3-connected net if three coordination sites are occupied by ligands that are not links of the network.

**27** M. O'Keeffe, B. G. Hyde, *Philos. Trans. R. Soc. London, Ser. A* **1980**, *295*, 553.

**28** An equivalent definition for *ring* is an *n*-membered circuit that represents the shortest possible path connecting *all* the $[n(n\text{-}1)/2]$ pairs of nodes belonging to that circuit.

**29** M. O'Keeffe, S. T. Hyde, *Zeolites* **1977**, *19*, 370.

**30** The program package TOPOS 4.0 with an advanced graphical interface (available at http://www.topos.ssu.samara.ru) was recently improved for the automatic determination and classification of the interpenetration degree, see Refs. [14] and [15]. The program is also able to analyze and simplify complex groups (including H-bonded networks) and to assign the topology of the resulting network according to the RCSR proposed symbol. As a result, a typical crystal structure of an organic, inorganic, or coordination compound may be processed in a few minutes from the .res or .cif files, to get a comprehensive description of the topology and of the interpenetration (if any). A detailed analysis of Voronoi-Dirichlet partition of crystal space and estimation of related crystallochemical parameters are also available in TOPOS. See, V. A. Blatov, *Cryst. Rev.* **2004**, *10*, 249.

**31** E. Flapan, *When Topology Meets Chemistry*, Cambridge University Press, Cambridge 2000.

**32** L. Carlucci, G. Ciani, D. M. Proserpio, *CrystEngComm* **2003**, *5*, 269.

**33** E. Koch, H. Sowa, *Acta Crystallogr., Sect. A* **2004**, *60*, 239; W. Fischer, *Acta Crystallogr., Sect. A* **2004**, *60*, 246.

**34** O. Delgado-Friedrichs, M. D. Foster, M. O'Keeffe, D. M. Proserpio, M. M. J. Treacy, O. M. Yaghi, *J. Solid State Chem.* **2005**, *178*, 2533.

**35** (a) O. Delgado-Friedrichs, M. O'Keeffe, *Acta Crystallogr., Sect. A* **2003**, *59*, 351; (b) O. Delgado-Friedrichs, *Nova Acta Leopold. NF*, **2003**, *88*, 39; (c) O. Delgado-Friedrichs, *Lecture Notes in Computer Science*, vol. 2912, Springer, Berlin, 2004, p.178; (d) O. Delgado-Friedrichs, *Discrete Comput. Geom.* **2005**, *33*, 67.

**36** (a) R. Robson, *J. Chem. Soc., Dalton Trans.* **2000**, 3735; (b) O. M. Yaghi, M. O'Keeffe, N. W. Ockwig, H. K. Chae, M. Eddaoudi, J. Kim, *Nature* **2003**, *423*, 705.

**37** G. Férey, *J. Solid State Chem.* **2000**, *152*, 37.

**38** (a) O. Delgado-Friedrichs, D. H. Huson, *Discr. Comput. Geom.* **1999**, *21*, 229; (b) O. Delgado-Friedrichs, A. W. M. Dress, D. H. Huson, J. Klinowsky, A. L. Mackay, *Nature* **1999**, *400*, 644; (c) M. O'Keeffe, *Nature*, **1999**, *400*, 617; (d) O. Delgado-Friedrichs, D. H. Huson, *Discr. Comput. Geom.* **2000**, *24*, 279.

**39** (a) O. Delgado Friedrichs, M. O'Keeffe, O. M. Yaghi, *Solid State Sci.* **2003**, *5*, 73; (b) O. Delgado Friedrichs, M. O'Keeffe, *Acta Crystallogr., Sect. A* **2005**, *61*, 358.

**40** Another application of the tiling theory is in the search for new zeolites, see the most recent results in: (a) M. D. Foster, A. Simperler, R. G. Bell, O. Delgado Friedrichs, F. A. Almeida Paz, J. Klinowski, *Nature Mater.* **2004**, *3*, 234; (b) A. Simperler, M. D. Foster, O. Delgado Friedrichs, R. G. Bell, F. A. Almeida Paz, J. Klinowski, *Acta Crystallogr., Sect. B* **2005**, *61*, 263.

**41** (a) S. Andersson, S. T. Hyde, H. G. von Schnering, *Z. Kristallogr.* **1984**, *168*, 1; (b) H. G. von Schnering, R. Nesper, *Angew. Chem. Int. Ed. Engl.* **1987**, *26*, 1059; (c) S. Andersson, S. T. Hyde, K. Larsson, S. Lidin, *Chem. Rev.* **1988**, *88*, 221; (d) S. T. Hyde, S. Andersson, K. Larsson, Z. Blum, T. Landh, S. Lidin, B. W. Ninham, *The Language of Shape: The Role of Curvature in Condensed Matter: Physics, Chemistry, and Biology*, Elsevier, Amsterdam, 1997; R. Nesper, S. Leoni, *ChemPhysChem 2001*, *2*, 413.

**42** (a) A. H. Schoen, *NASA Techn. Note*, 1970, *D-5541*; (b) W. Fischer, E. Koch, *Philos. Trans. R. Soc. London, Ser. A* **1996**, *354*, 2105; (c) C. Bonneau, O. Delgado Friedrichs, M. O'Keeffe, O. M. Yaghi, *Acta Crystallogr., Sect. A* **2004**, *60*, 517.

**43** (a) V. Robins, S. J. Ramsden, S. T. Hyde, *Eur. Phys. J. B*, **2004**, *39*, 365; (b) S. T. Hyde, A.-K. Larssen, T. Di Matteo, S. Ramsden, S. T. Hyde, *Aust. J. Chem.* **2003**, *56*, 981.

**44** S. R. Batten, R. Robson, in *Molecular Catenanes, Rotaxanes and Knots: A Journey Through the World of Molecular Topology*, J.-P. Sauvage, C. Dietrich-Buchecker (Eds.), Wiley-VCH, Weinheim, 1999, pp. 77–105.

**45** M. A. Withersby, A. J. Blake, N. R. Champness, P. A. Cooke, P. Hubberstey, M. Schröder, *J. Am. Chem. Soc.* **2000**, *122*, 4044.

**46** D. J. Price, S. R. Batten, B. Moubaraki, K. S. Murray, *Chem. Eur. J.* **2000**, *6*, 3186.

**47** M. O'Keeffe, *Z. Kristallogr.* **1991**, *196*, 21.

**48** (a) S. R Batten, *CrystEngComm.* **2001**, *3*, 67; (b) M. O'Keeffe, Nature **1998**, *392*, 879. However, taking into account our definition of self-catenation, for **icf** the catenated six- and ten-membered circuits are not both "shortest rings"; and for **itv** the catenated 8-rings are not "strong rings". It follows that a more detailed classification is possible considering also not only the "shortest rings" defining the VS but also any larger ring that may exist (as for **icf**), and the "strong rings" (as for **itv**) (see Ref. [21] for definition of all kind of rings).

**49** B. F. Abrahams, M. J. Hardie, B. F. Hoskins, R. Robson, E. E. Sutherland, *J. Chem. Soc., Chem. Commun.* **1994**, 1049.

**50** The same **rob** topology, together with a new self-catenated 6-connected net, has been recently reported for [Sc(4,4'-bipyridine-$N,N'$-dioxide)$_3$(ClO$_4$)$_3$]: D.-L. Long, R. J. Hill, A. J. Blake, N. R. Champness, P. Hubberstey, C. Wilson, M. Schröder, *Chem. Eur. J.* **2005**, *11*, 1384.

**51** B. F. Abrahamas, S. R. Batten, M. J. Grannas, H. Hamit, B. F. Hoskins, R. Robson, *Angew. Chem., In. Ed.*, **1999**, *38*, 1475.

**52** L. Carlucci, G. Ciani, D. M. Proserpio, S. Rizzato, *J. Chem. Soc., Dalton Trans.* **2000**, 3821.

**53** Self-catenation is observed not only in 3D nets but also in 2D nets: a few rare examples are known (see Ref. [3]) and two recent examples are: the 4-connected layers in a Cd coordination polymer (G. O. Lloyd, J. L. Atwood, L. J. Barbour, *Chem. Commun.* 2005, 1845) and the 6-connected layers in a supramolecular arrangement of a tetrazole derivative (A. T. Rizk, C. A. Kilner, M. A. Halcrow *CrystEngComm*, **2005**, *7*, 359).

**54** (a) J.-P. Sauvage, *Acc. Chem. Res.* **1998**, *31*, 611; (b) J. C. Chambron, C. O. Dietrich-Buchecker, J.-P. Sauvage, *Top. Curr. Chem.* **1993**, *165*, 131; (c) J.-P. Sauvage, *Acc. Chem. Res.* **1990**, *23*, 319; (d) J.-P. Sauvage, C. O. Dietrich-Buchecker, *Molecular Catenanes, Rotaxanes and Knots*, Wiley-VCH, Weinheim, 1999; (e) C. O. Dietrich-Buchecker, J.-P. Sauvage, *Angew. Chem. Int. Ed. Engl.* **1989**, *28*, 189; (f) D. B. Amabilino, J. F. Stoddart, *Chem. Rev.* **1995**, *95*, 2725; (g) F. M. Raymo, J. F. Stoddart, *Chem. Rev.* **1999**, *99*, 1643; (h) M. C. T. Fyfe, J. F. Stoddart, *Acc. Chem. Res.* **1997**, *30*, 393; (i) S. A. Nepogodiev, J. F. Stoddart, *Chem. Rev.* **1998**, *98*, 1959.

**55** (a) T. M. Reineke, M. Eddaoudi, D. M. Moler, M. O'Keeffe, O. M. Yaghi, *J. Am. Chem. Soc.* **2000**, *122*, 4843; (b) M. Kondo, M. Shimamura, S.-I. Noro, S. Minakoshi, A. Asami, K. Seki, S. Kitagawa, *Chem. Mater.* **2000**, *12*, 1288.

**56** S. R. Batten, *Curr. Opin. Solid State. Mater. Sci.* **2001**, *5*, 107.

**57** (a) J. Y. Lu, A. M. Babb, *Chem. Commun.* **2001**, 821; (b) Y.-C. Jiang, Y.-C. Lai, S.-L. Wang, K.-H. Lii, *Inorg. Chem.* **2001**, *40*, 5320; (c) D. M. Shin, I. S. Lee, Y. K. Cheung, M. S. Lah, *Chem. Commun.* **2003**, 1036.

**58** L. Carlucci, G. Ciani, D. M. Proserpio, *Chem. Commun.* **2004**, 380.

**59** Two-fold interpenetration in some prototypical nets has been discussed in terms of dual of the tiling of the net by O'Keeffe and coworkers (see Ref. [39a]). The dual of a tiling is obtained by putting new vertices in the center of the tiles and connecting such vertices by edges passing through common faces (each face of a tiling is common to two tiles) forming an interpenetrating net, so that all rings (faces) of the original structure are linked (catenated) to faces of the dual one and vice versa. When the tiling of the net and its dual are the same, we speak of self-dual nets (*e.g.* **dia**, **pcu**, **srs**, **fcu**, **cds**).

**60** (a) L. Carlucci, G. Ciani, P. Macchi, D. M. Proserpio, S. Rizzato, *Chem. Eur. J.* **1999**, *5*, 237; (b) L. Carlucci, G. Ciani, D. M. Proserpio, S. Rizzato, *Chem. Eur. J.* **2002**, *8*, 1520.

**61** H.-F. Zhu, J. Fan, T. Okamura, W.-Y. Sun, N. Ueyama, *Chem. Lett.* **2002**, 898.

**62** D. Venkataraman, S. Lee, J. S. Moore, P. Zhang, K. A. Hirsch, G. B. Gardner, A. C. Covey, C. L. Prentice, *Chem. Mater.* **1996**, *8*, 2030.

**63** D. S. Reddy, T. Dewa, K. Endo, Y. Aoyama, *Angew. Chem. Int. Ed. Engl.* **2000**, *39*, 4266.

**64** T. R. Shattock, P. Vishweshwar, Z. Wang, M. J. Zaworotko, *Cryst. Growth Des.* **2005**, *5*, 2046.

**65** (a) B. Chen, M. Eddaoudi, S. T. Hyde, M. O'Keeffe, O. M. Yaghi, *Science* **2001**, *291*, 1021; (b) J. L. C. Rowsell, O. M. Yaghi, *Angew. Chem. Int. Ed.* **2005**, *44*, 4670.

**66** See, e.g.: W. R. Wikoff, L. Liljas, R. L. Duda, H. Tsuruta, R. W. Hendrix, J. E. Johnson, *Science* **2000**, *289*, 2129; J. Chen, C. A. Rauch, J. A. White, P. T. Englund, N. R. Cozzarelli, *Cell* **1995**, *80*, 61.

**67** N. C. Seeman, P. S. Lukeman, *Rep. Prog. Phys.* **2005**, *68*, 237.

**68** See e.g.: A. J. Blake, N. R. Champness, A. Khlobystov, D. A. Lemenovskii, W.-S. Li, M. Schröder, *Chem. Commun.* **1997**, 2027; L. Carlucci, G. Ciani, D. M. Proserpio, *J. Chem. Soc., Dalton Trans.* **1999**, 1799.

**69** X.-L. Wang, C. Qin, E.-B. Wang, Y.-G. Li, Z.-M. Su, L. Xu, L. Carlucci, *Angew. Chem. Int. Ed. Engl.* **2005**, *44*, 5824.

**70** K. Biradha, A. Mondal, B. Moulton, M. J. Zaworotko, *J. Chem. Soc.,Dalton Trans.* **2000**, 3837.

**71** L. Carlucci, G. Ciani, D. M. Proserpio, *New J. Chem.* **1998**, 1319.

**72** L. Carlucci, G. Ciani, D. M. Proserpio, S. Rizzato, *CrystEngComm.* **2003**, *5*, 190.

**73** L. Carlucci, G. Ciani, M. Moret, D. M. Proserpio, S. Rizzato, *Angew. Chem. Int. Ed. Engl.* **2000**, *39*, 1506.

**74** (a) S. J. Cantrill, K. S. Chichak, A. J. Peters, J. F. Stoddart, *Acc. Chem. Res.* **2005**, *38*, 1; (b) K. S. Chichak, S. J. Cantrill,

A. R. Pease, S.-H. Chiu, G. W. Cave, J. L. Atwood, J. F. Stoddart, *Science* **2004**, *304*, 1308.

**75** C. Mao, W. Sun, N. C. Seeman, *Nature* **1997**, *386*, 137.

**76** D. B. Leznoff, B.-Y. Xue, R. J. Batchelor, F. W. B. Einstein, B. O. Patrick, *Inorg. Chem.* **2001**, *40*, 6026; M. P. Suh, H. J. Choi, S. M. So, B. M. Kim, *Inorg. Chem.* **2003**, *42*, 676; R. Liantonio, P. Metrangolo, T. Pilati, G. Resnati, *Cryst. Growth Des.* **2003**, *3*, 355.

**77** M. L. Tong, X.-M. Chen, B.-H. Ye, L.-N. Ji, *Angew. Chem. Int. Ed. Engl.* **1999**, *38*, 2237.

**78** S. Muthu, J. H. K. Yip, J. J. Vittal, *J. Chem. Soc., Dalton Trans.* **2002**, 4561.

**79** Not only poly-pseudo-rotaxanes but also polyrotaxanes should be considered trivial topological entanglements, extending the concepts developed for the related molecular entities. By means of "ideal" continuous deformations we could separate the components of any finite portion of a polyrotaxane. How-

ever, the extended motifs under examination can be considered infinite entities. Including this additional "boundary condition" a different topological description could result for such interweaving systems, but this is at present an almost completely unexplored area of chemical topology.

**80** J. W. Lee, S. Samal, N. Selvapalam, H.-J. Kim, K. Kim, *Acc. Chem. Res.* **2003**, *36*, 621.

**81** (a) C. J. Kuehl, F. M. Tabellion, A. M. Arif, P. J. Stang, *Organometallics* **2001**, *20*, 1956; (b) C. S. A. Fraser, M. C. Jennings, R. J. Puddephatt, *Chem. Commun.* **2001**, 1310.

**82** L. Carlucci, G. Ciani, D. M. Proserpio, *Cryst. Growth Des.* **2005**, *5*, 37.

**83** B. F. Hoskins, R. Robson, D. A. Slizys, *J. Am. Chem. Soc.* **1997**, *119*, 2952.

**84** C. Janiak, *Dalton Trans.* **2003**, 2781; J. S. Miller, *Adv. Mater.* **2001**, *13*, 525.

**85** J. S. Miller, T. E. Vos, W. W. Shum, *Adv. Mater.* **2005**, *17*, 2251.

# 2
# Design and Reactivity

## 2.1
### Prediction of Reactivity in Solid-state Chemistry
*Gerd Kaupp*

## 2.1.1
### Introduction

Organic solid-state reactions essentially started with photoreactions and many highly energetic molecular structures can only be built by that technique. Recently thermal, gas–solid, and solid–solid reactions have matured to the most sustainable and environmentally friendly syntheses. A recent review [1] collects more than 1000 waste-free reactions from virtually all fields of organic synthesis (including some inorganic syntheses, see Refs. [6–8] in Ref. [1]) with 100 % yield at quantitative conversion in stoichiometric runs, usually without any auxiliaries. Catalysis is also possible but mostly not necessary. The solid-state reactions have to be strictly differentiated from "solvent-free" reactions via the liquid phase (melt reactions) or reactions with a limited amount of solvent (kneading reactions) when the crystalline state is no longer profited from. Acceptable limiting cases are bulk thermal reactions and photoreactions in liquid suspension of insoluble crystals that do not involve the liquid as a reagent. Amorphous solids will behave atypically.

Unfortunately, there is a widespread habit of calling all solid-state reactions "topochemical reactions". However, such usage is highly misleading and confusing as it often implies a mechanistic proposal.

The term 'topochemistry' was introduced by Kohlschuetter in 1919 when he produced colloidal $Al_2O_3$ inside the boundaries of the original $KAl(SO_4)_2$ crystal [2]. It thus described a solid-state reaction without change in the outer crystal habit but without preserving the crystal structure or any relation to it. Rather, colloidal particles were formed from a previously well-organized crystal. These colloidal particles stuck together but did not form a new crystal. Schmidt and coworkers used the same expression in 1964 [3], when they studied single crystals of α- and β-cinnamic acid and obtained stereoselective product formation without preserving the outer shape of the crystals, because these formed surface structures and decomposed into microcrystals upon chemical conversion. This paper introduced

*Making Crystals by Design.* Edited by Dario Braga and Fabrizia Grepioni
Copyright © 2007 WILEY-VCH Verlag GmbH & Co. KGaA, Weinheim
ISBN: 978-3-527-31506-2

the so-called 'topochemical principle' on the basis of α- and ß-cinnamic acid, that did not withstand supermicroscopic scrutiny. Following these definitions and their challenge there was much effort to look for single-crystal-to-single-crystal reactions without crystal disintegration, but only a very few examples could be found [4]. These reactions appeared to only be possible if the shape of the product molecules did not significantly deviate from the starting molecules in the crystal [5]. As these conditions are almost never met, the overwhelming number of solid-state reactions proceed with crystal disintegration at low to medium conversion. But there are also thermal and photochemical reactions within crystals where the crystals maintain their outer shape due to formation of an amorphous phase [6] or due to a colorful surface reaction [7, 8]. This situation appears quite confusing, but a detailed scrutiny of the 'topochemical principle' (that unfortunately considers only the proximity of reacting centers in the starting crystal and claims 'minimal atomic and molecular movements') [3] clarifies the issue in the light of basic physical effects and nanoscopic studies [4, 9, 10]. The experimentally based new mechanism takes into account the local pressure upon chemical reaction in the bulk when the molecular shape changes. Pressure release must result in far-reaching anisotropic molecular migrations within the crystal, as detected by atomic force microscopy (AFM). Therefore, the crystal packing that enables such migrations along cleavage planes or along channels or to voids must be invoked. If the latter are missing in 3D-interlocked structures, no reaction occurs despite short distances. The new paradigm is known as the 'phase rebuilding mechanism' [4, 9, 10].

Crystal structures are available from the Cambridge Crystallographic Data Center (CCDC) and other databases. Importantly, crystal packing can be constructed from fractional coordinates and space groups with versatile software packets (e.g. Mercury, Schakal, Shelxl-97, etc.). A comprehensive classification of the different shapes and qualities of crystallographic cleavage planes and channels is provided in Sections 2.1.8 and 2.1.9 to demonstrate the predictive power of the experimental mechanistic approach. This leads to borderline assessments and reveals migrational selectivities when (interpenetrating) cleavage planes and channels occur in the same crystal. All of this is important for waste-free chemical syntheses and the stability of organic crystals (e.g. medicinal drugs) in an ambient atmosphere.

## 2.1.2
### Topochemistry and Topotaxy

Schmidt's topochemistry does not care whether the solid-state reactions proceed "topotactic" that means single-crystal to single-crystal without molecular movements beyond geometric relaxation. The crystal disintegration in the basic photo-dimerizations of α- and ß-cinnamic acid were mentioned in Ref. [3] but not considered important. It is therefore very confusing that the expression "topotactic" appears to be increasingly misused by several authors for all of their solid-state reactions. Also the use in a broader sense ("crystal orientations relating to the orientation in the starting crystal") is not helpful. These habits are particularly disturbing and should be discontinued, as they devalue the rare events where true topotaxy indeed occurs, when there is no geometric change within an accommodating lattice, so that no chemically induced local pressure occurs and no molecular migrations are necessary. Topotactic reactions are highly special and very rare. Several claims of topotaxy (in the true sense of the word) do not withstand experimental scrutiny, for example by atomic force microscopy (AFM), scanning near-field optical microscopy (SNOM), grazing incidence X-ray diffraction (GID), or simply by microscopic inspection (absence of crystal change or disintegration) and some rely on faulty kinetic interpretation (see Ref. [4]). Well-known examples that withstand AFM studies at the molecular level and other experimental tests are Nakanishi's photodimerization of 2-benzyl-5-benzylidenecyclopentanone [5] and the photo/thermal reversible [4 + 4]-cycloaddition/cycloreversion of [2.2]-(9,10)anthracenophane [9, 10]. Non-local X-ray diffraction studies are suitable for truly topotactic reactions, but not for the overwhelming majority of solid-state reactions with surface feature formation (i.e. anisotropic molecular migration) and crystal disintegration.

The topochemical misconception totally disregards the enormous local pressure in the crystal bulk if a chemical reaction occurs that changes the molecular geometry (for example flat molecules to 3D molecules). It does however admit disintegration of single crystals α-1 into powder of α-truxillic acid 2 [11] and surface features, as long as these do not relate to the crystal structure. It was therefore erroneously claimed that the surface features on photolyzed α-cinnamic acid 1 are at random ("no signs of orientation") [12], even though a look in a light microscope (before complete crystal disintegration) does show and would have shown, that these are strictly aligned along the cleavage plane direction. The basic claim of topochemistry was "minimal atomic and molecular movements" (not to speak of migrations) [3]. The reactions were supposed to occur in the interior of so-called "reaction cavities". From there sudden crystallization of product was assumed after a "solid solubility limit" was surpassed. Crystal packing has been totally disregarded. A 4.2 Å limiting distance for reactivity was defined, but distances of reacting centers in that range are quite common in crystals. It is not surprising that the disregard of the most important facts (immediate release of local pressure upon chemical reaction is necessary!) led to immediate or even long known failure of such claims. But due to its alleged simplicity (no need to invoke crystal packing) it found widespread use also in textbooks. All long known positive (reactivity at longer distance) and negative failures (non-reactivity at shorter distance than 4.2 Å) were either disregarded or termed irrelevant ("must be defect site reactions"). Even now some researchers look for various excuses for non-reactivity without invoking crystal packing, while denying the success with the crystal packing analysis (Section 2.1.11), despite 14 years of experience with AFM and the supporting techniques of SNOM, GID, nanoscratching, light microscopy, etc. All these techniques prove strict correlation of surface feature alignment or shape with the crystal packing. Analysis of the latter explains all of these pseudo-problems (Section 2.1.5) comprehensively and on a convincing experimental basis.

Unfortunately, the misconcept of minimal movements, in textbooks since 1964, severely hampered the development of waste-free environmentally benign gas–

solid and solid–solid reactions [1]. These reactions proceed particularly well but are not conceivable on the basis of "topochemistry" not "permitting" anisotropic molecular migrations within crystals.

### 2.1.3
### Far-reaching Molecular Migrations in Solid-state Reactions (AFM, GID, SNOM) and Experimental Solid-state Mechanism

The hardly understandable disregard of the enormous local pressure to the crystal lattice upon chemical reaction if the molecular shape changes, the disregard of crystal packing, the odd results with the "reaction cavity principle", and the numerous failures of topochemistry (Section 2.1.5) made urgent the replacement of the topochemical hypotheses by an experimentally founded solid-state mechanism. This was of the highest importance, as the lack of an acceptable mechanism hampered the advent of environmentally benign waste-free quantitative gas–solid reactions known since 1986 [13]. As X-ray crystal structure determinations had not opened a path for understanding the new synthetic possibilities (several renowned journals refused to publish the synthetic facts because of that) the first commercial AFM came into use in 1989 (although some STM already had an AFM addition). The AFM was used to reveal experimentally what is going on in photochemical, thermal, gas–solid and solid–solid reactions [14]. Clearly, anisotropic long-range molecular migrations were detected for all kinds of non-topotactic reactions that strictly correlated with the crystal packing. The sizes of the features above the surface were in the ranges of tens and hundreds of nm. Clearly, if changes occur at the surface that exceed the range of adsorption/desorption in the molecular range (1 nm range) this must derive from molecular migrations out of the bulk of the crystal, inasmuch as there is always a strict correlation to the crystal packing with indisputable face specificity. This applies to all of the four solid-state reaction types (gas–solid, solid–solid, thermal intracrystalline, photolytical intracrystalline). In gas–solid and solid–solid reactions linear reagents can enter the crystal bulk through small crystallographic channels or pores that are frequently available (Section 2.1.9), whereas larger reagents must choose wider channels or cleavage planes, if available, to enter the crystal bulk in the reaction zone. The crystal surface features develop gradually (phase rebuilding) and are followed by a sudden change (phase transformation into product phase) and finally by crystal disintegration with formation of a fresh surface [14]. These observations and conclusions from the AFM investigation were confirmed in all respects as bulk effects by GID [15]. Thus, the chemically induced local pressure is indeed released by anisotropic molecular migrations, in complete opposition to the topochemistry hypotheses. The crystal packing becomes most important for all reliable predictions, because it must offer "easy" ways (cleavage planes, channels, or voids) for the molecular migrations. If no such possibility exists in 3D-interlocked structures, no chemical reaction occurs, irrespective of the very close proximity and favorable orientations of the reactive centers [14]. Yet another point could be settled: the topochemistry proponents always claim "defect reaction" if their so-called "predictions" fail in thermo- and photochemistry, mostly without any proof or with highly laborious though inconclusive mechanistic studies, because any spectroscopic far-distance transfer of electronic energy does not immediately prove chemical reaction only at defect sites. Such luminescence processes are not efficient enough to explain ready solid-state photosynthesis that must therefore be a bulk reaction (Section 2.1.5). This topic could also be experimentally settled by chemical contrast studies with SNOM that indicated uniform reaction all over the surface, irrespective of feature sites, for example in the most famous example of the "topochemically forbidden" ($d = 6.038$ Å) anthracene **14** photodimerization [9, 16], removing all speculation in terms of reaction only at defect sites (it most likely also occurs there!). On the other hand, SNOM is also the method of choice if island mechanisms around nucleation sites in certain types of gas–solid reactions have to be proven by the chemical contrast at the island. That situation occurs if the crystal packing impedes reaction or the exit of migrating molecules at particular faces. Still another nanoscopic technique, nanoscratching, reveals that long-range molecular migrations can also be induced by mechanical pressure [17]. This is a highly desired support for far-reaching anisotropic molecular migrations with strict correlation to the crystal structure.

The systematic classification of the feature shapes revealed 12 basic forms that are named after well-known objects, see Table 2.1.1. Subdivisions occur and further geometric types await detection. The basic shapes may occur in pure form or mixed and merged. Importantly, they relate to the crystal packing. For example, if a cleavage plane ends on the surface, parallel floes will occur, or flat lying molecules on a surface will lead to random volcanoes, etc. An exhaustive collection of these feature types and subtypes can be found in Ref. [18].

**Table 2.1.1** Gas–solid and solid–solid reactions.

| 12 basic geometric types of surface features |
| --- |
| flat covers |
| volcanoes |
| pyramids |
| craters |
| egg-trays (volcanoes and craters) |
| floes |
| fissures |
| heights and valleys |
| islands |
| moving zones |
| pool basins |
| bricks |

The type of surface features is highly characteristic and reproducible, as it depends on the crystal packing (subdivisions depend on the geometric shape of the product

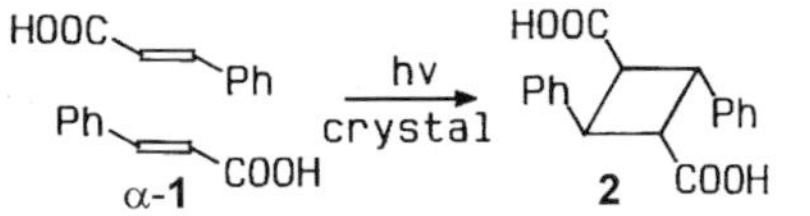

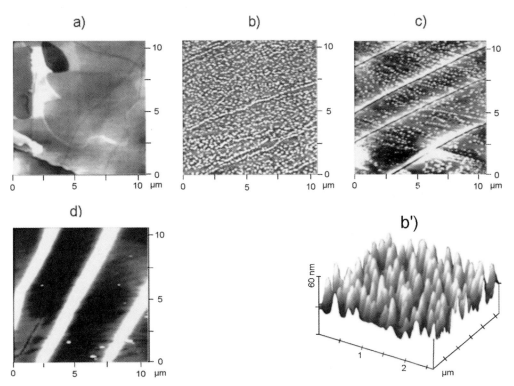

Scheme 2.1.1

molecules). This by necessity implies face selectivity (Section 2.1.4). As expected, the shapes of the observed surface features are characteristically dependent on the crystal face, if the crystal packing is different under these. For example the photo-dimerization of anthracene **14** provides chainy hills along molecular steps on its (001) face and random craters on its (110) face [19]; the 9-chloroanthracene photo-dimerization forms parallel pyramidal floes on the main face and random volcanoes on its long side face [10]; the ß-cinnamic acid photodimerization leads to craters on the (100) face and volcanoes on the (010) face [20] (Section 2.1.4). The differences in feature formation on different faces of the same crystal and on different polymorphs of the same compound have been amply demonstrated in

**Figure 2.1.1** AFM surfaces (11 μm) on (010) of α-cinnamic acid **1** after 365 nm (6.0 mW cm⁻²) irradiation: (a) fresh, (b) and (b') after 30 min (b' is a 3 μm scan), (c) after 2 times 45 min with a delay of 40 min, (d) after 90 min continuous irradiation; the orientation of the crystal varied slightly but the *c*-axis was always cut by the parallel trenches and ridges at an angle of 40° (crystallographic cleavage plane direction: 39°); the Z-scale covers 10 nm in (a) and 100 nm in (b), (c) and (d).

numerous publications. For example, the addition of gaseous bromine to α-cinnamic acid **1** on (010) leads to "egg-tray" features, while ß-cinnamic acid gives moving zones with a sharp front on (100) and flat lying scales on (010). These features strictly correlate with the crystal packing [16].

If crystal structures are available, their packing can be used to predict reactivity or non-reactivity. Distances of reactive centers are less important, but it is self-evident that close distances and favorable orientation can be advantageous in most cases (see exceptions in Section 2.1.6) and that far distances with obstruction in between will be detrimental. But it is essential that molecules can migrate along "easy paths" and these must be available for reactivity.

The experimental proof for anisotropic far-reaching molecular migrations within crystals by AFM analysis is particularly important with the former standard of topochemistry: the α-cinnamic acid **1** photochemistry to give predominantly α-truxillic acid **2** (Scheme 2.1.1).

It has been known since 1992 [20] that submicroscopic cracks and hills form along the cleavage plane direction on the major (010) face of **1**. Figure 2.1.1 shows the AFM results at lower conversion with monochromatic irradiation (365 nm). [21, 22].

The AFM images in Fig. 2.1.1 show that two processes occur with far reaching molecular migrations. The continuous growth of the volcanoes in (b) and beyond

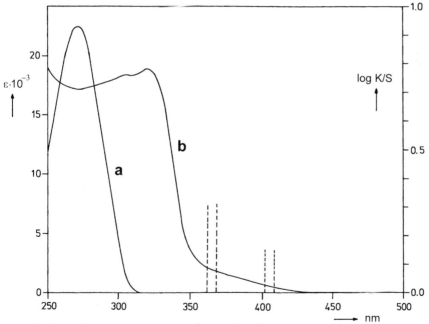

**Figure 2.1.2** (a) UV absorption spectrum of cinnamic acid **1** in methanol and (b) diffuse reflection spectrum of α-cinnamic acid **1** powder, exhibiting considerable shift of the absorption to long wavelengths with tailing up to 440 nm.

(90 % light penetration through about 1300 molecular layers at 365 nm) is followed by their disappearance at the expense of the formation of trenches and ridges that is complete in (d). The first process is to be called "phase rebuilding" as the product phase is not yet formed, whereas the second process that occurs suddenly at higher intensities [at low intensities this cannot be very abrupt and (c) is an intermediate state] is phase transformation (separation of the product phase) [14, 21]. The direction of the parallel ridges and trenches corresponds to the cleavage direction on that face. Similar irradiation at 405 nm (90 % light penetration through about 4400 layers at 405 nm) increases the number of volcanoes (craters in between) but the attainment of trenches and ridges takes very long, also due to the lower intensity.

The diffuse reflection spectrum of powdered α-cinnamic acid **1** reflects its solid-state absorption spectrum, which extends into the visible, in contrast to the solution spectrum (Fig. 2.1.2) [20]. It also indicates the wavelength range in the irradiations (365 and 405 nm) and helps in the judgment of light penetration when compared with the absorptivity in methanol (the molecules under (010) of the single crystal are inclined by ± 32°).

The crystal packing of α-cinnamic acid **1** (Fig. 2.1.3) exhibits two cleavage directions. Therefore, molecules can migrate and the photodimerization can proceed. The shapes of the features are easily understood on the basis of the crystal packing. The diagonal (10−1) cleavage planes between the single layers and the extended molecules lie flat in their rows. They end at the probed overwhelming (010) face. Therefore, upward materials transport to form the initial volcanoes (Fig. 2.1.1(b) and (b')) is expected, as is trenching and hill formation along the cleavage plane direction (Fig. 2.1.1(c) and (d)). The differences between (c) and (d) from different experiments show that the anisotropic migrations have a finite rate. The interpenetrating (010) cleavage plane is also used for molecular migrations, as the exit at the thin side faces (that could not be probed by AFM due to alignment

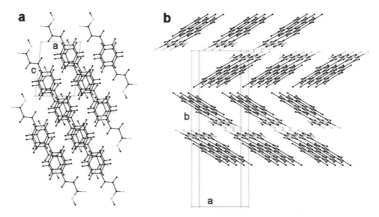

**Figure 2.1.3** Packing diagrams of α-cinnamic acid **1** (P2$_1$/n) [24]:
(a) on (010) showing the diagonal (10−1) cleavage plane and (b) on the (001) side face showing the (010) cleavage plane.

**a**          **b**

**Figure 2.1.4** Micrographs of α-cinnamic acid **1** crystals from benzene on (010) (200-fold in the microscope): (a) fully transparent after 6 months storage under argon in the dark; (b) same crop crystal that turned dull after 6 months exposure to filtered daylight under argon, exhibiting cracks along the (10–1) cleavage plane direction, crystallites on (010), and particularly crystallites on the side faces.

problems) becomes visibly opaque during photolysis (seen in Fig. 2.1.4(b)). The molecules are not very steep on the (010) cleavage plane (32°) and this facilitates their migration. SNOM measurements indicate chemical uniformity at any stage of the photoreaction on (010) (no chemical contrast at the features in Figs. 2.1.1 and 2.1.5) [21, 23].

As the paradigm change between minimal atomic and molecular movements and necessary requirement for far-reaching anisotropic molecular migrations for releasing local pressure came unexpectedly for topochemistry proponents we studied the molecular migrations also by giving a very thin crystal (40–60 μm depth) an exceedingly long time for any alleged relaxation to occur. Six months irradiation of α-**1** with filtered daylight (Pyrex and double glass window, argon atmosphere) that passed the whole colorless crystal almost unabsorbed (Fig. 2.1.2) provided 30% conversion into **2**. The crystal became dull and started to show disintegration by cracks along the cleavage direction on (010) under the microscope. Furthermore, there was formation of crystallites, particularly at the side faces (001) and (100) where molecules migrate in the (010) cleavage plane (Fig. 2.1.4) [21].

AFM investigation of the long-term irradiated crystal on (010) (between the crystallites) showed images as depicted in Fig. 2.1.5. Clearly, fence-like walls (typically 40 nm high) in the cleavage direction are formed by far-reaching anisotropic molecular migration [21]. The thorough light penetration with minimal absorption slows down the dimer formation and lattice guided feature formation. But this does not avoid the lattice guided phase rebuilding with anisotropic molecular migrations, because the enormous local pressure that is created by geometrical mismatch in the bulk of the crystal upon photodimerization has to be immediately released. Clearly, the crystal cannot take any chance to behave like a topochemical or topotactical system, even if it is given the longest possible time for any alleged relaxation in a so-called "reaction cavity".

We have to stress this self-evident fact, because a still frequently cited paper [25] spread confusion with claims of a so-called "quantitative single-crystal-to-single-crystal conversion" in the α-cinnamic acid **1** photodimerization to give **2** upon polychromatic irradiation with wavelengths > 345 and > 360 nm. These obviously contained shorter wavelength light than the experiments in Figs. 2.1.1 and 2.1.5. Nevertheless, these authors called their experiment "tail irradiation" but did not appreciate the published absorption spectrum [20] (Fig. 2.1.2) and claimed to have saved "topochemistry". Unfortunately, topochemistry proponents still keep on citing this paper [25] while disregarding the content of the experimentally challenging papers [4, 9, 14, 21, 22] (the immediate and repeated attempt to challenge paper [25] in *J. Am. Chem. Soc.* was blocked). Importantly, it is not possible to achieve a conversion of higher than 30% without crystal disintegration, even with very thin

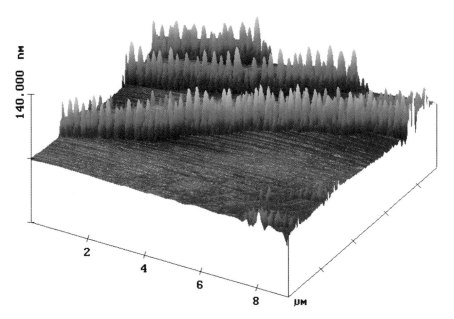

**Figure 2.1.5** 9 μm AFM surface of α-cinnamic acid **1** on (010) (between the crystallites) after 6 months of daylight exposure in a Pyrex vessel under argon through two additional window glass plates showing fence-like features along the cleavage plane direction.

crystals at extremely long wavelengths or with lower conversion upon monochromatic irradiations with shorter wavelength light (405 nm, 365 nm). Therefore, the claimed X-ray data up to so-called "100% conversion" [25] can at best be extrapolations. Equally severe is the data challenge: the reported X-ray data (both Supplementary Material for [25] and CIF in the Cambridge Crystallographic Database deposit) do not withstand standard scrutiny and do therefore not correspond to a natural structure. Due to the importance of the subject these data are analyzed in Fig. 2.1.6. It is easily seen that the data cannot in any way represent the packing of a real crystal species. There are severe C–H-to-C–H interactions of 2.033 Å (85% of the van der Waals radii). Furthermore, even labile polymorphs of carboxylic acids never disdain the high bond energy inherent in planar cyclic hydrogen bond dimers if they are not hindered from doing so (for example **70**, below). But Fig. 2.1.6 exhibits chair-like hydrogen bond dimer interactions. There is no reason whatsoever for the 68 pm displacement of the planes in such eight-membered rings of the carboxylic acid dimers as seen in Fig. 2.1.6. These eight-membered rings are not enforced to be non-planar: they calculate from the claimed data to go up and down alternatively in adjacent rows, but in reality these would choose to assume the stable planar rings for energetic reasons. Clearly, the claimed X-ray data [25] are invalid. Not unexpectedly, "tail irradiation" does not alter the three-step solid-state mechanism with long-range anisotropic molecular migrations, if the original lattice cannot accommodate the product molecules. This has been amply experimentally proved in Figs. 2.1.1, 2.1.4, and 2.1.5.

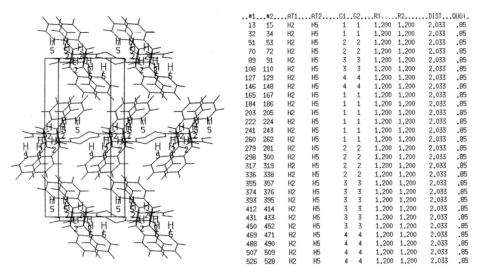

| ..#1 | .#2 | .AT1 | .AT2 | .G1 | .G2 | .R1 | .R2 | .DIST | .QUOI. |
|---|---|---|---|---|---|---|---|---|---|
| 13 | 15 | H2 | H5 | 1 | 1 | 1.200 | 1.200 | 2.033 | .85 |
| 32 | 34 | H2 | H5 | 1 | 1 | 1.200 | 1.200 | 2.033 | .85 |
| 51 | 53 | H2 | H5 | 2 | 2 | 1.200 | 1.200 | 2.033 | .85 |
| 70 | 72 | H2 | H5 | 2 | 2 | 1.200 | 1.200 | 2.033 | .85 |
| 89 | 91 | H2 | H5 | 3 | 3 | 1.200 | 1.200 | 2.033 | .85 |
| 108 | 110 | H2 | H5 | 3 | 3 | 1.200 | 1.200 | 2.033 | .85 |
| 127 | 129 | H2 | H5 | 4 | 4 | 1.200 | 1.200 | 2.033 | .85 |
| 146 | 148 | H2 | H5 | 4 | 4 | 1.200 | 1.200 | 2.033 | .85 |
| 165 | 167 | H2 | H5 | 1 | 1 | 1.200 | 1.200 | 2.033 | .85 |
| 184 | 186 | H2 | H5 | 1 | 1 | 1.200 | 1.200 | 2.033 | .85 |
| 203 | 205 | H2 | H5 | 1 | 1 | 1.200 | 1.200 | 2.033 | .85 |
| 222 | 224 | H2 | H5 | 1 | 1 | 1.200 | 1.200 | 2.033 | .85 |
| 241 | 243 | H2 | H5 | 1 | 1 | 1.200 | 1.200 | 2.033 | .85 |
| 260 | 262 | H2 | H5 | 1 | 1 | 1.200 | 1.200 | 2.033 | .85 |
| 279 | 281 | H2 | H5 | 2 | 2 | 1.200 | 1.200 | 2.033 | .85 |
| 298 | 300 | H2 | H5 | 2 | 2 | 1.200 | 1.200 | 2.033 | .85 |
| 317 | 319 | H2 | H5 | 2 | 2 | 1.200 | 1.200 | 2.033 | .85 |
| 336 | 338 | H2 | H5 | 2 | 2 | 1.200 | 1.200 | 2.033 | .85 |
| 355 | 357 | H2 | H5 | 3 | 3 | 1.200 | 1.200 | 2.033 | .85 |
| 374 | 376 | H2 | H5 | 3 | 3 | 1.200 | 1.200 | 2.033 | .85 |
| 393 | 395 | H2 | H5 | 3 | 3 | 1.200 | 1.200 | 2.033 | .85 |
| 412 | 414 | H2 | H5 | 3 | 3 | 1.200 | 1.200 | 2.033 | .85 |
| 431 | 433 | H2 | H5 | 3 | 3 | 1.200 | 1.200 | 2.033 | .85 |
| 450 | 452 | H2 | H5 | 3 | 3 | 1.200 | 1.200 | 2.033 | .85 |
| 469 | 471 | H2 | H5 | 4 | 4 | 1.200 | 1.200 | 2.033 | .85 |
| 488 | 490 | H2 | H5 | 4 | 4 | 1.200 | 1.200 | 2.033 | .85 |
| 507 | 509 | H2 | H5 | 4 | 4 | 1.200 | 1.200 | 2.033 | .85 |
| 526 | 528 | H2 | H5 | 4 | 4 | 1.200 | 1.200 | 2.033 | .85 |

**Figure 2.1.6** Trial plot of the resulting crystal packing as calculated from the published fractional coordinates and space group based on a 0.3–0.7 mm long crystal of α-cinnamic acid **1** after "tail irradiation" to give a claimed [25] "100% conversion" to α-truxillic acid **2**.

The three-step solid-state mechanism (phase rebuilding, phase transformation, crystal disintegration) in thermal and photochemical reactions has been shown to work well if all three essential steps succeed [9]. The first step (phase rebuilding) is structure dependent. If there is no possibility for molecular migration one has to crystallize a suitable polymorph. The stages two (phase transformation) and three (disintegration) can be engineered by the reaction conditions if the former are impeded, but this is not often necessary. For example, the additions of bromine and chlorine to *trans*-stilbene {cleavage plane (001), small channels along [010] and [100]} were known to be incomplete but could be brought to completion by previous milling of the crystals to < 1 μm grain size [26]. Suitable techniques are collected in Ref. [9]. It is very fortunate that the three-step mechanism also applies to gas–solid and solid–solid reactions that cannot be foreseen by topochemistry, severely hampering their development as already mentioned. This may be demonstrated with the quantitative gas–solid cascade reaction of benzohydrazide **3** with cyanogen bromide to give 2-amino-5-phenyl-1,3,4-oxadiazole hydrobromide **4**

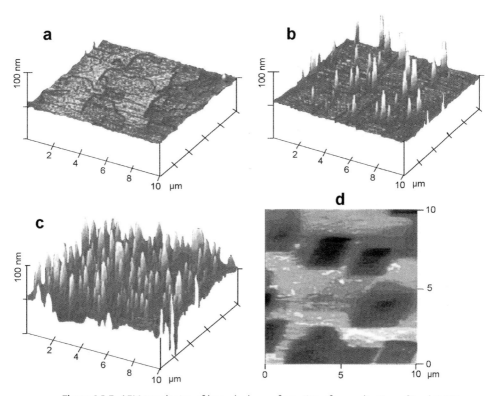

**Figure 2.1.7** AFM topologies of benzohydrazide **3** on the (100) face: (a) flat fresh surface with molecular terraces, (b) after application of 0.2 ml BrCN vapor; (c) after application of 0.4 ml BrCN, (d) phase transformation after application of 1 ml BrCN, showing rhombic "pool basins" of 60–100 nm depth at the expense of the initial gradual growth of "volcanoes"; the Z-scale in (d) is 400 nm [27].

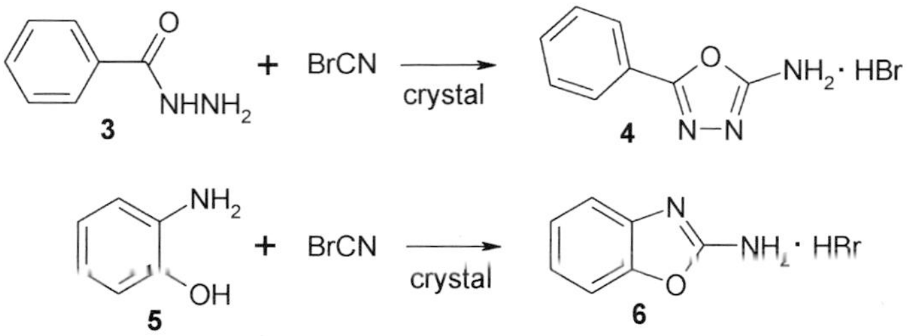

Scheme 2.1.2

(Scheme 2.1.2, Fig. 2.1.7) [27]. It can be nicely seen that there is an initially gradual growth of random volcanoes (phase rebuilding) that are suddenly replaced by the sharp edged rhombic "pool basins" with depths of 60–100 nm (phase transformation) (Fig. 2.1.7). Shortly thereafter, the reacted zone disintegrates from the crystal with formation of a fresh surface for a new reaction cycle etc. so that a complete reaction with 100% yield becomes possible in preparative runs.

The original crystal **3** packs in multiple hydrogen-bridged double layers that are parallel to (100) with very steep (71° in two directions) molecules [27]. The clear signs of an induction period for the penetration of the double layers additionally to the migrations in the (100) cleavage direction is well understood.

Rectangular "pool basins" are obtained upon phase transformation (second stage) if *o*-aminophenol (**5**) crystals (001) react with gaseous BrCN to give 2-aminobenzoxazole (**6**) in a cascade reaction (Scheme 2.1.2, Figure 2.1.8). Disintegration of the reacted zone follows. Again a double layer with steep molecules (73° but now arranged diagonally across at ± 45°) with the parallel (001) cleavage plane characterizes the crystal packing of **5** [27].

Sharp regular edges are not always obtained. The diazotization of 4-aminobenzoic acid (**7**) on (101) with nitrogen dioxide (Scheme 2.1.3) is a typical example for a very large increase in the surface features upon the sudden phase transition that follows gradual phase rebuilding (Fig. 2.1.9). Preparatively, this is a waste-free quantitative synthesis of the solid diazonium nitrate **8** in a complicated multistep cascade reaction [9, 28].

The molecular migrations are detected by AFM on the (101) face of a single crystal in Fig. 2.1.9. The initial surface is very smooth. Upon application of $NO_2$ gas there is continuous growth of volcano-type features in the starting period, as shown in the phase rebuilt face from an image series until suddenly a very large change (*Z*-scale change from 200 to 600 nm) occurs. Shortly after that the reacted zone

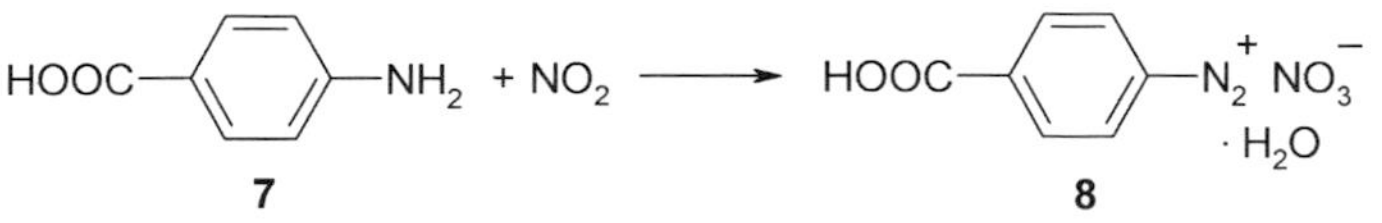

Scheme 2.1.3

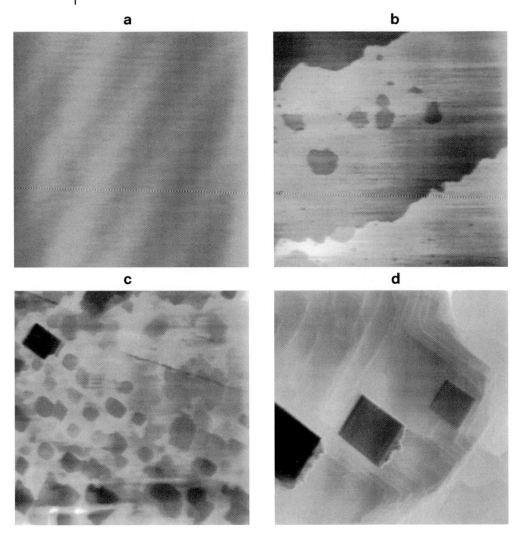

**Figure 2.1.8** 10 μm AFM topologies on (001) of 2-aminophenol
**5**: (a) fresh, (b), (c) and (d) after successive treatments with
BrCN vapor; the Z-scale for (a), (b), (c), (d) is 10, 10, 30, 100 nm,
respectively [27].

disintegrates from the crystal and forms a fresh surface to start again, and so on, in
preparative runs. Clearly, the three-step solid-state mechanism of phase rebuilding,
phase transformation and crystal disintegration occurs, as has been shown for
numerous other systems [9, 14].

In still further cases the phase transformation can flatten out the features from
the phase rebuilding stage. A typical example is the solid-state reaction of thiourea
(monolayer structure) on (001) with phenacylbromide [(100)-cleavage plane] giving

starting face          phase rebuilt face

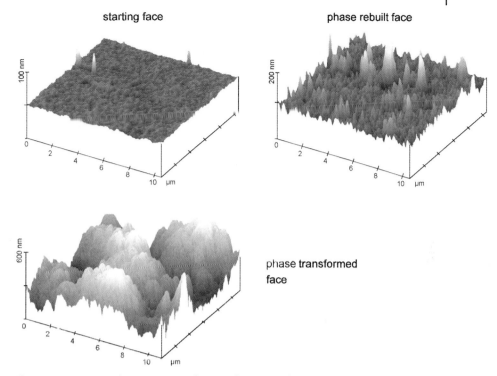

phase transformed
face

**Figure 2.1.9** AFM topologies on (101) of p-aminobenzoic acid **7**
(P2₁/n) [30]: (a) rather flat fresh surface, (b) after one application
of NO₂ gas, (c) after three applications of NO₂ gas.

the hydrobromide of 2-amino-4-phenylthiazole [29]. It is, however, always necessary that the product phase disintegrates from the crystal and thus creates a fresh
surface. If it does not disintegrate, as in several cases of flat cover formation, the
chemical reaction stops and the crystal is protected from further reaction by
passivation. Several cases are discussed in Ref. [18]. This type of stopped reactivity
can be circumvented by milling or salt formation. The chemical mechanism of
solid-state reactions is independent from the solid-state mechanism. This is already
clear from the multistep cascade transformations in Figs. 2.1.7–2.1.9 that follow
the same three-step solid-state mechanism as the one-step chemical reactions. For
predictions of solid-state reactivity of chemically possible reactions one has to
analyze the crystal packing of the starting material for the possibility of molecular
migrations in the same way as with thermal and photochemical reactions in
crystals. This does not exclude that in unusual cases an unreactive metastable
crystal structure might become reactive by a spontaneous phase transition to a
favorable structure, if catalyzed by the reagent. No such cases have come to mind
yet. We do not treat here the case of mixtures of polymorphs. Another complication
is described in Section 2.1.5 where the initial thermolabile photoproduct cyclo-
reverts and thus gives rise to an unexpected product upon further photolysis.

2.1.4
**Face Selectivity of Reactivity**

We have to distinguish face selectivity for reactivity and face selectivity for the shape of the features. AFM easily detects at the molecular level if certain faces are unreactive. Typical examples are given in the solid-state reactions of Scheme 2.1.4.

The reaction of benzimidazole (**9**) and cyanogen chloride gas gives *N*-cyanobenzimidazole (**10**) and HCl [27]. The plain (100) face of **9** is not attacked by ClCN, because the functional groups are hidden under shielding benzo-groups. However, the nitrogen atoms are freely available at the (010) and (001) side-faces. These are available at molecular steps on the (100)-face as in Fig. 2.1.10(a). It is clearly shown in Fig. 2.1.10(b) that the chemical reaction occurs only at the steps with formation of volcano type features up to > 300 nm. In ground material **9** all faces are available and a 100% yield of **10** is obtained [27].

Another example is the solid–solid pinacol rerrangement of benzopinacol **11** catalyzed by *p*-toluenesulfonic acid monohydrate polycrystals **12** to give benzopinacolone **13** (Fig. 2.1.11). The reaction proceeds on (100) of **11** where the crystal packing has the hydroxy-Hs down (crater formation) but not on (001) where the hydroxy-Hs are up. The catalyzing protonation occurs only on (100) where the free electron pairs of the hydroxy group are available. This is a highly remarkable result that has also been imaged at a *Z*-scale of 10 nm and at 0.1 mm to the acid crystal in

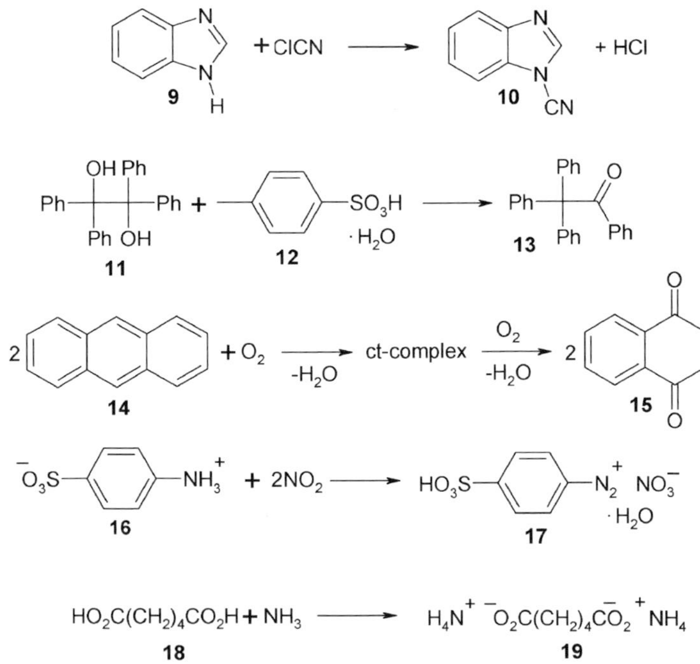

Scheme 2.1.4

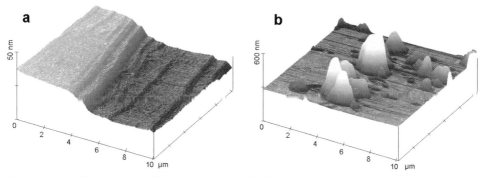

**Figure 2.1.10** AFM topologies of benzimidazole **9** on the (100)-face at a step site corresponding to the (001)-face where the functional groups are available: (a) before reaction, (b) same site after application of diluted ClCN gas, showing formation of high features at the step sites [27].

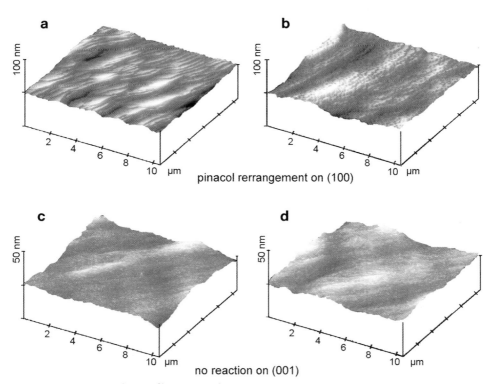

pinacol rerrangement on (100)

no reaction on (001)

**Figure 2.1.11** AFM topologies of benzopinacol **11**: (a) on (100), (c) on (001), (b) and (d) after placing a small polycrystal of *p*-toluenesulfonic acid monohydrate on the respective face, heating to 50 °C for 12 h and scanning at 0.5 mm distance from the acid polycrystal. The initial sites could not be precisely found again after the 12 h heating off AFM stage.

Ref. [31]. The craters on the (100) face and thus the anisotropic molecular migrations can be measured at up to 2 mm distance from the acid polycrystal.

Also self-protection of crystals against autoxidation in air is a molecular phenomenon that can be evaluated by AFM and SNOM [18]. For example, anthracene **14** crystallizes in thin scales that are stable against oxygen in air for at least two years, because their overwhelming (001) and their (110) faces hide both and one, respectively, of the reactive 9,10-positions in the outermost layer, so that no anthraquinone **15** is formed. However, the {110} faces of specially crystallized prisms of anthracene **14** with the same crystal structure tend to exhibit craters down to 400 nm in depth. Their slopes present both the 9- and 10-positions of **14** for autoxidation to give first the charge transfer complex of **14** and **15** and finally anthraquinone **15**. These packing effects of **14** are most sensibly unraveled with SNOM in combination with AFM (Fig. 2.1.12) [14, 23, 32]. AFM indicates tiny walls at the crater rims that might be indicative of reaction, but only SNOM (reflection back to the fiber mode) [18] exhibits chemical contrast that is basic in the flat parts (unreacted **14**), dark in the depths (light absorbing ct-complex **14·15**), and bright at the slopes and rims **15** (Fig. 2.1.12 (b)).

Further prominent examples for face selective reactivity (but for different reasons) are encountered with sulfanilic acid monohydrate **16** and 4-aminobenzoic acid **7**. They do not diazotize on their (010) face with $NO_2$ or on the (001) face with NOCl, respectively, where molecules cannot exit due to infinite hydrogen-bonded strings. Conversely, at slopes on (010) of **16** or on (100) of **7**, where the hydrogen-bonded amino groups of the strings are freely available, reaction occurs [18, 32, 33].

The reaction of carboxylic acids and phenols with ammonia and amines was studied by Curtin, Paul, Miller, and coworkers using X-ray data, light microscopy, and solid-state NMR [34, 35]. They found anisotropic behavior under the light microscope, but frequently could not correlate to what they tried to "predict" on the basis of topochemistry. The discussion of that failure was postponed and not

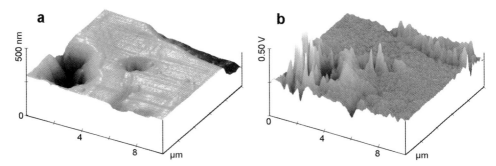

**Figure 2.1.12** (a) AFM on {110} of anthracene **14** with some craters, the slopes of which rarely exceed 20° (but 30° is also observed) after pre-exposure to ambient clean air for several weeks; (b) simultaneous SNOM image that indicates three types of contrast: normal at the flat sites, bright (positive) at the sites of anthraquinone **15**, and dark (negative) at the sites of the intermediate charge transfer complex (**14·15**).

taken up again, even after the advent of the phase rebuilding mechanism with its analysis of crystal packing. For example, the "extraordinary behavior" of adipic acid **18** upon forming its bis-ammonium salt **19** could not be explained. The answer has been provided in Refs. [14, 28] and is very clear-cut: AFM reveals that the reaction of adipic acid **18** with ammonia on (010) gradually forms "egg-tray" features and phase transforms into 1.6 μm high mountains that shortly thereafter disintegrate. The (100) surface of **18** also has carboxylic acid groups freely available and random volcanoes form, up to 50 nm height, (not seen under a light microscope) upon reaction with ammonia gas. However, these features do not change furter, even at 120-fold application of ammonia. The crystal packing reveals infinite strings of the hydrogen-bonded diacid under (100). Therefore, molecules cannot migrate out over that face once the irregularities at the hydrated surface are settled by salt formation and the reaction stops. Conversely, on the (010)-face the strings are hit by ammonia from their sides and the molecules can migrate upon reaction by releasing the imposed local pressure [14, 28]. The latter and the packing analysis are not taken into account by topochemistry as already pointed out in Section 2.1.2.

## 2.1.5
### Some of the Important Failures of Topochemistry and Their Remedy by the Experimental Mechanism

It is not surprising, that disregard of the local pressure and thus long-range anisotropic molecular migrations results in severe failures of topochemical "predictions" by claiming minimal atomic and molecular movements and by claiming a maximal distance of 4.2 Å for the reacting centers. It has been known since 1905 that anthracene crystals **14** undergo [4+4] photodimerization [36], but the distance is > 6 Å and the molecules are parallel displaced in the crystal [10, 14, 16, 37]. Also α-benzylidene butyrolactone **20** photodimerizes (3.666 Å, parallel double bonds) [38, 39], but both reactions exhibit long-range molecular migrations that are strictly correlated to the molecular packing. Conversely, the double bonds in 2-benzylidene cyclopentanone **21** do not photodimerize (4.123 Å, parallel double bonds) (Scheme 2.1.5). The parallel distances in **20** and **21** have been termed favorable for "topochemistry" but **21** is not reactive in the solid state. These are clear failures of topochemistry.

The crystal packing analysis clearly shows, that there is nothing special with these examples. Anthracene **14** has a layered structure with a (100) cleavage plane

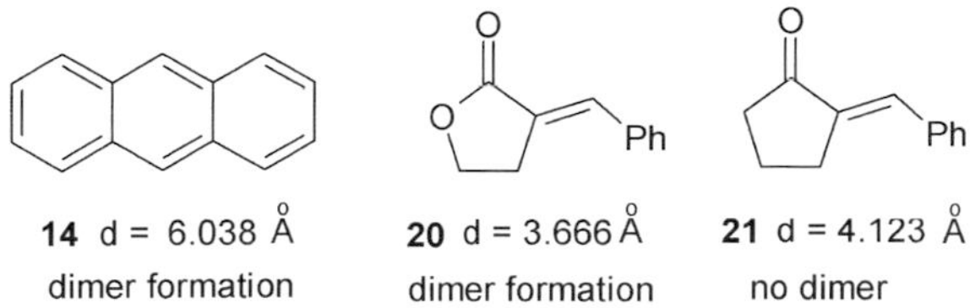

**14** d = 6.038 Å     **20** d = 3.666 Å     **21** d = 4.123 Å
dimer formation      dimer formation      no dimer

**Scheme 2.1.5**

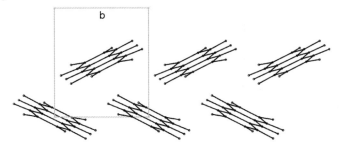

**Figure 2.1.13** Crystal packing diagram of anthracene **14** (P2$_1$/a) along [001]; the cleavage plane is (100).

(Fig. 2.1.13) and the molecules can slip within the horizontal rows and migrate (further plots in Ref. [14]). The lactone **20** exhibits layers in three directions with crossing cleavage planes (101) and (001) (Fig. 2.1.14). Furthermore it exhibits channels along [001] and [100]. The dimer bond formation is as indicated in Fig. 2.1.14, and molecules can use the cleavage planes for anisotropic migration, also with the support of the channels. Conversely, the crystal packing of benzylidene cyclopentanone **21** (Fig. 2.1.15) is so much 3D-interlocked that the anticipated dimer molecules could not use the small channels (neither those along [100]) for migration because they would hook together. Thus, the crystal packing impedes photodimerization of **21**.

There are many more failures of topochemistry, which one may not become aware from textbooks. The molecular formulae of some of these are shown in Fig. 2.1.16 [9]. For example, the first row solids with very short parallel double bonds are not photoreactive, while the second row solids with longer distances and the three solids with very skew arrangements form photodimers. Furthermore, in the last case three photocycloadducts arise in comparable yield rather than one and

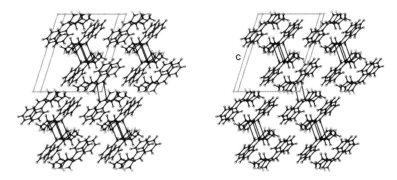

**Figure 2.1.14** Stereoscopic view of the crystal packing diagram of *trans*-benzylidene butyrolactone **20** (P2$_1$/n) on (010) but turned around Y by 10° for a better view; the dimer bond interactions are indicated by long bonds.

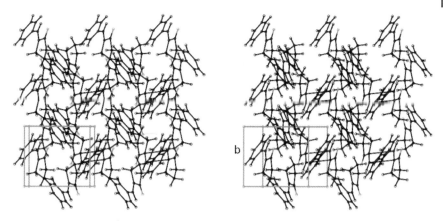

**Figure 2.1.15** Stereoscopic view of the crystal packing diagram of *trans*-2-benzylidene cyclopentanone **21** (P2₁/n) on (001); the anticipated but inactive dimer bond interactions (eight of them in the model) are indicated by long bonds.

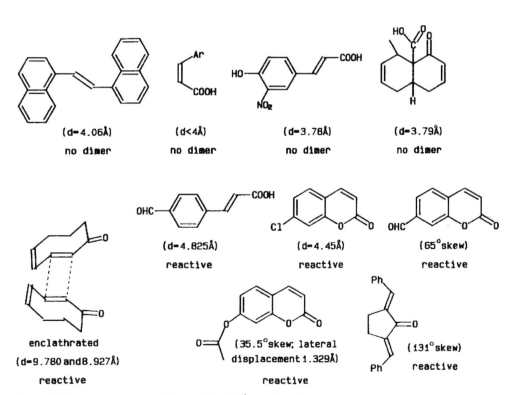

**Figure 2.1.16.** Some prominent failures of the 4.2 Å criterion of Schmidt's "topochemistry".

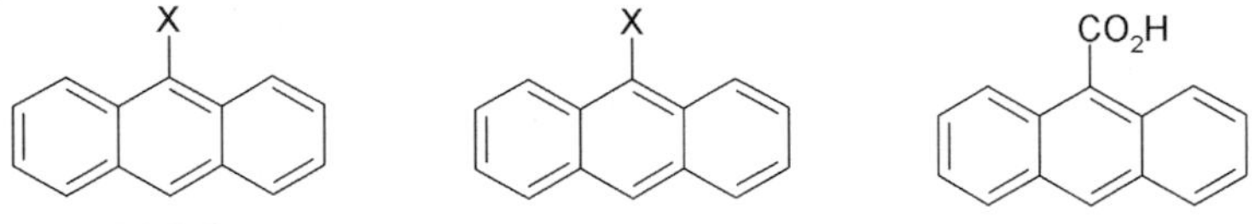

**22** P$2_1 2_1 2_1$         **23** P$2_1$/a         **24** P1

a: X = CHO: d = 3.93      a: X = CO$_2$Me: d = 4.15      d = 3.88
b: X = CN: d = 4.29       b: X = Br: d = 4.1
c: X = Cl: d = 4.0

packing: head to head     packing: head to head     packing: head to head
product: head to tail     unreactive                unreactive

**Scheme 2.1.6**

there are further examples for multiproduct solid-state photochemistry [39]. Interestingly, all of these and further examples [40] with known crystal structure data can be successfully analyzed and predicted by their crystal packing. The crystal packing clearly reveals whether or not molecules can undergo anisotropic migrations within the crystal. AFM studies provide the experimental evidence for anisotropic face selective molecular migrations or relevant facts concerning phase transformation and crystal disintegration if required. Several of these studies have already been performed. The common habit of topochemistry proponents to term all reactions that do not fit their hypothesis as "defect site reactions" and to immediately accept all reactions that would formally fit is neither scientific nor helpful for prediction of reactivity.

Very severe drawbacks and failures of topochemistry exist in the photodimerization of substituted anthracenes [41] (Scheme 2.1.6).

The monosubstituted anthracenes **22–24** with so-called ß-packing (head-to head arrangement) are either "unreactive" (**23 a,b** and **24**) or they form head to tail products if their crystals are irradiated (**22a–c**) that is opposite to the packing arrangement. The failure of head to head dimerization was termed a "topochemical abnormality" [41]. No satisfactory explanation for the formation of only the head to tail photodimers from **22a–c** has been given by topochemistry proponents for the failure of their "predictions", as all six examples (**22–24**) shown (and four further unreactive anthracene derivatives with ß-packing) seemed to be excellently arranged for the photoreaction in the crystals to form head to head [4+4]-dimers. Therefore, the three reactions with the "wrong" stereochemical outcome were termed "defect site reactions" by citing some papers that described excimer emission according to the Förster energy transfer mechanism (compare the same excuse with anthracene **14** and the exclusion of such a claim by SNOM [16], see also Fig. 2.1.12). Unfortunately, the many unreactive examples were left undiscussed and their X-ray data were not disclosed. All examples in the solid-state photodimerization of anthracene derivatives that seemed to fit to the topochemistry hypothesis were immediately considered as granted without further proof.

This totally unacceptable situation required an experimental approach. Clear answers were found from solution photolyses [42] and AFM studies [14, 37]. Firstly, it has been shown in solution that head to head [4+4]-photoadducts can be formed but that these are thermolabile. For example, 9-methylanthracene in benzene yields the *syn*- and *anti*-[4+4]-dimers in a 1:1.5 ratio at 40 °C when irradiated with 350 nm light. The *syn*-dimer was thermally cycloreverted upon heating in benzene. Clearly, the substituents in the above six anthracene derivatives **22–24** decrease the stability of their *syn*-dimers even further (due to radical stabilization effects) and that is the reason why these are not found in solution photodimerizations at room temperature. As the solid-state photolyses were performed at or above room temperature it must be concluded that the intended head to head photodimers are in fact formed upon photolysis but rapidly cyclorevert because they are not stable. This is the straightforward explanation for the three "unreactive" cases **23 a** and **b** and **24** (if they should happen to exhibit cleavage planes, but their crystal data are only guessed and not reported). Low-temperature photolyses of the $P2_1/n$ polymorph of **24** [43] [*syn*-packing, distance 3.897 Å, bilayers under (100), zigzag channel under (010)] and 9-bromo-10-methylanthracene $\{(P2_12_12_1)$ [44], syn-packing, C(Br)···C(Br)' distance 3.959 Å, (100) cleavage plane$\}$ should be tried.

Thermolability is also expected for the *syn*-[4+4] dimers of the compounds **22a–c** that give the "wrong" *anti*-stereoisomers upon solid-state photodimerization. We performed crystal packing analyses as the AFM studies on 9-H- **14**, 9-methyl-, 9-chloro- **22 c**, and 9-cyano-anthracene **22 b** all exhibit long-range anisotropic molecular migrations in phase rebuilding, phase transformation, and disintegration [37], irrespective of the topochemistry "predictions". The possibility for molecular migrations is evidenced for 9-cyanoanthracene (**22 b**) in Fig. 2.1.17 that indicates packing in monolayers with a comparably poor cleavage plane in between [45], but the adjacent vertical channels give enhanced free space to help the migrations.

How can it happen that the *anti*-[4+4]dimer forms instead of non-reaction due to the thermally unstable *syn*-[4+4]dimer? There are several possible explanations. But geometric constraints exclude rotation of the molecules around their long axis

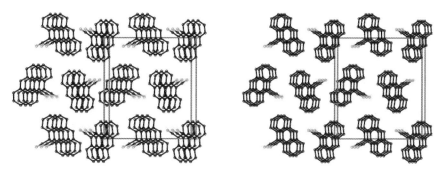

**Figure 2.1.17** Stereoscopic view of the crystal packing of 9-cyanoanthracene **22 b** on (001) but rotated around Y by 10° and without hydrogens for a better view.

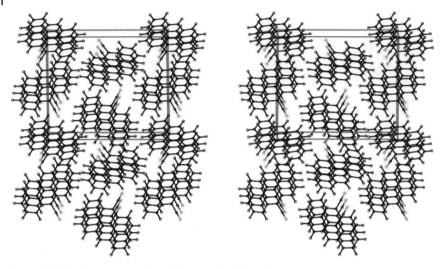

**Figure 2.1.18.** Stereoscopic view of the crystal packing of 9-cyanoanthracene **22 b** ($P2_12_12_1$) on (011) showing two central molecules that are surrounded from all sides with the possibility to rotate within its molecular plane.

within rows or migrations of molecules between neighboring rows for creating the suitable arrangement for *anti*-dimerization (Figs. 2.1.17 and 2.1.18). Clearly, the formation of head to tail dimers must be explained by initial formation of the *syn*-[4+4]-dimer. After this very likely event there exist two possibilities that cannot yet be distinguished. If the *syn*-dimer has a sufficient lifetime for anisotropic migration before its thermal reversion it will rebuild the phase and apparently form *anti*-pairs that allow formation of the stable *anti*-dimer upon further photolysis. If however the lifetime of the *syn*-dimer is too short for the time consuming migration it will decompose more or less at its original site via the most stable diradical [46] and lose energy by rotating one of the planar rings within the molecular plane. Such processes are well feasible within the crystal as long as the substituents are within the molecular plane (CHO, CN) or comparably small (Cl) but not with large (Br, $CO_2CH_3$) or hydrogen-bridged ($CO_2H$) substituents. This is shown in Fig. 2.1.18: a circle can circumscribe the flat molecules without severely interacting with the neighboring molecules. Thus, the crystal packing seen under (011) allows such proposal (Fig. 2.1.18). After such local creation of *anti*-pairs these can photodimerize to the stable *anti*-[4+4]-dimer, again with far-reaching anisotropic molecular migrations etc. Further detailed AFM studies in combination with synchrotron nanobeam diffraction and low temperature experiments will be required for a more detailed mapping of the occurring processes. It should be noted here that **23 a**, but as its $P2_1/c$ polymorph that crystallizes with close *anti*-pairs (so-called α–packing), provides the *anti*-[4+4]-dimer upon photolysis [41]. The necessary migrational ability is given by interpenetrating channels along [100], [010], and [001], that can be most easily found by a packing analysis.

Another subject that cannot be handled by a "minimal atomic and molecular movement" hypothesis concerns the numerous $E \rightarrow Z$, $Z \rightarrow E$, and *s-trans*→*s-cis* (**32**, **55**, **63**, below) isomerizations within crystals. Their direction and occurrence can only be predicted on the basis of molecular migrations along cleavage planes and channels or to voids (**63**, below). In many cases, the geometric double bond isomerizations are unidirectional if only one of the stereomers exhibits a crystal structure ready for molecular migrations. This has been exhaustively analyzed for all known systems with complete crystal structure data and already reported in Refs. [4, 6, 10, 14, 47–49]. The local pressures (geometric changes) are particularly high for these interesting reactions and far-reaching molecular migrations always correspond with the crystal packing and the three-step solid-state mechanism. For example, it has been rightfully predicted on the basis of the crystal packing that solid (*E*)-2-chloro-cinnamic acid (with cleavage plane for migrations) undergoes photoisomerization to the (*Z*)-compound [50], while the closely related solid (*Z*)-2-methylcinnamic acid (three-dimensionally interwoven molecules that cannot migrate) does not isomerize [4]. It was also studied in which cases a rotational cooperative mechanism might be envisioned (both ends in opposite direction by 45° to reach a transition state) [4, 10, 14, 48] and when the space conserving "hula-twist" mechanism (simultaneous double bond and adjacent single bond isomerization within a plane while only one C–H unit undergoes out-of-plane translocation) [51] or bicycle pedal movements are the only thinkable choices for the isomerization [4, 6, 47, 49].

## 2.1.6
## Molecular Migrations in the Absence of Severe Local Pressure

A particular situation is encountered if a solid-state chemical reaction leads to shrinking while the original crystal lattice is able to accommodate the product shape. In that case molecular migrations are not enforced by enormous local pressure. Such reactions are rare, as they require special conditions of the crystal packing, but they occur in linear dimerizations and polymerizations, where extended molecules lead to extended oligomers that do not severely disturb the original crystal structure. A number of crystalline 1,1-diarylethenes and 2-vinyl-naphthalene undergo acid catalyzed head to tail linear dimerization with *cis*-selectivity/specificity (Scheme 2.1.7) [52]. The crystal structure of **25** is known [53] (Fig. 2.1.19). It has the closest C1–C2' / C2–C1' distances for the head-to-tail interaction at 4.380 and 4.417 Å. The catalyzing proton can enter from the (100) face, but no addition of HCl occurs.

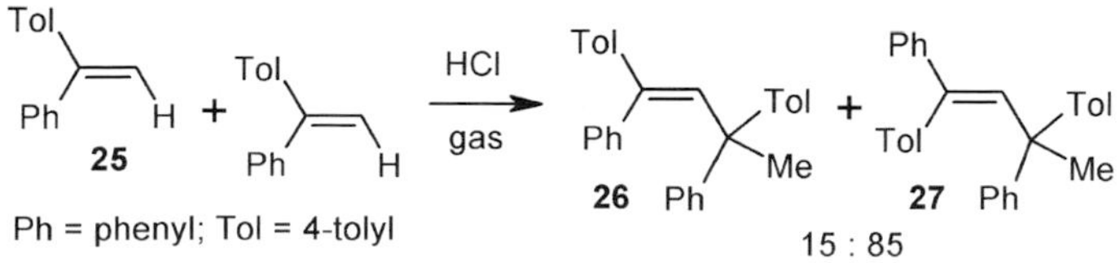

Ph = phenyl; Tol = 4-tolyl

**Scheme 2.1.7**

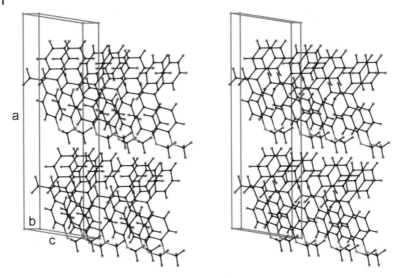

**Figure 2.1.19** Stereoscopic representation of the crystal packing of **25** (Cc) [53] on (010) but turned around Y by 10° for a better view showing layers of steep molecules (62°) that interlock in the cleavage plane, which is not suitable for migration.

The reason for the non-addition of HCl is the lack of suitable cleavage planes for molecular migration [26]. Figure 2.1.19 shows that there is considerable inter-locking in the cleavage plane and that the main axis of the molecules on it is very steep (62°). It remains linear head to tail dimerization with hydrogen migration (100% yield of **26** + **27**). In that case the dimer molecules transmit the catalyzing proton to the next molecule in the crystal (despite the short distances between C1 and C2' no polymerization continues in the original lattice). The linear dimers **26** and **27** do not need to use the unsuitable cleavage plane for migration but can be accommodated in the original structure, as they create space in the right direction and participate in the shrinking. AFM shows that no change occurs on the (100) surface of **25**, except for the formation of large craters without walls around the rims [26]. This also shows that the catalytic reaction proceeds around the sites of initial reaction (proton transfer to the next monomer). Local melting has to be efficiently impeded for complete reaction at −25 °C in this case. The *cis*-selectivity can be explained by easier bond libration when the larger anisyl group turns inward while forming the double bond. Unfortunately, no crystal structure is available for 2-vinylnaphthalene (powder data are available) [54] where *cis*-specificity is observed in the corresponding linear dimerization [52].

Thermal polymerizations of vinylic crystalline monomers giving isotactic or syndiotactic crystalline polymers are rare. More frequent are photolytical and radiolytical studies. The crystallographic packing requirements are most stringent. Clearly, the polymer chains cannot exit the crystal. They must therefore be able to accommodate with the monomer structure. Despite enormous efforts only a few suitable systems could be found for stereoselective polymerization, even though numerous systems have been found with parallel stacks at very short distance. Nevertheless atactic polymers were mostly obtained. This is a striking failure of the "topochemical" hypothesis and it is hard to understand why the necessary accommodation of the intended polymer with the monomer structure was not addressed [55]. Later, 1,3 dienes were found that gave stereoregular polymers and the crystal structures of these monomers were collected [56]. The latter indicated a close to 5 Å wide and 30–60° inclined stacking of the extended $E,E$ and $Z,Z$-dienes with respect to the polymerization axis. The importance of the 5 Å stacking distance was emphasized, it was found close to the fiber period of fully stretched polymer chains, and a translational problem was mentioned, but this was not mechanistically interpreted beyond "topochemistry". Interestingly, six dienes were listed with distances between the reacting centers of 5.3–5.7 Å that polymerized whereas ten systems with distances of the corresponding centers of only 4.2–3.5 Å did not polymerize. These were taken from a study of 14 reacting (such as **28** → **29**, Scheme 2.1.8) and 39 non-reacting crystalline diene structures [56]. Clearly, the favorable packing for reactivity could have been easily predicted (45° inclination at 5 Å stacking is good in most cases, it depends on bond lengths and bond angles) [52], because the polymer must accommodate with the monomer lattice, but "topochemistry" provides no basis for such prediction. Rather, it is decisive that the stacking width and the length of the polymer do not significantly differ from the monomer stacks, which is essential, as the polymer chain cannot migrate [52, 56a]. Clearly, crystal-packing analysis is appropriate, but not distance of reacting centers assessment. Thus, "topochemistry" again fails in its predictions. Some of these diene polymerizations might be truly topotactic if the polymer tacticity corresponds to the monomer arrangement and if there is a perfect fit of the periods so that no monomer migrations are necessary. But already a small mismatch adds up to large distances of monomer migration at considerable degrees of polymerization. AFM measurements would be able to distinguish different situations on the front face with molecular precision [56a].

Another claim of a thermal and photochemical homogeneous (or topotactic) polymerization of the diacetylene **30 a** (Scheme 2.1.9) is questionable. The claim had to be modified several times, it was admitted later that the polymorph required crystal solvent (e.g. 0.5 dioxane) [57], and **30 a** was no longer discussed in the review of the same research group [58] that dealt with **30 b** and listed 32 further diacetylenes. First order reaction was claimed in order to try to substantiate "topochemical" behavior [58]. However, inspection of the published kinetic curves indicates zero order up to 90% conversion after an induction period instead of first order for

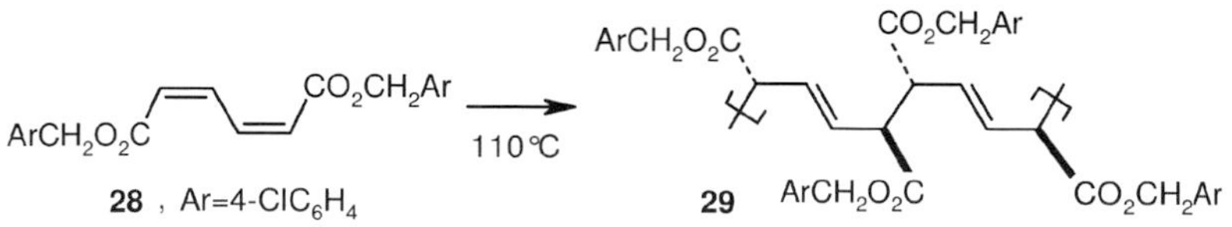

**Scheme 2.1.8**

30      **a**: R=CONHPh     31

**b**: R=SO$_2$(4-Tol)             **Scheme 2.1.9**

the thermal reaction. The *trans-tactic* crystalline polymer **31 b** was reported to have little defects. Most of numerous related diacetylenes with "suitably" short distances of the reactive centers and equal packing type failed to undergo thermal polymerization. This is again clear evidence that the continuously invoked topochemistry fails to "predict" reactivity. Rather the isolated successful results rest on the particular packing, as in the case of **30 b**. As shown in the formula scheme, there must be columnar stacking with an angle of close to 45° between the diacetylene moieties and the stacking axis at a stacking distance of close to 5 Å. This particular packing achieves a period of the monomer stacking that closely corresponds with the polymer fiber period. It is this condition that makes the polymer (that cannot escape the crystal) accommodate to the monomer structure. However, there remains a mismatch of the periods of –3.9 % (polymer shorter) [52, 56 a]. Therefore, there must be a lot of far-reaching anisotropic migrations of monomer molecules along the stacking axis upon polymerization in the stacking column (e.g. by ten stacking units at a degree of polymerization of 260). This should be easily detected by AFM in the form of deep craters on the (010) face. However, an AFM investigation did not scan this more important face but only reported "ridges along the polymerization axis" on (100) of thermalized **30 b** [59]. Clearly, predictions require detailed crystal packing analysis also with this type of molecular migration for obvious reasons that escaped attention on the basis of "topochemistry". It is to be predicted that a moderate negative mismatch in the periods will not be detrimental as monomer molecules can almost pressurelessly migrate to the growing polymer string within the favorable stack, largely without destroying the delicate arrangement. At increasing distances switches to neighboring stacks producing defects in the *trans-tactic* structure are to be expected. However, too much negative mismatch will soon destroy the favorable stacking. Conversely, positive mismatch must impede the *trans-tactic* polymerization. The pressure that would build up by the alleged polymer string provoked monomer migrations with the aid of cleavage planes or channels. But such phase rebuilding would locally destroy the delicate stacking for the *trans-tactic* polymerization. The stacking periods and angles of the non-reacting monomers are known. Periods of alleged polymers can be judged from standard bond lengths and angles or via *ab initio* calculations. Therefore, predictions of the mismatch type and thus reactivity for these and new systems are easily available from the experimental "phase rebuilding mechanism" with its detailed crystal packing analyses but not with the static distance considerations of the topochemical misconception. Much useless work for decades on self-generated pseudoproblems could have been easily avoided [see 4, 14, 26, 52, 56a].

## 2.1.7
## Multiple Cleavage Planes

The occurrence of interpenetrating cleavage planes increases the reactivity in solid-state reactions. Perhaps the most famous example is α-cinnamic acid with two cleavage planes, as depicted in Fig. 2.1.3. The quality of cleavage planes for enabling molecular migrations can vary considerably. Points of concern are whether they separate mono- or bi-layers, the degree of residual interlocking, their polarity, and the steepness of the molecules (the flatter the better). The most direct and promising approach for rating cleavage planes is the search for systems with various different cleavage directions and determination by AFM which ones are preferred for molecular migration at the expense of others. A suitable system is realized in the crystals of *meso*-3,4-dibromo-1,6-diphenyl-1,6 bis(*p*-tolyl]-1,2,4,5-hexatetraene **32**, which thermally isomerizes to give the bismethylene cyclobutene **34** undoubtedly via *cisoid* **33** (Scheme 2.1.10) [8]. Figure 2.1.20 shows the crystal packing of **32** in three directions and indicates three different cleavage planes by straight lines that cut the major face (010) at (a) 65°, (b) 61°, and (c) 83°. The first extends along *c* in the [*a* + *b*] direction, the second along the [*a* + *b*] and [*a* + *c*] directions, and the third along *b* in the [*a* + *c*] direction. All three cleavage directions can be executed by touching the plates of **32** with a needle on the overwhelming (010) face.

When judging the qualities of the three cleavage planes in Fig. 2.1.20 with respect to their migrational abilities we do not find much difference in the inter-locking between the adjacent layers or the tilt angle of the molecular axis with respect to the cleavage direction. However, there are significant differences in the Br$\cdots$Br distances in (a) with $x = 6.226$ Å, (b) $y = 3.940$ Å, and (c) $z = 3.940$ Å. Clearly, the space consuming internal rotation or a bicycle-pedal-like movement (to

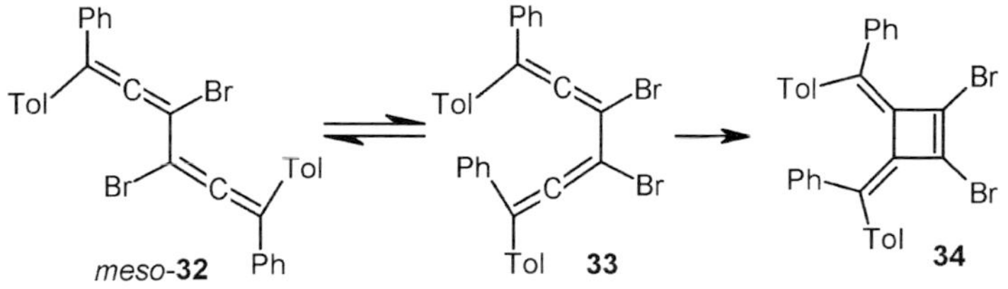

Scheme 2.1.10

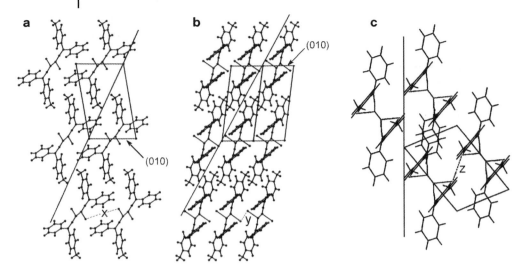

**Figure 2.1.20** Crystal packing of *meso-32* (P-1): (a) along [001] showing the direction of the skew (1−10) cleavage plane ($x$ = 6.226 Å); (b) along [101] showing the direction of the skew (-111) cleavage plane ($y$ = 3.940 Å); (c) along [010] showing the almost vertical (-101) cleavage plane cutting the (010)-surface diagonally ($z$ = 3.940 Å).

give **33**; a hula-twist-like process has been excluded by the overall stereochemistry [49]) and cyclization (to give **34**) should be more favorable in the proximity of the cleavage plane where the staple distance to the next molecules is largest. It is therefore predicted that the thermochemical processes and the pressure releasing molecular migrations occur at the (1−10) cleavage plane of **32**. Experimental verification of this prediction is provided by AFM investigation on its (010) surface, where all of the three cleavage directions end [8]. The exiting molecules from the bulk of the crystal align exclusively along the *c*-axis, the direction of the (1−10) cleavage plane. The other cleavage planes are not used. We have selectivity in the strict correlation of the molecular migrations with the crystal structure. These facts are of high importance, because they show, that the quality of cleavage planes for far-reaching molecular migrations might require scrutiny in such a way that their occurrence alone does not guarantee reactivity in unfavorable situations. The discovery of such borderlines is a matter of continuing active research. But the availability of cleavage planes (alternatively channels or voids) is an absolute necessity for non-topotactic reactions unless the effective volume of the products shrinks (Section 2.1.6).

It should be remembered here that cleavage planes are not the only means to facilitate far-reaching molecular migrations. Thus, the racemate of **32** undergoes the analogous solid-state reaction with migrations in channels [d/l-**55** (P2$_1$/n) in Scheme 2.1.11, Fig. 2.1.42, below].

2.1.8
**Various Types of Cleavage Planes**

Many solid-state reactions of all types rely on cleavage planes [1]. For example α-cinnamic acid **1** (Fig. 2.1.3) or anthracene **14** (Fig.2.1.13) with slightly inclined planar molecules in double or single layers are reactive in both solid-state photolysis and bromination [16]. The molecular packing of **35** is even more favorable for solid-state reactivity at its varied functionality (Fig. 2.1.21) as the molecules lie flat and as there is additional free space between the aromatic rings. The molecules **35** adopt the zwitterionic form with hydrogen bonds exclusively within the double layers.

Some further interesting systems may be predicted as solid-state reactive at all of the functionalities that are present, as there appear to be no restrictions with

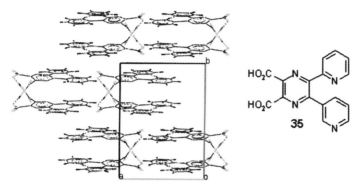

**Figure 2.1.21** Crystal packing of **35** (P2₁/c) [60] along [001] showing the (010) cleavage plane; hydrogen bonds are indicated.

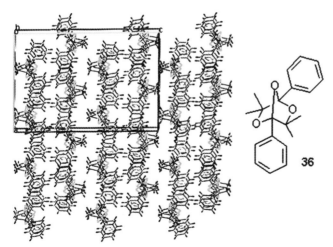

**Figure 2.1.22** Crystal packing of **36** (P2₁/c) [62] along [010] (slightly turned around X) showing the (001) cleavage plane with minor interlocking.

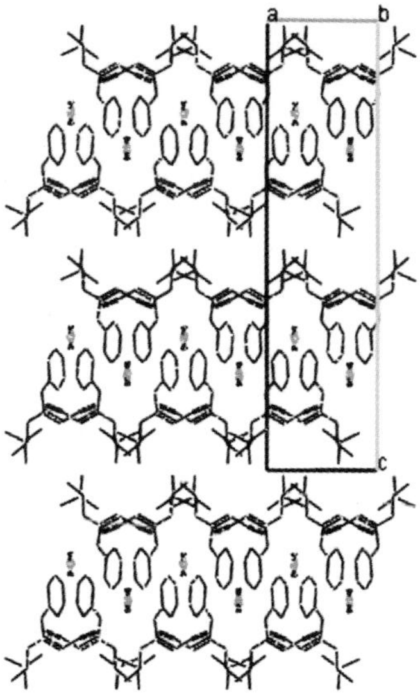

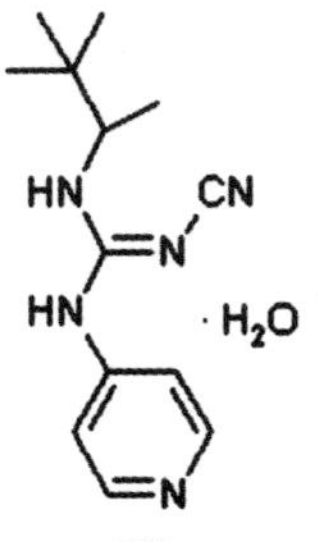

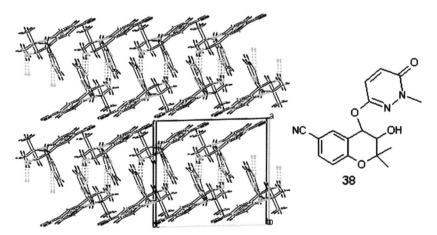

**37**

**Figure 2.1.23** Crystal packing of **37** (Pbca) [63] along [010] showing the nonpolar (001) cleavage plane with minor interlocking; the water molecules reside within the double layers.

respect to reaction types [1]. For example, the acetal functions and the phenyl groups of **36** should be reactive in suitable gas–solid and solid–solid reactions due to the cleavage plane between the double layers. The crystal packing under (010) has the main molecular axis of the 3D-molecules in the cleavage plane direction (favorable) (Fig. 2.1.22). The slight interlocking within the cleavage plane should not be detrimental for molecular migrations that are pushed by pressure as locally imposed by change in molecular shape upon chemical reaction. A comparable degree of interlocking is found in the (100) cleavage plane of the solid-state reactive

**38**

**Figure 2.1.24** Crystal packing of **38** (P2$_1$/c) [63] along [010] showing (100) cleavage plane with minor interlocking; hydrogen bonds are indicated.

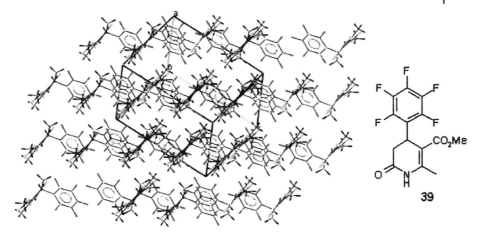

**Figure 2.1.25** Crystal packing of **39** (P-1) [65] approximately on (110) showing the cleavage plane between monolayers; hydrogen bonds are indicated.

phenacylbromide [29]. Interpenetrating although narrow channels along [001] and [100] in the $P2_12_12_1$ structure [61] also help.

A similar type of cleavage plane is found in the crystal packing of **37**. The crystal water is not in the cleavage plane but it occupies the bottleneck interconnected polar channels. Solid-state reactivity is predicted due to the nonpolar (001) cleavage plane (Fig. 2.1.23).

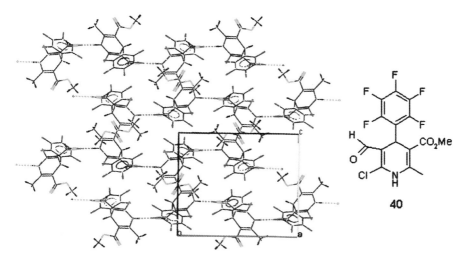

**Figure 2.1.26** Crystal packing of **40** ($P2_1/c$) [65] along [100] showing the (010) cleavage plane with weak hydrogen bonds between the adjacent alternating monolayers; hydrogen bonds are indicated.

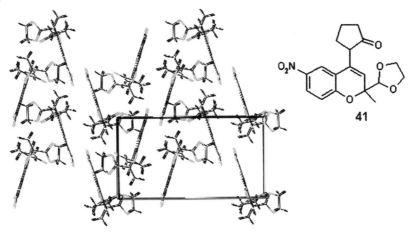

**Figure 2.1.27** Crystal packing of **41** (P2$_1$2$_1$2$_1$) [66] along [100] showing the (010) cleavage plane.

Waste-free solid-state syntheses might be particularly valuable in synthesis or derivatization of pharmaceuticals. For example, the potassium-channel activator **38** exhibits a cleavage plane that is reminiscent of the (010) cleavage plane of α-cinnamic acid **1** (Fig. 2.1.3) even though the hydrogen bond dimers in the double layers consist of 3D-molecules (Fig. 2.1.24). Clearly, solid-state reactivity is predicted and important chemical selectivities are expected.

The calcium channel modulators **39** and **40** might be suitable candidates for solid-state reactions. The skew cleavage plane between monolayers approximately under (110) in **39** (Fig. 2.1.25) appears very suitable for release of pressure by

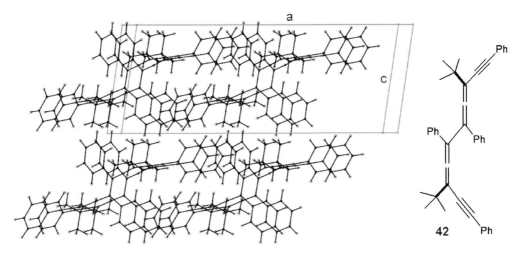

**Figure 2.1.28** Crystal packing of **42** (C2) on (010) but rotated around Y by 10° for viewing 8 molecules in the model; the very strong interlocking between the layers is impeding bulk reactions.

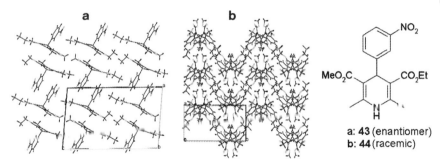

**Figure 2.1.29** Crystal packing of (a) **43** (P2₁) [67] along [010] and (b) **44** (P2₁/c) [68] along [001] showing 3D interlocking but no cleavage planes for molecular migrations in both cases.

molecular migrations and there are several functionalities that might be reacted. On the other hand, the monolayers in **40** are connected by hydrogen bonds between NH and aldehyde groups (Fig. 2.1.26). It is hard to predict if these relatively weak hydrogen bonds (distances N···O: 2.010, N–H: 0.860, H···O: 2.066 Å) can obstruct chemical reactivity, but there are also small channels along [001] in the structure that should be helpful. So it is worth trying with various reagents.

The (010) cleavage plane of the potassium channel activator **41** also appears suitable for solid-state reactivity. The 3D-molecules should have a chance to migrate (Fig. 2.1.27).

The bis-diallene **42** exhibits a monolayered structure (Fig. 2.1.28) [7]. The cleavage plane along the *a*-axis is highly interlocked. Clearly this cleavage plane does not allow molecular migrations, and thermolysis of **42** at 140–150 °C leads to a cyclization reaction only at the outer and inner surfaces with molecular migrations and phase transformation but not disintegration of the product phase in that case. This has been clearly shown by the AFM investigation [7]. None of the other cleavage planes discussed in this chapter has such a high degree of interlocking.

No cleavage planes are present in the crystals of the strongly 3D-interlocked calcium channel antagonist nitrendipine, both in the enantiomer **43** (P2₁) where channels along [100] are too small (not shown) and in the racemate **44** (P2₁/c) (Fig. 2.1.29). It is therefore not expected, that the crystals of nitrendipine (**43** or **44**) will autoxidize in air (except for a surface layer) or undergo other solid-state reactions.

## 2.1.9
## Channels

Channels in organic crystals are not uncommon. Evidently, such channels are genuine self-assembled nanotubes throughout the crystal in parallel well-ordered arrangement, the technical use of which awaits urgent exploration beyond solid-state chemistry. Channels in non-layered and often 3D-interlocked crystals are characteristically different from cleavage planes in layered structures, but there are

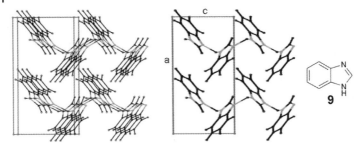

**Figure 2.1.30** Stereoscopic representation of the crystal packing of benzimidazole **9** (Pna2₁) [69] along [010] turned around Y by 2° for a better view, showing vertical stacks linked by hydrogen bonds forming heavily interlocked horizontal "bilayers" and almost square channels along [010].

also bottlenecked channels that resemble cleavage planes (for example Figs. 2.1.38–2.1.40 below). The differentiation may be useful, if the channel parts exhibit particularly wide diameters as compared to the bottleneck. Large enough channels can also be used for molecular migrations when local pressure due to chemical reaction has to be released. The first examples were revealed by AFM [10]. Thus, photochemical di-π-methane rearrangements of dibenzo-7,8-dicarbethoxybarrelene and dicyano-(1-cyclohexenyl)-(3-oxocyclopent-1-enyl)methane exhibit vertical channels under (001) and (010), respectively, and their photolysis leads to both steep sharp hills and steep sharp craters in both cases [10].

Channels can have very different shapes. Their cross-section can be large and small, round, oval (egg shape), elliptical, drop-like, square, rectangular, rhombic, hexagonal, sickle-shaped, two-chamber like, and "irregular". They may be intramolecular in stacked macrocycles or zeolitically interpenetrated by other channels.

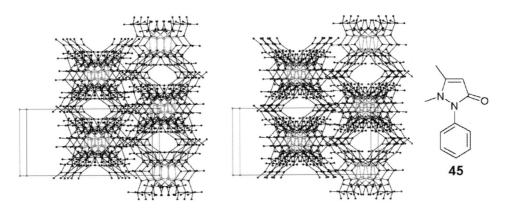

**Figure 2.1.31** Stereoscopic representation of the crystal packing of antipyrine **45** (C2/c) [71] along [001] showing elliptical channels that do not involve the nitrogen atoms, the projected plane calculates (-1 0 2.3); the (010) face is on top of the image.

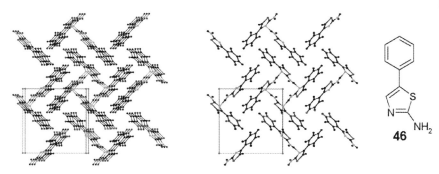

**Figure 2.1.32** Stereoscopic representation of the crystal packing of **46** (P4₃) [73] including hydrogen bonds, showing both square polar and rectangular nonpolar channels along [001].

A square channel cross-section is available in crystalline benzimidazole **9**. The well developed channels are depicted in Fig. 2.1.30. Therefore, the crystals of benzimidazole **9** have been predicted to be reactive in solid-state reactions, as the pressure induced by chemical reaction can be released by molecular migrations in the parallel channels along [010]. This has already been verified by the reactions with ClCN (Fig. 2.1.10) [27], triphenylchloromethane (alkylation) [29] and para-formaldehyde (to give 1-hydroxymethyl-benzimidazole) [1]. These are both gas–solid and solid–solid waste-free stoichiometric reactions without any solvent. Similarly, the waste-free solid-solid reaction of L-proline with paraformaldehyde to give L-N-hydroxymethyl proline [1] profits from irregular channels along [001] and [010]. Further solid-state reactions can be envisaged. Of course, solid-state secondary amine condensations with paraformaldehyde are not restricted to channel structures but occur also with layered structures (exhibiting cleavage planes). An example is the waste-free production of 1-hydroxymethylimidazole by stoichiometric milling of imidazole and paraformaldehyde at 0 °C [1].

The heterocyclic pharmacon antipyrine **45** exhibits highly interlocked double-layers with (001) cleavage planes. Particularly impressive are beautiful elliptical

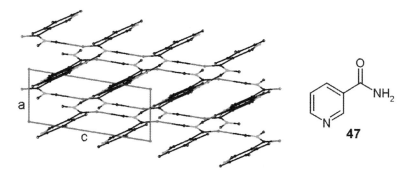

**Figure 2.1.33** Rhombic channels of nicotinic-acid amide **47** (P2₁/c) [74] along [010].

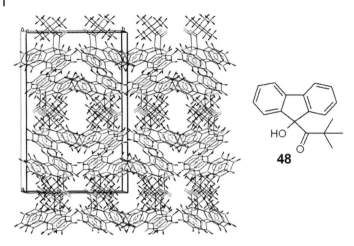

**Figure 2.1.34** Stretched elliptical and sickle-shaped channel structures of 9-hydroxy-9-(tert-butylcarbonyl)fluorene **48** (Pbcn) [76] along [001] but slightly turned around Y; hydrogen bonds are indicated.

channels along [001] (Fig. 2.1.31) that predict solid-state reactivity. This could be verified by reaction with bromine vapor which provides a quantitative yield of 4-bromoantipyrine hydrobromide [70].

The compound 4-phenyl-2-aminothiazol (**46**) exhibits a highly interlocked structure, but it can be seen in Fig. 2.1.32 that distinct polar square and unpolar rectangular channels occur. The high synthetic versatility of this compound [72] makes it a promising target for waste-free solid-state reactions.

Very prominent rhombic channels are seen in Fig. 2.1.33 along [010] of nicotinic-acid amide (vitamin B3) **47** that also exhibits smaller channels along [2–10] and [201], which should make it a good candidate for solid-state reactions.

The remarkable solid-state reactivity of phenylboronic acid [75] relies on polar rhombic channels that increase after breakage of the hydrogen bonds upon reaction. The reaction partners mannitol or *myo*-inositol also exhibit channels (along [001] or along [010] and [100]).

Further promising structures can be found in the Cambridge Crystallographic Database. Most of them include guest molecules and are only persistent in their presence, which may not always wipe out their solid-state reactivity. There are also so-called "zeolitic" structures without included guests. For example, 9-hydroxy-9-(tert-butylcarbonyl)fluorene **48** exhibits prominent stretched elliptical and sickle-shaped channels along [001] (Fig. 2.1.34). Numerous solid-state reactions can be envisaged. Real zeolitic structures exhibit connected channels in different directions though (for example **49**, **16**, **51**, **54**, **59**, **60** and cholesterol below).

Prominent "double-sickle" channels along [001] cross the "hanging drop" channels along [100] in the crystal of dibenzobarrelene (**49**) (Fig. 2.1.35) that also exhibits an interlocked (100) cleavage plane. Thus, solid-state reactivity of **49** is firmly predicted.

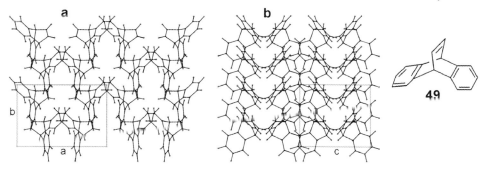

Figure 2.1.35 Double sickle (a) along [001] and hanging drop channels (b) along [100] of **49** (C2) [77].

Oval (egg-shape) channels occur along [001] of sulfanilic acid monohydrate **16** that exhibits a layered structure with hydrogen bond connection of the layers (Fig. 2.1.36) and additional rectangular channels along [100] and [010] ("zeolitic" inter-penetration). Its quantitative gas–solid diazotization has been studied prepara-tively, by AFM, and GID [15, 18, 32, 33]. Also the diazotization of water-free crystalline sulfanilic acid (P2$_1$/c) [78] profits from prominent channels along [001] [33], whereas 4-nitroaniline has only a flat channel along [001] and its diazotization profits from its (101) cleavage plane [79].

Very large hexagonal channels are known from hydrido-tris(bis-trimethylsilyl)a-mido-uranium(IV) **50** (Fig. 2.1.37).

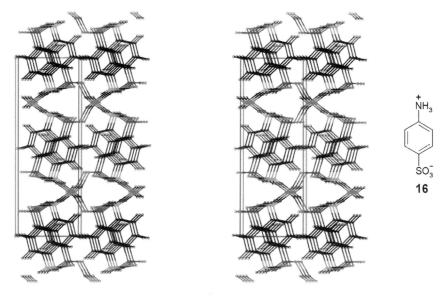

Figure 2.1.36 Stereoscopic representation of the crystal packing of **16** (P2$_1$/c) [80] along [001] including all hydrogen bonds, showing oval channels.

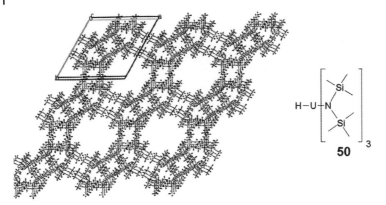

**Figure 2.1.37** Large hexagonal channels in the structure of hydrido-tris(bis-trimethylsilyl)amido-uranium(IV) (**50**) (P-3₁c) [81] along [001]; the uranium atoms are not shown, the structure is two-fold disordered in the Z-direction at the positions 2/3, 1/3, 0.310 and 2/3, 1/3, 0.190.

Very large channels are known from numerous porous MOFs (metal–organic frameworks). For example, exceptionally high capacities for storage of carbon dioxide at room temperature and pressures up to 45 bar have been claimed for some of these [82].

An interesting case of cyclic bottleneck-connected channels that are interpenetrated by cleavage planes in two directions is found for the disulfide **51** (Fig. 2.1.38). This sponge-like solid may be solid-state reactive, even though it contains intractably mobile molecules of tetrahydrofuran, dichloromethane, or other solvents, depending on the crystallization.

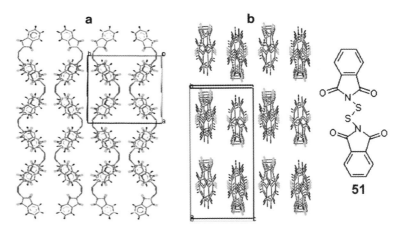

**Figure 2.1.38** Crystal packing of **51** (C2/c) [83]; (a) along [001] showing bottlenecked cyclic channels; (b) interpenetrating cleavage planes along [100]; these crystals contain intractably mobile molecules of solvents, for example tetrahydrofuran.

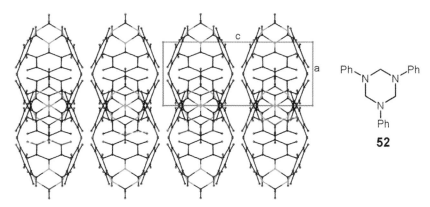

Figure 2.1.39 Bottlenecked rhombic channels along [010] of
52 (Pbcm) [84].

The rhombic bottlenecked channels of 1,3,5-triphenylhexahydrotriazin **52** along
[010] do not contain solvents (Fig. 2.1.39). They are closer to a genuine cleavage
plane and they enable the solid-state formation of *N*-phenylmethylenimine hydrochloride by reaction with HCl gas [1].

The calcium channel blocker **53** exhibits an irregular bottlenecked channel
structure that may also be similar to a cleavage plane (Fig. 2.1.40). This solid

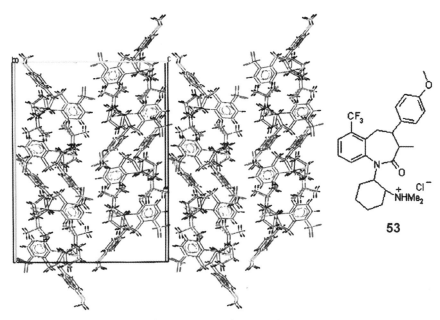

Figure 2.1.40 Crystal packing of **53** (Pca2₁) [85] along [010] but
slightly rotated around Y showing bottlenecked channels and
additionally smaller channels along [010]; hydrogen bonds are
indicated.

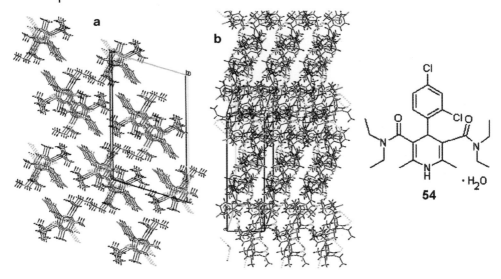

**Figure 2.1.41** Crystal packing of **54** (P2₁/n) [86] showing (a) "two-chamber" channels along [010] and (b) interpenetrating flat rectangular channels along about [5.2 1 1]; hydrogen bonds are indicated.

should be suitable for solid-state reactions. The smaller channels in the neighborhood are certainly helpful for the molecular migrations upon solid-state reactions.

The "two-chamber" channels in Fig. 2.1.41(a) interpenetrate the rectangular channels of the calcium channel antagonist crystals of **54** in Fig. 2.1.41(b). Such "zeolitic" quality could be suitable for molecular migrations (and thus solid-state chemical reactivity) even if these channels appear rather small.

Channels can be very irregular but that should not prevent them from enabling solid-state reactivity. A typical example is the stereospecific cyclization of the bisallene *rac*-**55** (P2₁/n) to give both the in,in- **57** and the out,out-cyclobutene **58** (Scheme 2.1.11). These are isomers of **34** as formulated for the solid-state thermolysis of *meso*-**32** (Section 2.1.7) [8]. Upon heating the crystals of *rac*-**55** to 130 °C grooves and later heights (about 100 nm high) and valleys align along the diagonal on the (-101) face between the natural (011) and (01–1) faces in the [100] direction. The sudden phase transformation leads to very high features in the μm range and cracks that are

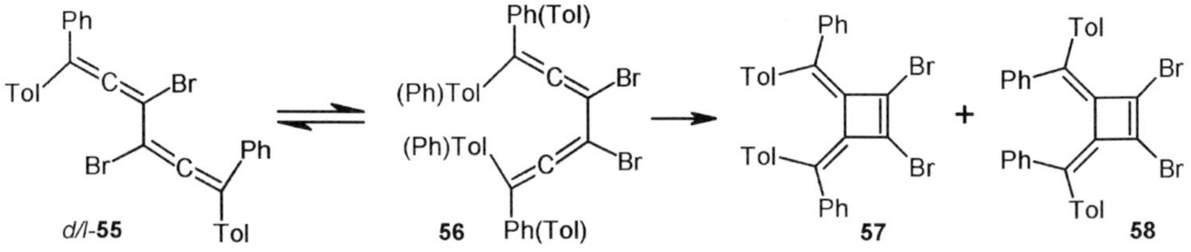

**Scheme 2.1.11**

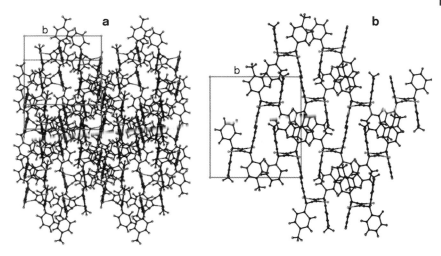

**Figure 2.1.42** Crystal packing of *d/l*-**55** (P2₁/n); (a) on (-101); (b) on (3.4 0-1) showing channels along [100] that end with skew cut on (-101) and give there rise to surface features in the AFM topology after isomerization [8] that align in the direction of the channels.

too deep for AFM investigation [8]. The crystal packing of *d/l*-**55** (Fig. 2.1.42) exhibits no cleavage planes but channels in the [100]-direction (Fig. 2.1.42(b)). These enable *s-trans* to *s-cis* isomerization to give **56** followed by cyclizations to give the products **57** and **58** with molecular migrations that end on the (-101) surface.

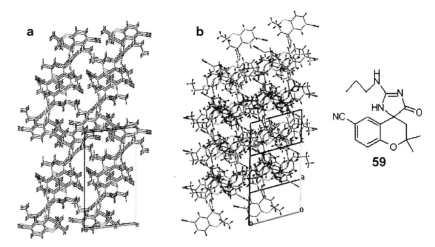

**Figure 2.1.43** Crystal packing of **59** (P-1) [87]; (a) along [100] but slightly rotated around Y; (b) along [-201]; irregular two-chamber and elliptical channel systems interpenetrate in this organic "zeolite"; hydrogen bonds are indicated.

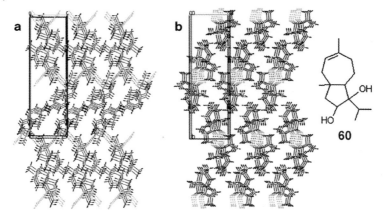

**Figure 2.1.44** Crystal packing of **60** (P2₁2₁2₁) [88]; (a) along [100] but slightly turned around X, (b) along [001] but slightly turned around Y; two irregularly formed channel systems of favorable size interpenetrate in a "zeolitic" manner; hydrogen bonds are indicated.

An example of a drug with potassium channel activity **59** exhibiting very irregular "one-" and "two-chamber" channels is given in Fig. 2.1.43. The "zeolitic" interpenetration of moderate size elliptical channels is very interesting and should enable solid-state reactivity.

The crystals of compound **60** from Arthrinium phaeospermum (**60** acting as a Maζ-K channel modulator) also exhibit irregularly formed channels of considerable size that interpenetrate and thus form an organic "zeolite" (Fig. 2.1.44). The size of

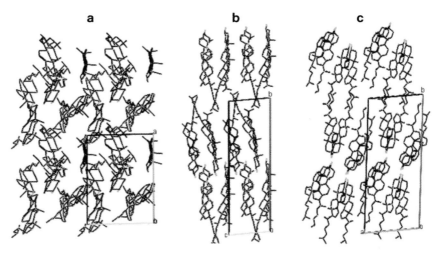

**Figure 2.1.45** Irregular channels through cholesterol (P1) [89]; (a) along [010], (b) along [100], (c) along [001]; H-atoms omitted for clarity.

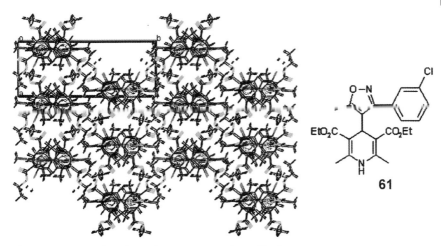

**Figure 2.1.46** Crystal packing of **61** (P2₁/c) [92] along [001] but slightly turned around X exhibiting small channels; the disorder of the methyl group from one ester substituent and hydrogen bonds are indicated.

these channels promises solid-state reactivity both at the hydroxy groups and at the double bond.

A very typical "zeolitic" organic material is anhydrous cholesterol. Its crystals are penetrated by channels along the three crystallographic directions *a*, *b*, and *c* (Fig. 2.1.45). It is therefore a good reagent for gas–solid and solid–solid reactions. This has already been verified by quantitative reactions with bromine [1] and oxalic acid [1].

The pure 5α,6ß-dibromide of cholesterol is quantitatively and waste-free formed upon reaction with the calculated quantity of bromine vapor [90]. Co-milling of cholesterol with a stoichiometric quantity of oxalic acid at room temperature gives the 1:1-complex, as is clearly indicated by characteristic changes in the IR spectra and the uniformity of the obtained crystals [1]. It is also possible to directly esterify crystalline cholesterol with oxalic acid. The diester quantitatively forms upon co-milling in a 2:1 ratio at 90 °C for 1 h without intermediate melting. Furthermore, briefly (10 min) co-milled components at room temperature can be heated for 1 h at 100 °C for a waste-free quantitative solid-state di-esterification. Interestingly, also the milled 1:1 complex provides only the diester and unreacted oxalic acid at 90 °C. Therefore, the intermediate monoester must be much more reactive in the solid state than the oxalic acid [1]. As the esterification produces water, there could be a change in the crystal structure of cholesterol into that of cholesterol monohydrate (P1) [91], which exhibits a cleavage plane and thus also enables solid-state reactivity. Further studies are necessary to clarify if such an autocatalytic process is actually responsible for this reaction.

Of course, there are also channels through 3D-interlocked structures that appear too small in diameter to enable solid-state reactivity. The calcium channel blocker **61** exhibits distinct channels along [001] (Fig. 2.1.46), but these might be too small or a limiting case for solid-state reactivity.

Small channels are certainly important in gas–solid reactions for permitting access of linear reagents to the bulk of crystals. For example, bromine often penetrates visibly into crystals, sometimes reversibly without reaction. Striking examples are reported in Ref. [1]. The *exo-* ($P2_12_12_1$) and *endo-* ($P2_12_12_1$) Diels-Alder adducts of cyclopentadiene and maleic anhydride exhibit small channels along [010] and [001], respectively. They also exhibit cleavage planes [intersecting (-110) with (110) and less prominent (0-12), respectively]. They therefore add bromine gas to their reactive C=C-double bond in gas–solid reactions. Conversely, the Diels-Alder adduct of 1,3-cyclohexadiene and maleic anhydride ($P2_1/c$) crystallizes 3D-interlocked and exhibits small channels along a, b, c (for additional voids see p. 134). If dilute bromine vapor up to 0.4 equivalents is applied to these crystals, the bromine is included without addition to the C=C-double bond, as there is no possibility for anisotropic molecular migration of the expected dibromide product in that lattice. The brown bromine can be evaporated at 20–50 °C without adding to the double bond. Upon application of concentrated or excess bromine the crystals melt at room temperature and this has to be avoided [1].

Macrocycles can form big intramolecular stacked channels such as the hexagonal ones of the α-cyclodextrin **62** (cyclohexa-amylose potassiumacetate hydrate) [93] (Fig. 2.1.47). These channels are formed as straight stacks of the voids inside the macrocycle. The smaller interconnecting channels between the macrocycles are filled with potassium acetate and water. This is a very favorable situation for channel formation as the molecules are interconnected by multiple hydrogen bonds straight and parallel behind each other (Fig. 2.1.47). Numerous solid-state reactions can be envisioned for **62**. However, these are to be classified as reactions at the inner surface of the aligned large rings. The latter are firmly fixed in their

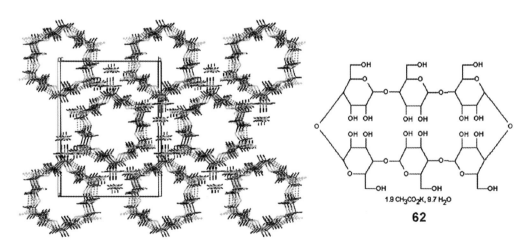

1.9 CH$_3$CO$_2$K, 9.7 H$_2$O

**62**

**Figure 2.1.47** Huge channel structure (including hydrogen bonds) of cyclohexa-amylose potassiumacetate hydrate (**62**) ($P2_12_12$) [93] along [001] (slight rotation around Y to visualize three layers) with potassium acetate and water in the channels between the macrocycles.

stacking. They will not leave before phase transformation and disintegration or such reactions could stay topotactic. Less interesting would be reactions with ring fragmentation.

The occurrence of macrocycles in cyclodextrins must not necessarily form channels in the crystals. The rings may arrange skew or displaced in various directions and obstruct each other. Crystal structures of water-free cyclodextrins are not available in the Cambridge Crystallographic Database. Cyclohexa-amylose hydrate (7.57 H$_2$O) [94] exhibits only small irregular channels due to skew displaced alignment of the molecular staples with the water molecules inside the rings, while other hydrates do not exhibit any molecular channels at all. Also in cyclohepta-amylose hydrate (12 H$_2$O) [95] channels are filled with water and they are small due to skew arrangement. In cycloocta-amylose hydrate (14 H$_2$O) irregular molecular channels occur that include the water. In cyclodeca-amylose hydrate (23.5 H$_2$O) [96] large channels are again present but with water molecules inside and the same is true for tetradeca-amylose hydrate (29.7 H$_2$O).

Numerous channel structures of host–guest compounds are known that owe their channel structure to the inclusion. Filled channels are not the subjects of this section, but displacement of guests by reagents followed by solid-state reaction might also be envisioned in channels that are engineered by inclusion.

## 2.1.10
### Closed Voids

The occurrence of significant closed voids (cavities) in three-dimensionally interlocked crystals is also able to remove chemically imposed pressure by accommodation of the product molecules and thus allows solid-state reactivity. These circumstances appear to be rare and might be easily overlooked. AFM measurements are not able to analyze processes with migrations to closed voids, as no features are occurring at the surface. The outer crystal shape might not change in such a situation, but nevertheless: such reactions must not be topotactic. Rather, amorphous product phases are most likely and this can be shown by loss of X-ray reflection intensity, in particular if the voids to be filled are the consequence of large and space demanding molecules that form the lattice [6]. A typical example is the 1:1-complex of the very spacious cis-stilbene **63** with acetone. The molecules pack in monolayers that are in the (100) and (010) planes, but they interlock considerably (Fig. 2.1.48). All hydrogen bonds are within the host–guest pairs. Very spacious molecules have difficulties with migration. They can, however, be reactive in the solid-state if their crystal structure contains nearby voids that can be used for accommodation of the geometrically changed molecules. Fig. 2.1.48 shows polar voids that are filled with acetone and nonpolar empty voids of similar size. Photolysis of crystals of the 1:1-complex of cis-3,3'-bis(diphenylhydroxymethyl)stilbene **63** gives the trans-isomer of the spacious stilbene derivative, as the local pressure can be released by short-range molecular migration. Internal rotation of the huge groups is not necessary in the crystal confinement, as the space conserving "hula-twist" or "bicycle pedal" mechanisms [51, 97] can be used.

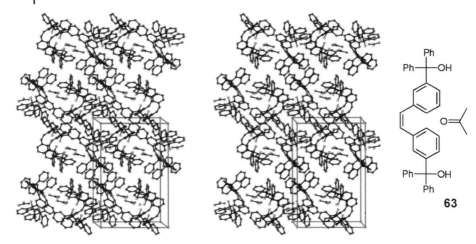

**Figure 2.1.48** Stereoscopic representation of the crystal packing
of **63** (P2₁/n) [6] on (001) but rotated around Y by 10° and
hydrogens omitted for a better view; the polar voids are filled with
acetone unlike the empty nonpolar voids.

However, a positive proof for hula-twist could not be obtained by X-ray diffraction, as
the product phase became amorphous. The external shape of the crystal did not
change at the microscopic level, but AFM indicated some loss of acetone on (100) by
efflorescence, forming a protective cover which can be correlated with the crystal
packing. Further molecular migrations on other faces were not detected with the
molecular sensitivity of AFM [6]. The crystal stayed clear transparent, but a topotactic
conversion is excluded if a crystalline phase becomes an amorphous product phase.

Other closed voids (cavities) have been found (in addition to small channels
along a, b, and c) in the crystals of *endo*-bicyclo[2.2.2]oct-5-ene-2,3-dicarboxyanhy-
dride (P2₁/c), the Diels-Alder adduct of 1,3-cyclohexadiene and maleic anhydride
[1]. The voids extend as tall three-axis ellipsoids under the skew faces of about
(01{10}) and (118). These crystals include up to 0.4 equivalents of diluted bromine
vapor without melting. No addition of bromine to the double bond occurs, as no
large channels and no cleavage planes are available for anisotropic migrations in
the 3D-interlocked structure. The bromine can be evaporated in a vacuum at 20–
50 °C without having attacked the double bond of the host [1]. It can enter into the
crystal and exit via the small channels. Numerous crystallographic closed voids are
to be expected in various crystals.

## 2.1.11
### Interpretation of Some Recent Literature Data

A series of crystalline arylidene oxindoles (**64** and **65**) has been irradiated with the
aim of photodimerization [98]. It was found that the (*Z*)-derivatives **65 a,b** were
photostable, the (*E*)-derivative **64 a** did not photodimerize (polymer mixtures were

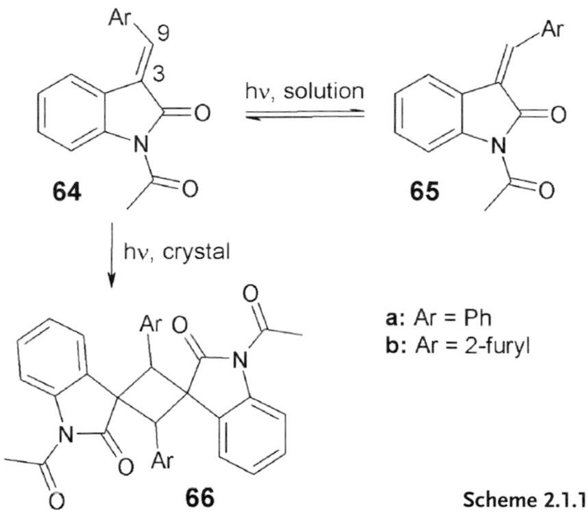

**64**   hv, solution   **65**

hv, crystal

Ar O

N

**66**

a: Ar = Ph
b: Ar = 2-furyl

Scheme 2.1.12

unspecifically formed very slowly in low yield), and only the (E)-derivative **64 b** effectively gave the head-to-tail dimer **66 b** (Scheme 2.1.12). This was said to be "in full accord with the topochemical principles" as the crystal structure of **64 b** was of the "α-type", but obviously despite the fact that the non-dimerizing **64 a** also exhibited the α-type structure. Furthermore, it was pretended that AFM studies (which include strict crystal packing correlation, Sections 2.1.3 and 2.1.4) would not be able to explain the non-reactivities [98]. Such statements disregard the ample evidence to the contrary (for example Sections 2.1.3–2.1.5) and are not tenable.

From the X-ray structural data [98] we calculate the distances (in Å) of inter-molecular double bond centers in possible dimer pairs of **64**, and **65**. They come out as follows:

**64 a**: C3–C3' = 3.988; C9–C9' = 4.216; C3–C9' = C3'–C9 = 3.876

**64 b**: C3–C3' = 3.795; C9–C9' = 3.705; C3–C9' = C3'–C9 = 3.498

**65 a**: not available

**65 b**: C3–C3' = C9–C9' = 4.585; C3–C9' = 3.951; C3'–C9 = 5.478

These values clearly prove that the distances of potentially reactive centers are unsuitable for predictions of reactivity. **64 a** (P2$_1$/c) and **64 b** (P2/n) have distances that are very close and also **65 b** (P2$_1$/c) has distances that are not out of reach for dimerization, as several examples with longer distances are reactive (Section 2.1.5). However, only crystalline **64 b** undergoes photodimerization, while crystalline **64 a** does not photodimerize, and crystalline **65 b** is photostable. Explanation of the different reactivities of **64 a** and **64 b** has been attempted with a difference of 2 kcal mol$^{-1}$ in cohesion energy of the packing or with a slight twist of the phenyl ring in **64 a** that was claimed to be responsible for "reactions only at defects" [98].

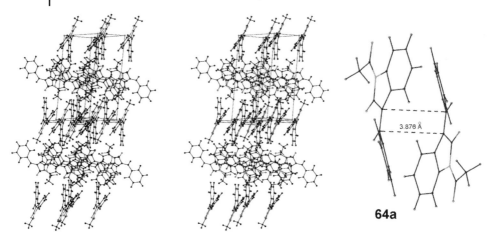

**Figure 2.1.49** Stereoscopic representation of the crystal packing of **64 a** on (110) showing layers that very strongly interlock and a pair of the closely arranged molecules that cannot photodimerize due to lack of migrational capability of the crystal.

Such minute energetic or steric differences are not at all convincing in photochemistry with a huge excess energy for deactivation.

Clearly, the crystal packing must be considered in order to understand these experimental data. As the geometry of the expected photodimers changes consid-

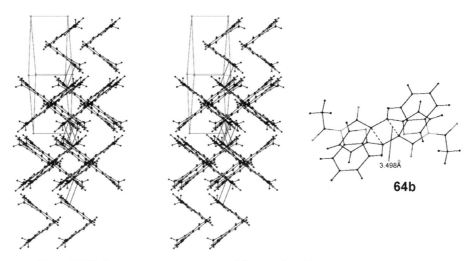

**Figure 2.1.50** Stereoscopic representation of the crystal packing of **64 b** on (-102) showing layers separated by cleavage planes next to the sickle-shaped channels and a pair of the closely arranged molecules that photodimerize due to the migrational capability of the crystal; the dimer interactions are indicated with long bonds in the dimer pair and in the crystal model.

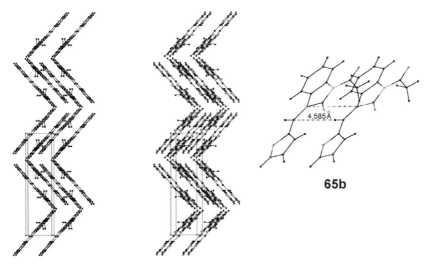

**65b**

**Figure 2.1.51** Stereoscopic representation of the crystal packing of **65 b** on (001) showing layers of steep molecules (48°, unfavorable) that slightly interlock in the cleavage plane and a pair of the closely arranged molecules that does not dimerize due to impeded migrational aptitude of the crystal.

erably with respect to the starting pairs of closely aligned molecules, it is essential that the pressure imposed on the lattice be immediately released by molecular migrations. The packing analysis (Figs. 2.1.49–2.1.51) shows that molecular migrations are only possible with **64 b** (Fig. 2.1.50), but not with **64 a** and **65 b**, in accordance with the experimental result. The crystal packing of **64 a** in Fig. 2.1.49 indicates a layered structure, but there is severe interlocking between the layers and the sharp (angle: 109°) zigzagged planes along [001] separating interwoven double layers is also not suitable. Therefore, the "expected" photoreaction cannot occur.

The packing of **64 b** is different even though the geometry of the starting pairs of closely aligned molecules is not very different from **64 a**. The horizontal cleavage planes and sickle-shaped channels are clearly seen in Fig. 2.1.50 and the molecules in the layers are not very steep (32°), which renders the crystal suitable for molecular migrations immediately after the single photodimerization events. Interlocked cleavageplanes under (010) may also help. As reported [98] an efficient reaction ensues.

The molecules of **65 b** (with $d = 4.585$ Å) are also suitably arranged for intended photodimerisation, because there are several reactive systems with even larger distance and less suitable non-parallel arrangement (Sections 2.1.5). The crystal packing in Fig. 2.1.51 clearly shows a cleavage plane with slight interlocking. However, the molecules are very steep on it (± 48°) and channels along [010] are too small, which impedes the sliding into the cleavage plane for migration. Apparently, the three-dimensional intended dimer molecules would not have a chance to leave their sites with a more significant turn into the direction of migration for releasing pressure. No reaction occurs.

Apparently, the quality of cleavage planes (Sections 2.1.7 and 2.1.8) plays its role here and it is to be assumed that a slope limit might be found between 32 and 48° for these typical examples, as we start with almost flat molecules in both **64 b** and **65 b**. Clearly, more examples have to be analyzed in order to definitely assess the borderline. Unfortunately, most of the early claims with cinnamic acid derivatives and their so-called α-, ß-, and γ-structures were not substantiated with published X-ray data and these were not available upon request. Finally, the steepness of molecules on cleavage planes is less important for $Z \to E$ isomerizations (Section 2.1.5). It should therefore be checked whether **65 b** undergoes as yet undetected solid-state photoisomerization at a suitable wavelength to give **64 b**.

## 2.1.12
### Applications in Addition to Solid-state Syntheses

The very high efficiency of gas–solid reactions [1] poses problems for the storage of crystals with suitable cleavage planes, channels, or voids in ambient atmosphere, as these are generally reactive, depending on their functional groups. Relevant reactive gases are also active at high dilution. These include water, carbon dioxide, acids, ammonia, amines, sulfur dioxide, nitrous oxides, ketones and other volatile organic solvents. The problems in storage stability might be most severe with pharmaceuticals that must retain the utmost purity upon storage and release. It

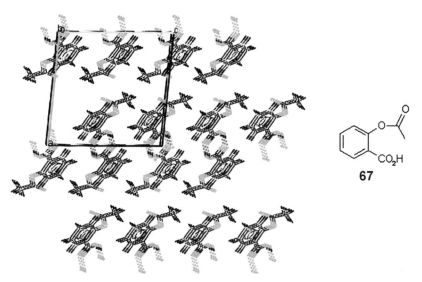

**67**

**Figure 2.1.52** Crystal packing of aspirin **67** (P2$_1$/c) [99] along [010] but slightly rotated around X and Y; the hydrogen bonds are only within the double layers and an excellent cleavage plane separates the layers.

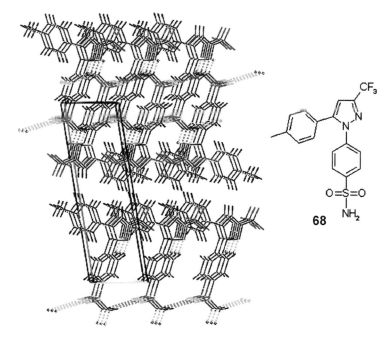

**Figure 2.1.53** Crystal packing of Celecoxib® (**68**) (P-1) (Ref. [100]) along [100] but slightly rotated around Y exhibiting hydrogen bonds within double layers and nonpolar cleavage plane in between with some interlocking.

should therefore be good practice to take care of the crystal packing of solid pharmaceuticals that will be exposed to ambient atmosphere. We select here some current and widely used analgesics, antirheumatics, and antiphlogistics that are mostly applied in tablet form. Their reactivity is judged according to the crystal packing.

It is well known, that acetyl salicylic acid (Aspirin®) **67** hydrolyses by the inevitable water content upon storage and that the liberated acetic acid (gas phase) catalyzes the gas-solid reaction. The crystal packing in Fig. 2.1.52 indicates double layers with ideal cleavage plane in between.

The drugs Celecoxib® (**68**) and Rofecoxib® (Vioxx, no longer on the market) **69** have been widely used as COX–II inhibitors against rheumatics. Their crystal packings are very different. **68** forms double layers with slightly interlocked cleavage plane between them but molecular migrations of the steep nonplanar molecules should be blocked due to difficulties in leaving the double layers (Fig. 2.1.53). Also the small irregular channels should not help much. Solid-state reactivity is thus not to be expected.

The crystal structure of **69** is 3D-interlocked leaving only rectangular channels that appear too narrow (Fig. 2.1.54) to enable molecular migrations and thus no solid-state reactivity is to be expected for this drug either. While these judgments

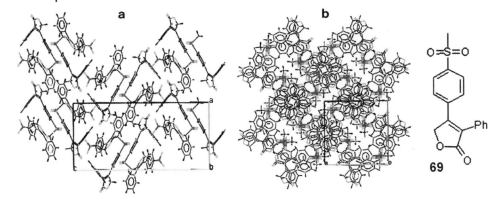

**a**        **b**

Figure 2.1.54 Crystal packing of Rofecoxib® (Vioxx) (**69**) (P4$_1$2$_1$2)
(Ref. [101]) along [100] (a) and along [001] (b), indicating heavily
interlocked packing and tall rectangular channels that appear too
small for allowing chemical reactivity in the solid-state.

require experimental testing the crystal qualities would appear well suited for air-stable solid drugs.

The antiarthritic drug naproxen® (*S*-**70**) has a fully interlocked structure, but it exhibits prominent irregular channels to both sides of an infinite linear hydrogen bond array of the carboxylic groups (Fig. 2.1.55). This should make it sensitive to exposure to proper atmospheric pollutants. Indeed, crystalline **70** reacts with gaseous ammonia, methylamine, or diluted bromine without melting [70].

Ibuprofen (*R/S*-**71**) is one of the most used antirheumatic drugs. Its crystal structure exhibits hydrogen bond connected double layers with a nonpolar smooth zigzag

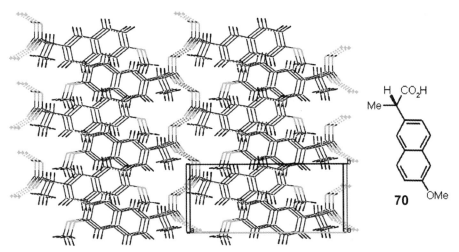

Figure 2.1.55 Crystal packing of (*S*)-**70** (P2$_1$) (Ref. [102]) along
[001] but slightly turned around Y for viewing the stacks, exhibiting
prominent irregular channels; hydrogen bonds are indicated.

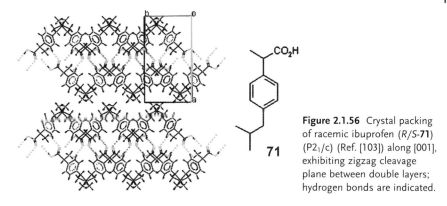

71

**Figure 2.1.56** Crystal packing of racemic ibuprofen (*R/S*-**71**) (P2$_1$/c) (Ref. [103]) along [001], exhibiting zigzag cleavage plane between double layers; hydrogen bonds are indicated.

cleavage plane between them (Fig. 2.1.56). The interlocking in the cleavage plane appears significant but molecular migrations and thus gas–solid reactions might not be totally impeded. It forms the ammonium salt with gaseous ammonia, which overcomes that hindrance, but it does not react with diluted bromine or nitrogen dioxide gas. Evidently, molecular migrations are stopped in that situation [70].

The more selective drug (*S*)-(+)-ibuprofen **72** does not exhibit cleavage planes between the layers that are linked by very strong hydrogen bonds but the nonpolar irregular channels might offer some reactivity potential (Fig. 2.1.57). Again this is a limiting case. While gaseous ammonia and methylamine give the salts in the gas–solid reaction, diluted bromine gas does not react with **72** in the solid state [70]. It should be noted here that excess bromine or nitrogen dioxide gas tend to liquefy the relatively low melting drugs **70**, **71**, and **72** and destroy these in the melt, which has to be avoided.

Three polymorphs of the most prominent antirheumatic drug diclofenac **73** have been described at room temperature. If these are long-term stable or metastable at

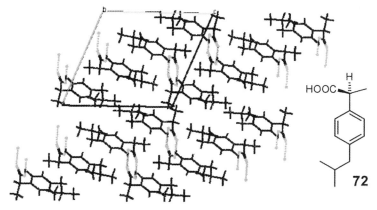

72

**Figure 2.1.57** Crystal packing of (*S*)-(+)-ibuprofen (**72**) (P2$_1$) (Ref. [104]) along [010], without cleavage plane between hydrogen bond connected layers but with irregular channels of moderate size; hydrogen bonds are indicated.

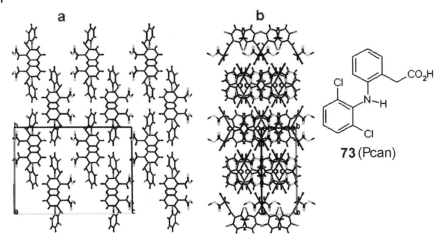

**Figure 2.1.58** Crystal packing of **73** (Pcan) (Ref. [105]); (a) along [100] exhibiting elliptical channels besides not hydrogen bonded carboxylic acid groups and (b) along [001] with interlocked cleavage plane.

room temperature and technically available, it can be easily judged, which polymorph is most suitable for storage and for tablets in ambient atmosphere.

The Pcan polymorph is unusual, as it does not exhibit hydrogen bonds between the carboxylic acid groups (distance O–H···O=C is 3.293 Å) (Fig. 2.1.58(a)). This is a double layer structure with interlocked (001) cleavage plane (b) and highly polar zone (a) with elliptical channels. Such packing appears to endow reactivity to atmospheric constituents. It should immediately hydrate in moist air.

The packing of the C2/c polymorph of diclofenac **73** exhibits regular acid dimer hydrogen bonds and rhombic channels along [001] (Fig. 2.1.59), which should endow it with solid-state reactivity.

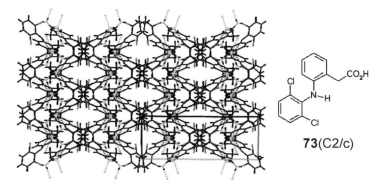

**Figure 2.1.59** Crystal packing of **73** (C2/c) (Ref. [106]) along [001] exhibiting prominent rhombic channels; hydrogen bonds are indicated.

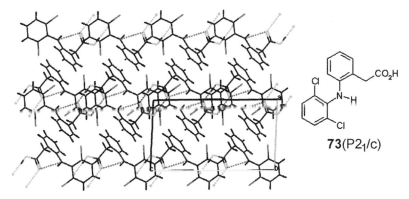

**73(P2₁/c)**

**Figure 2.1.60** Crystal packing of **73** (P2₁/c) (Ref. [106]) along [010] exhibiting 3D-interlocked structure without significant channels; hydrogen bonds are indicated.

Finally, the P2₁/c polymorph of **73** (Ref. [106]) with regular acid dimer hydrogen bonds is best suited for storage and exposure to ambient atmosphere, as it has a 3D-interlocked packing without cleavage planes (Fig. 2.1.60). The irregular channels appear too small with respect to the molecular size to be significant. This polymorph is therefore expected to stay unchanged in ambient atmosphere or against purposeful gas–solid or solid–solid reactions as long as it does not undergo (catalyzed) phase transformations.

Crystals of the analgesic and antipyretic paracetamol (4-hydroxyacetanilide, **74**) (P2₁/a) [107] exhibit comparatively small channels. However, these might appear large enough to take up water vapor followed by phase transformation to the monohydrate of **74** (P2₁/n) [108] with larger channels that should be more favorable for solid-state reactivity. Such hydration might also occur with **67** or **70–73** (hydrate structures not known in CCDC) and change their properties, as do complexations of solid drugs (for example with caffeine, etc.). Clearly, the preparations for the administration forms of the actual pills can change the prediction of atmospheric stability according to different crystal structures that must be separately studied and analyzed.

## 2.1.13
### Conclusions and Outlook

This chapter dealt with molecular crystals but not with infinitely covalent crystals or polymers, where mechanochemistry (or tribochemistry) occurs upon milling. Covalent bonds are broken in mechanochemistry of crystals and that produces surface plasmas, which can, for example, be used for the mineralization of all organic materials (including TCDD and methane) [109, 110]. An experimentally confirmed prediction is that three-dimensional infinitely covalent crystals (such as $SiO_2$, $Fe_2O_3$, glass, ceramics, etc.) are most efficient. Mechanochemical reactions have to be strictly distinguished from pure surface reactions and from solid-state reactions in molecular crystals where no covalent bonds are broken by mechanical interaction, unless a Bridgman's anvil is used.

The "topochemical principle" or in short the so-called "topochemistry" for molecular solid-state reactions has no physical basis, as it neglects the necessity for immediate removal of local pressure by molecular migrations along cleavage planes or channels or into voids of the crystal. Therefore, so-called "predictions" on the basis of "minimum amount of atomic and molecular movements" failed. Some apparent "success" was trivial, as there are high chances that near planar molecules crystallize at short distance, which in most cases (exceptions in Section 2.1.6) cannot disfavor the reactivity, but this is self-evident irrespective of the correct solid-state mechanism. The disproval of "topochemistry" in various instances is crucial. The most important drawback of topochemistry misconception for synthesis and environment has been its inability to "predict" or handle particular versatile gas–solid and solid–solid reactions. As these are by far the most environmentally benign techniques of synthesis with now more than 1000 published 100% yield waste-free reactions [1], the high acceptance of the seemingly simple "topochemistry" was highly counterproductive. For example, in the 1980 s referees and editors of a renowned journal rejected important papers that reported the first of the waste-free 100% yield gas–solid reactions despite the direct comparison with the wasteful low yield solution counterparts. The dictum was: "We do not need gas–solid reactions, the same reactions also occur in solution". Thereafter, new products that can not be prepared by any other technique suffered from similar reports. These and all the other waste-free solid-state syntheses could be mechanistically understood after the application of AFM to the reacting crystals became possible. It clearly revealed that molecules must anisotropically migrate far distances within the crystal for immediate release of local pressure (or less frequently for geometric reasons, as discussed in Section 2.1.6) in all types of solid-state reactions (thermal, photochemical/radiolytical, gas–solid and solid–solid). Only the very few genuine topotactic reactions [4, 10] do not develop either local pressure or voids (Section 2.1.6) and therefore they do not exhibit far reaching molecular migrations. Clearly, the crystal packing has to be considered if predictions of reactivity are desired. If no migration is possible, no reaction can occur. Crystal packing analysis may appear more involved than just looking at distances of reaction centers. However, the experimentally based effort of crystal packing analysis is the only successful way. Consequently, SNOM and GID and nanoscratching were also applied to multiply support for the AFM results and their obvious interpretation, because the experimentally founded phase rebuilding mechanism was a change in paradigm.

This chapter collects various appearances of cleavage planes, channels, and voids in organic crystals and further (including some inorganic) examples are reported in Ref. [111]. It also tries to point out borderlines for reactivity on the basis of experimental material. Packing analysis is well possible with generally available and easy to use software such as Shelxl-97, Schakal, Mercury from CCDC, and other software packages. The crystal packing analysis for cleavage planes, channels, or voids is the most important step in the prediction of reactivity, as molecules must be able to anisotropically migrate in the phase rebuilding step, given a certain polymorph. However, reactivity also depends on the reagent, unimpeded phase transformation, detachment of the product phase, and there must be no local melting. Thus, AFM is still essential if a positive reactivity prediction should fail in a first experimental trial, to find the reasons and engineer ways around these possible obstacles in step two or step three of the experimental solid-state mechanism. Such obstacles are fortunately not very frequent. Examples of the engineering have been published, both for facilitating phase transformation and disintegration.

Of course, there are still limiting cases for the judgment of migrational ability in the cases of difficult molecular shape and size, less favorable cleavage planes, short diameter channels, multiple short diameter channels, and small voids. These limiting cases are of the highest interest and merit detailed study with the aid of local techniques such as AFM.

Further improvement of the knowledge about anisotropic migrations within molecular crystals can also be obtained by their purely mechanical induction using nanoscratching and nanoindentation, techniques that have been comprehensibly treated [18]. Conversely, averaging (non-local) techniques such as formal kinetics, IR, UV/Vis, solid-state NMR, time-dependent X-ray diffraction, etc. are not very helpful in detecting the solid-state mechanism of non-topotactic reactions that are by far the rule. Previous kinetic plots have been mistreated several times (e.g. kinetic first order claimed for undoubtedly zero order data, as "first order kinetics" was obscurely thought to "prove topotactic behavior"). Formal kinetics does not tell anything about the local events in solid-state reactions that can only be studied with nanotechniques, GID, and emerging nanobeam X-ray diffraction. But experimental evidence shows that well-behaved gas–solid and solid–solid reactions tend to occur with kinetic near zero order and thus come to rapid completion with 100% yield in stoichiometric reactions without excess of reagent. The three-step solid-state mechanism creates a fresh surface by disintegration, and crystals that become smaller and smaller react more rapidly than the initially larger ones. Very small (micronized) crystals even tend to react explosively in gas–solid reactions, a fact that has to be accounted for when running these. It must be stated again: topotactic reactions are very rare exceptions that must be secured by AFM with molecular precision and they must provide a product crystal rather than an amorphous object with the shape of the original crystal.

The systematic elucidation of solid-state reactions both in synthetic and mechanistic terms opened the important field of waste-free sustainable chemical production with many more applications for environmental protection that can now be safely predicted.

# References

**1** G. Kaupp, *Top. Curr. Chem.* **2005**, *254*, 95–183; http://www.springerlink.com/index/10.1007/b100997.

**2** V. Kohlschütter, *Z. Anorg. Allg. Chem.* **1919**, *105*, 1–25.

**3** M. D. Cohen, G. M. J. Schmidt, *J. Chem. Soc.* **1964**, 1996–2000.

**4** Review: G. Kaupp, *Curr. Opin. Solid State Mater. Sci.* **2002**, *6*, 131–138.

**5** H. Nakanishi, W. Jones, J. M. Thomas, M. B. Hursthouse, M. Motevalli, *J. Phys. Chem.* **1981**, *85*, 3636–3642.

**6** K. Tanaka, T. Hiratsuka, S. Ohba, M. R. Naimi-Jamal, G. Kaupp, *J. Phys. Org. Chem.*, **2003**, *16*, 905–912.

**7** G. Kaupp, J. Schmeyers, M. Kato, K. Tanaka, N. Harata, F. Toda, *J. Phys. Org. Chem.*, **2001**, *14*, 444–452.

**8** G. Kaupp, J. Schmeyers, M. Kato, K. Tanaka, F. Toda, *J. Phys. Org. Chem.*, **2002**, *15*, 148–153.

**9** G. Kaupp, *CrystEngComm* **2003**, *5*, 117–133.

**10** G. Kaupp, *Adv. Photochem.* **1995**, *19*, 119–177, with 16 color tables.

**11** M. D. Cohen, G. M. J. Schmidt, F. I. Sonntag, *J. Chem. Soc.* **1964**, 2000–2013.

**12** G. M. J. Schmidt, *J. Chem. Soc.* **1964**, 2014–2021.

**13** G. Kaupp, D. Matthies, *Chem. Ber.* **1986**, *119*, 2387–2392.

**14** G. Kaupp, *Supermicroscopy in Supramolecular Chemistry: AFM, SNOM, and SXM*, in Comprehensive Supramolecular Chemistry, Vol. 8, J. E. D. Davies (Ed.), Elsevier, Oxford, **1996**, pp. 381–423 + 21 color plates.

**15** A. Herrmann, G. Kaupp, T Geue, U. Pietsch, *Mol. Cryst. Liq. Cryst.* **1997**, *293*, 261–275.

**16** G. Kaupp, *Chem. Unserer Zeit* **1997**, *31*, 129–139.

**17** G. Kaupp, M. R. Naimi-Jamal, *CrystChemComm.* **2005**, *7*, 402–410.

**18** G. Kaupp, *Atomic Force Microscopy, Scanning Nearfield Optical Microscopy, Nanoindentation and Nanoscratching – Applications to Rough and Natural Surfaces*, in Springer Series NanoScience and Technology, Springer Verlag GmbH, Heidelberg, **2006**.

**19** G. Kaupp, M. Plagmann, *J. Photochem. Photobiol. A: Chem.* **1994**, *80*, 399–407.

**20** G. Kaupp, *Angew. Chem. Int. Ed. Engl.* **1992**, *31*, 592–594.

**21** G. Kaupp, M. Haak, *Mol. Cryst. Liq. Cryst.* **1998**, *313*, 193–198.

**22** G. Kaupp, *Int. J. Photoenergy* **2001**, *3*, 55–62.

**23** G. Kaupp, A. Herrmann, M. Haak, *J. Vac. Sci. Technol. B* **1997**, *15*, 1521–1526.

**24** D. A. Wierda, T. L. Feng, A. R. Barron, *Acta Crystallogr., Sect.C* **1989**, *45*, 338–339.

**25** V. Enkelmann, G. Wegner, K. Novak, K. B. Wagener, *J. Am. Chem. Soc.* **1993**, *115*, 10390–10391.

**26** G. Kaupp, A. Kuse, *Mol. Cryst. Liq. Cryst.* **1998**, *313*, 361–366.

**27** G. Kaupp, J. Schmeyers, J. Boy, *Chem. Eur. J.* **1998**, *4*, 2467–2474.

**28** G. Kaupp, J. Schmeyers, M. Haak, T. Marquardt, A. Herrmann, *Mol. Cryst. Liq. Cryst.* **1996**, *276*, 315–337.

**29** G. Kaupp, J. Schmeyers, J. Boy, *J. Prakt. Chem.* **2000**, *342*, 269–280.

**30** T. F. Lai, R. E. Marsh, *Acta Crystallogr.* **1967**, *22*, 885–893.

**31** G. Kaupp, M. Haak, F. Toda, *J. Phys. Org. Chem.* **1995**, *8*, 545–551.

**32** G. Kaupp, A. Herrmann, *J. Phys. Org. Chem.* **1997**, *10*, 675–679.

**33** G. Kaupp, A. Herrmann, J. Schmeyers, *Chem. Eur. J.* **2002**, *8*, 1395–1406.

**34** R. S. Miller, D. Y. Curtin, I. C. Paul, *J. Am. Chem. Soc.* **1974**, *96*, 6340–6349.

**35** L. G. Butler, D. G. Cory, K. M. Dooley, J. B. Miller, A. N. Garroway, *J. Am. Chem. Soc.* **1992**, *114*, 125–135.

**36** R. Luther, F. Weigert, *Z. Phys. Chem. (Leipzig)* **1905**, *51*, 297–328.

**37** G. Kaupp, *Angew. Chem. Int. Ed. Engl.* **1992**, *31*, 595–597.

**38** G. Kaupp, E. Jostkleigrewe, H. J. Hermann, *Angew. Chem. Int. Ed. Engl.* **1982**, *21*, 435–436.

**39** G. Kaupp, *J. Microsc.* **1994**, *174*, 15–22.

**40** G. Kaupp, Cyclobutane Synthesis in the Solid Phase, in *CRC Handbook of Organic Photochemistry and Photobiology*, W. Horspool, P. S. Song (Eds.), CRC Press Boca Raton, **1995**, pp. 50–63.

41 E. Heller, G. M. J. Schmidt, *Isr. J. Chem.* **1971**, *9*, 449–462.

42 G. Kaupp, E. Teufel, *Chem. Ber.* **1980**, *113*, 3669–3674.

43 L. J. Fitzgerald, R. E. Gerkin, *Acta Crystallogr., Sect. C* **1997**, *53*, 71–73.

44 T. R. Welberry, R. D. G. Jones, J. Epstein, *Acta Crystallogr., Sect. B* **1982**, *38*, 1518–1525.

45 G. Kaupp, GIT *Fachzeitschrift für das Laboratorium* **1993**, *37*, 581–586.

46 G. Kaupp, *Liebigs Ann. Chem.* **1973**, 844–878.

47 G. Kaupp, J. Schmeyers, *J. Photochem. Photobiol. B: Biology* **2000**, *59*, 15–19.

48 G. Kaupp, M. Haak, *Angew. Chem. Int. Ed. Engl.* **1996**, *35*, 2774–2777.

49 G. Kaupp, *Photochem. Photobiol.* **2002**, *76*, 590–595.

50 T. C. James, *J. Chem. Soc.* **1911**, *99*, 1620–1627.

51 R. S. H. Liu, *Acc. Chem. Res.* **2001**, *34*, 555–562.

52 G. Kaupp, Stereoselective thermal solid-state reactions in *Topics in Stereochemistry*, Vol. *25*, Ch. 9, Wiley, **2006**, 303–354.

53 Y. Y. Wei, B. Tinant, J. P. Declerc, M. Van Meerssche, *Acta Crystallogr., Sect. C,* **1987**, *43*, 86–89.

54 L. J. E. Hofer, W. C. Peebles, *Anal. Chem.* **1951**, *23*, 690–695.

55 L. Di, B. M. Foxman, *Supramol. Chem.* **2001**, *13*, 163–174.

56 A Matsumoto, *Polymer J.* **2003**, *15*, 93–121.

56a G. Kaupp, *Lecture at the PACIFICHEM.* **2005**, *ORGN 10 (236)*, 75, Honolulu, 15.12.2005.

57 E. Hädicke, E. C. Mez, C. H. Krauch, G. Wegner, J. Kaiser, *Angew. Chem. Int. Ed. Engl.* **1971**, *10*, 266–267.

58 V. Enkelmann., *Adv. Polym. Sci.* **1984**, *83*, 91–136.

59 T.Yaji, K. Izcemi, S. Isoda, *Appl. Surf. Sci.* **2002**, *188*, 519–523.

60 M. Alfonso, Y. Wang, H. Stoeckli-Evans, *Acta Crystallogr., Sect. C* **2001**, *57*, 1184–1188.

61 M. P. Gupta, S. M. Prasad, *Acta Crystallogr., Sect. B* **1971**, *27*, 1649–1653.

62 P. D. Robinson, C. Y. Meyers, V. M. Kolb, J. L. Tunnell, J. D. Ferrara, *Acta Crystallogr., Sect. C* **1992**, *48*, 1033–1036.

63 B. Pirotte, P. de Tullio, B. Masereel, M. Schynts, J. Delarge, L. Dupont, *Helv. Chim. Acta* **1993**, *76*, 1311–1318.

64 R. Gericke, J. Harting, I. Lues, C. Schittenhelm, *J. Med. Chem.* **1991**, *34*, 3074–3085.

65 M. Suarez, E. Salfran, E. Ochoa, Y. Verdecia, L. Alba, N. Martin, C. Seoane, R. Martinez-Alvarez, H. N. de Armas, N. M. Blaton, O. M. Peeters, C. J. De Ranter, *J. Heterocycl. Chem.* **2003**, *40*, 269–275.

66 W. Shin, C. H. Chae, S. E. Yoo, *Acta Crystallogr., Sect. C* **1996**, *52*, 2058–2060.

67 A. Burger, J. M. Rollinger, P. Bruggeller, *J. Pharm. Sci.* **1997**, *86*, 674–679.

68 D. A. Langs, P. D. Strong, D. J. Triggle, *J. Computer-Aided Mol. Des.* **1990**, *4*, 215–230.

69 A. Escande, J. L. Galigne, *Acta Crystallogr., Sect. B* **1974**, *30*, 1647–1648.

70 G. Kaupp, M. R. Naimi-Jamal, unpublished results.

71 T. P. Singh, M. Vijayan, *Acta Crystallogr., Sect. B* **1973**, *29*, 714–720.

72 M. A. Metwally, E. Abdel-latif, F. A. Amer, G. Kaupp, *J. Sulfur Chem.* **2004**, *25*, 63–85.

73 O. Au-Alvarez, R. C. Peterson, A. A. Crespo, Y. R. Esteva, H. M. Alvarez, A. M. P. Stiven, R. P. Hernandez, *Acta Crystallogr., Sect. C* **1999**, *55*, 821–823.

74 Y. Miwa, T. Mizuno, K. Tsuchida, T. Taga, Y. Iwata, *Acta Crystallogr., Sect. B* **1999**, *55*, 78–84.

75 G. Kaupp, M. R. Naimi-Jamal, V. Stepanenko, *Chem. Eur. J.* **2003**, *9*, 4156–4160.

76 C. Y. Meyers, J. L. Tunnell, P. D. Robinson, D. H. Hua, S. Saha, *Acta Crystallogr., Sect. C* **1992**, *48*, 1815–1818.

77 J. Trotter, F. C. Wireko, *Acta Crystallogr., Sect. C* **1990**, *46*, 103–106.

78 J. N. Low, C. Glidewell, *Acta Crystallogr., Sect. C* **2002**, *58*, o209–o211.

79 G. Kaupp, A. Herrmann, *J. Prakt. Chem.* **1997**, *339*, 256–260.

80 A. F. M. Rae, E. N. Maslen, *Acta Crystallogr.* **1962**, *15*, 1285–1291.

81 R. A. Andersen, A. Zalkin, D. H. Templeton, *Inorg. Chem.* **1981**, *20*, 622–623.

82 A. R. Millward, O. M. Yaghi, *J. Am. Chem. Soc.* **2005**, *127*, 17998–17999.

83 D. M. M. Farrell, C. Glidewell, J. N. Low, J. M. S. Skakle, C. M. Zakavia, *Acta Crystallogr., Sect. B* **2002**, *58*, 289–299.

84 A. G. Giumanini, G. Verardo, L. Randaccio, N. Bresciani-Pahor, P. Traldi, *J. Prakt. Chem.- Chem. Ztg.* **1985**, *327*, 739–740.

85 S. D. Kimball, J. T. Hunt, J. C. Barrish, J. Das, D. M. Floyd, M. W. Lago, V. G. Lee, S. H. Spergel, S. Moreland, S. A. Hedberg, J. Z. Gougoutas, M. F. Malley, W. F. Lau, *Bioorg. Med. Chem.* **1993**, *1*, 285–307.

86 S. Mehdi, K. Ravikumar, *Acta Crystallogr., Sect. C*, **1992**, *48*, 1627–1630.

87 R. C. Gadwood, B. V. Kamdar, L. A. C. Dubray, M. L. Wolfe, M. P. Smith, W. Watt, S. A. Mizsak, V. E. Groppi, *J. Med. Chem.* **1993**, *36*, 1480–1487.

88 J. G. Ondeyka, R. G. Ball, M. L. Garcia, A. W. Dombrowski, G. Sabnis, G. J. Kaczorowski, D. L. Zink, G. F. Bills, M. A. Goetz, W. A. Schmalhofer, S. B. Singh, *Bioorg. Med. Chem. Lett.* **1995**, *5*, 733–734.

89 H. S. Shieh, L. G. Hoard, C. E. Nordman, *Acta Crystallogr., Sect. B* **1981**, *37*, 1538–1543.

90 G. Kaupp, C. Seep, *Angew. Chem. Int. Ed. Engl.* **1988**, *27*, 1511–1512.

91 B. M. Craven, *Acta Crystallogr., Sect. B* **1979**, *35*, 1123–1128.

92 G. W. Zamponi, S. C. Stotz, R. J. Staples, T. M. Andro, J. K. Nelson, V. Hulubei, A. Blumenfeld, N. R. Natale, *J. Med. Chem.* **2003**, *46*, 87–96.

93 A. Hybl, R. E. Rundle, D. E. Williams, *J. Am. Chem. Soc.* **1965**, *87*, 2779–2788.

94 K. K. Chacko, W. Saenger, *J. Am. Chem. Soc.* **1981**, *103*, 1708–1715.

95 K. Lindner, W. Saenger, *Carbohydr. Res.* **1982**, *99*, 103–115.

96 J. Jacob, K. Gessler, D. Hoffmann, H. Sanbe, K. Koizumi, S. M. Smith, T. Takaha, W. Saenger, *Carbohydr. Res.* **1999**, *322*, 228–246.

97 R. S. H. Liu, G. S. Hammond, *Proc. Natl. Acad. Sci. USA* **2000**, *97*, 11153–11158.

98 M. Milanesio, D. Viterbo, A. Albini, E. Fasani, R. Bianchi, M. Barzaghi, *J. Org. Chem.* **2000**, *65*, 3416–3425.

99 C. C. Wilson, *New J. Chem.* **2002**, *26*, 1733–1739.

100 R. Vasu Dev, K. S. Rekha, K. Vyas, S. B. Mohanti, P. R. Kumar, G. O. Reddy, *Acta Crystallogr., Sect. C* **1999**, *55*, IUC9900161.

101 K. S. Rekha, K. Vyas, C. M. H. Raju, B. Chandrashekar, G. O. Reddy, *Acta Crystallogr., Sect. C* **2000**, *56*, e68.

102 Y. B. Kim, H. J. Song, I. Y. Park, *Arch. Pharm. Res.* **1987**, *10*, 232–238.

103 J. F. McConnell, *Cryst. Struct. Commun.* **1974**, *3*, 73–75.

104 L. K. Hansen, G. L. Perlovich, A. Bauer-Brandl, *Acta Crystallogr., Sect. E* **2003**, *59*, o1357–o1358.

105 N. Jaiboon, K. Yos-In, S. Ruangchaithaweesuk, N. Chaichit, R. Thutivoranath, K. Siritaedmukul, S. Hannongbua, *Anal. Sci.* **2001**, *17*, 1465–1466.

106 C. Castellari, S. Ottani, *Acta Crystallogr., Sect. C* **1997**, *53*, 794–797.

107 M. Haisa, S. Kashino, R. Kawai, H. Maeda, *Acta Crystallogr., Sect. B* **1976**, *32*, 1326–1328.

108 A. Parkin, S. Parsons, C. R. Pulham, *Acta Crystallogr., Sect. E* **2002**, *58*, o1345–o1347.

109 G. Kaupp, M. R. Naimi-Jamal, H. Ren, H. Zoz, *Process-Worldwide* **2003**, *4*, 24–27; http://vmg01.dnsalias.net/vmg/process-worldwide/download/102327/Long-Version.doc

110 G. Kaupp, H. Zoz, Ger. Offen. **2004**, 5 pp., DE 10261204; *Chem. Abs.* **2004**, 549456.

111 G. Kaupp, Solid-State Reactivity/Gas-Solid and Solid-Solid Reactions, in *Encyclopedia of Supramolecular Chemistry* DOI : 10.1081/E-ESMC-120024356, **2005**, pp.1–13.

## 2.2
## Making Crystals by Reacting Crystals
*Fumio Toda*

### 2.2.1
### Introduction

Crushed grapes give wine by fermentation, but dried grapes do not result in wine. Although milk turns sour and shaking of milk gives cheese, dried milk can be kept unaltered. Similarly dried meat can be stored for a long time, whereas meat soup rapidly putrefies on standing. By observation of these phenomena, one can see that conversion of one material into another one occurs in the liquid state but not in the solid state. One of the most famous ancient philosophers in Greece, Aristotle, summarized these observations by "No Corpora nisi Fluida", which means "No reaction occurs in the absence of solvent". Such philosophies had a big influence on the evolution of the modern sciences in Europe, and this provides one historical reason why most organic reactions have been studied in solution.

Nevertheless, it is remarkable that chemists still carry out their reactions in solution, even when a special reason for the use of solvent cannot be found. We have found that many reactions proceed efficiently in the solid state. Indeed, in many cases, solid-state organic reaction occurs more efficiently and more selectively than does its solution counterpart, since molecules in a crystal are arranged tightly and regularly. In other words, very high concentration and selective reactions can be performed in the solid state. Furthermore, the solid-state reaction has many advantages: reduced pollution, low costs, and simplicity in process and handling. These factors are especially important in industry. Very recently, it was disclosed that the solid-state reaction can be accelerated in the presence of a small amount of solvent vapor. This is described in Section 2.2.2.

When greater selectivity is required in the solid-state reaction, host–guest chemistry techniques can be applied efficaciously. Reaction in the solid state of the guest compound as its inclusion complex crystal with a chiral host can give an optically active reaction product. Various host compounds have been designed by us to follow this simple principle.

The occurrence of efficient solid-state reactions shows that the molecules reacting are able to move freely in the solid state. In fact, host–guest inclusion complexation can occur by simply mixing and grinding both crystals in the solid state. Surprisingly, solid-state complex crystallization even occurs selectively. For example, mixing and grinding racemic guest and optically active host in the solid state gives an inclusion complex crystal involving just one enantiomer of the guest with the host, and from which the optically active guest can be obtained. Such efficient chiral recognition has been observed in many inclusion crystals, and efficient optical resolution has been achieved by using this phenomenon.

In an inclusion crystal formed by mixing crystals of guest and host compounds, molecules of the former are ordered regularly, therefore, photoreaction of the guest in the inclusion crystal proceeds efficiently and stereoselectively to give one stereoisomer in a pure state. When the guest molecules are not ordered at appropriate positions for efficient reaction, the molecular ordering can be transferred into a favorable one for the reaction via a phase transition. In an inclusion crystal formed by mixing of a prochiral guest crystal and a chiral host crystal, molecules of the former are arranged in a chiral form and such chirality becomes permanent upon solid-state reaction. In some cases, an optically active product results from mixing and grinding an achiral guest with a chiral host, followed by irradiation. This shows that movement and chiral arrangement of achiral molecules can occur by mixing and grinding host and guest crystals in the solid state. This is also described in Section 2.2.2.

In some special cases, prochiral molecules are arranged in a chiral form within the crystal without using any chiral source. Once again this chirality can become permanent through solid-state photoreaction. This chirality in the chiral crystal can easily be transformed into racemic or opposite chirality in the solid state by heating or contact with solvent vapor. Very interestingly, the inclusion complex crystal of racemic 2,2'-dihydroxy-1,1'-binaphthyl and tetramethylammonium chloride is converted to a conglomerate complex crystal by heating or contact with solvent vapor. These phase transitions in crystals are described in Section 2.2.3.

Section 2.3.4 deals with a seeding process in the solid state. Racemic 2,2'-dihydroxy-1,1'-binaphthyl forms two different kinds of inclusion complex crystals with ether in the host:guest ratio, namely, 3:1 and 1:2. Selective inclusion complexation by contact of the racemic host crystal with ether vapor in the solid state can be controlled by adding a seed crystal.

## 2.2.2
### Thermal Solid-state Reactions

Various thermal crystal–crystal reactions and reactions in crystals have been reported [1]. In this section, some typical examples of these reactions are described. Recently, it was found that solid-state reactions are accelerated by a small amount of solvent vapor. This finding is important and some examples are also described.

### 2.2.2.1   Crystal–Crystal Reactions

Some Bayer-Villiger oxidations of ketones with *m*-chloroperbenzoic acid **1** were found to proceed much faster in the solid state than in solution [2]. The oxidations were carried out at room temperature with a mixture of powdered ketone and 2 mol equiv. of powdered **1**. When the reaction time was longer than 1 day, the reaction mixture was ground once a day with an agate mortar and pestle. The excess of **1** was decomposed with aqueous $NaHSO_4$, and the product was taken up in ether. From the ether solution the product was isolated by column chromatography (Table 2.2.1). For comparison, the oxidation was also carried out in $CHCl_3$. Yields obtained from a solution of the ketone and 2 mol equiv. of **1** in $CHCl_3$ at room temperature are also summarized in Table 2.2.1. In some cases there is no marked difference between reactions in the solid state and in $CHCl_3$. However, there is a big difference depending on the kind of ketone.

**Table 2.2.1** Yields of Baeyer-Villiger oxidation products in the solid and in CHCl₃[a]

$$R^1COR^2 \xrightarrow[\quad 1 \quad]{m\text{-}ClC_6H_4CO_3H} R^1CO_2R^2$$

| ketone | reaction time | product | yield (%) solid state | yield (%) CHCl₃[b] |
|---|---|---|---|---|
| tBu–cyclohexanone | 30 min | tBu–(7-membered lactone) | 95 | 94 |
| Br–C₆H₄–C(O)–Me | 5 days | Br–C₆H₄–C(O)–O–Me | 64 | 50 |
| Ph–C(O)–CH₂–Ph | 24 h | Ph–C(O)–O–CH₂–Ph | 97 | 46 |
| Ph–C(O)–Ph | 24 h | Ph–C(O)–O–Ph | 85 | 13 |
| Ph–C(O)–C₆H₄–Me | 24 h | Ph–C(O)–O–C₆H₄–Me | 50 | 12 |
| Ph–C(O)–C₆H₄(o-Me) | 4 days | Ph–C(O)–CH₂–C₆H₄(o-Me) and Ph–O–C(O)–C₆H₄(o-Me) (1:1) | 39 | 6 |

ᵃ Molar ratio of ketone and (1) is 1:2. ᵇ The reaction was carried out with 1 g of ketone in 50 ml of CHCl₃.

We have shown that molecules move quite easily in the solid state [3]. Namely, some host–guest complex crystals can be formed by mixing crystalline host and guest compounds in the solid state or by keeping a mixture of powdered host and guest at room temperature. Therefore, the ease of oxidation in the solid state is not unexpected.

When the oxidation reaction of a prochiral guest is carried out in inclusion crystals with a chiral host, enantioselective oxidation of the guest can be expected. For example, enantioselective oxidation of β-ionone **2** to optically active epoxide **3** was accomplished by reacting a 1:1 inclusion complex **5** of **2** and the chiral host compound (**4 c**) [4] with **1** in the solid state. In this reaction it was found that a small amount of water accelerated the reaction. Keeping a mixture of finely pow-

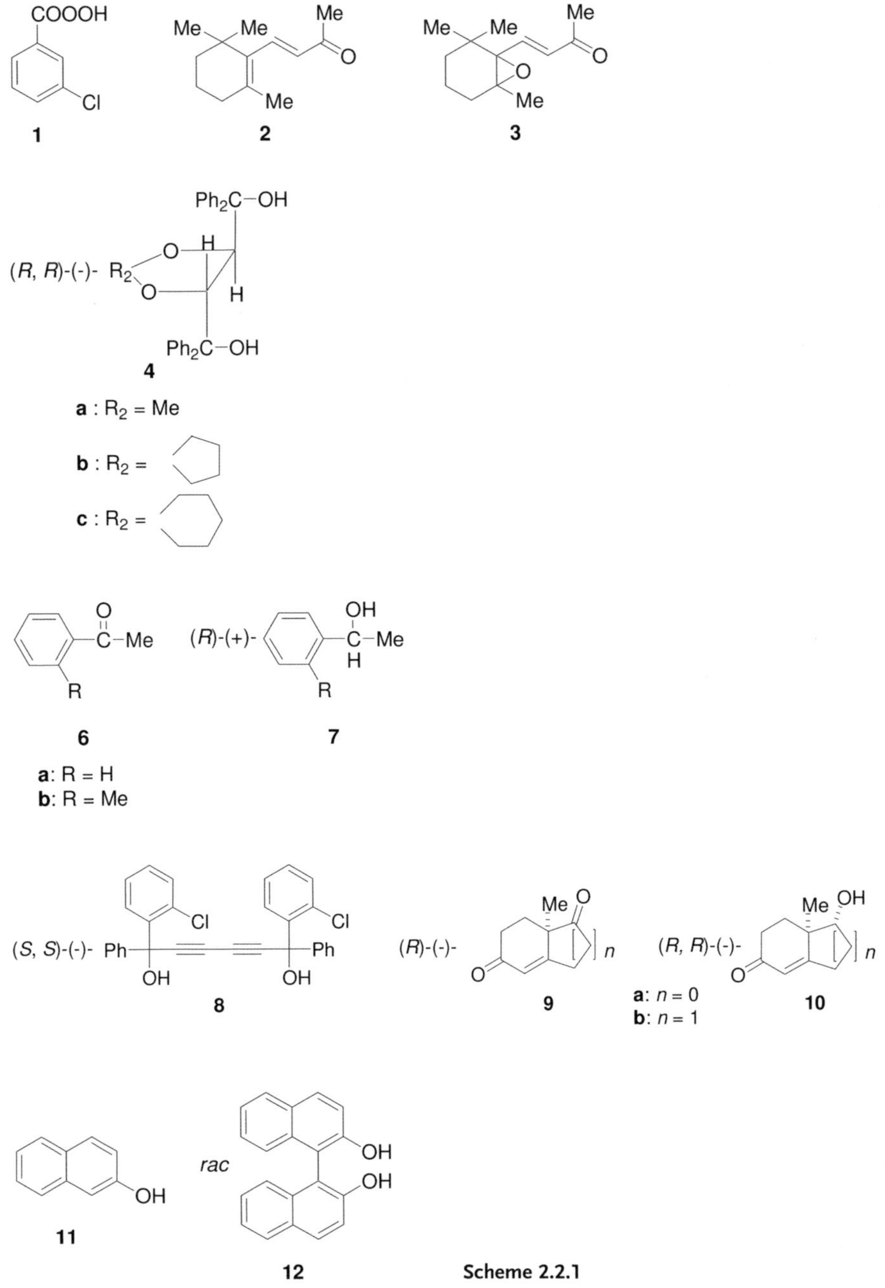

**a** : R$_2$ = Me

**b** : R$_2$ =

**c** : R$_2$ =

**6**

**7**

**a**: R = H
**b**: R = Me

**8**

**9**

**a**: n = 0
**b**: n = 1

**10**

**11**

rac

**12**

**Scheme 2.2.1**

dered **5** (1.7 g), **1** (1.24 g), and water (5 ml) at room temperature for 3 days gave **3** of 54% ee (0.41 g, 81% yield) [5].The X-ray single crystal structure of **5** has been determined [5].

Reduction of ketone by $NaBH_4$ can also be carried out in the solid state, although the reaction proceeds slowly. When a mixture of the ketones and a 10-fold molar amount of $NaBH_4$ was kept at room temperature with occasional mixing and grinding using an agate mortar and pestle for 5 days, the corresponding alcohols were obtained [6]. These reactions were also accelerated by a small amount of water [7].

Enantioselective reduction of ketones was also accomplished by carrying out the reaction in its inclusion complex crystal with a chiral host. For example, treatment of inclusion complexes of **6 a** and **6 b** with the chiral host (**8**) [8] by a $BH_3$-ethylenediamine complex in the solid state gave the chiral alcohols, **7 a** of 44% ee (96%) and **7 b** of 59% ee (57%), respectively, in the yield indicated [9]. Treatment of 1:1 inclusion complexes [9] of **9 a** and **4 a** with $NaBH_4$ in the solid state for 3 days gave $(R,R)$-$(-)$-**10 a** of 100% ee in 54% yield [6]. Similar reduction of a 1:1 complex of **9 b** and **4 a** with $NaBH_4$ in the solid state gave **10 b** of 100% ee in 55% yield. The enone moiety of **9** is masked by forming a hydrogen bond with the hydroxy group of **4 a** [10], so that the other carbonyl group is reduced selectively [9].

Coupling reactions of phenol derivatives proceed very efficiently in the solid state. This method is especially useful for the preparation of *rac*-2,2'-dihydroxy-1,1'-bi- naphthyl **12** by an oxidative coupling of 2-hydroxynaphthalene **11**. For example, when a mixture of finely powdered **11** and 2 mol equiv. of $FeCl_3 \cdot 6H_2O$ is kept at 50 °C for 2 h, **12** was obtained in 95% yield [11].

An enantioselective Wittig reaction was accomplished by the crystal–crystal reaction of a 1:1 complex of **4 b** and **13 a** or **13 b** with **14**, which gives the corresponding $(-)$-olefin **15 a** of 42% ee (73%) or **15 b** of 73% ee (73%), respectively, in the yield indicated [12]. An enantioselective Michael addition reaction was also achieved in an inclusion complex with a chiral host compound. Treatment of a 1:1 complex of **4 c** and **16** with 2-mercaptopyridine **17** in the solid state gave $(-)$-**18** of 80% ee in 51% yield [13].

## 2.2.2.2 Reactions in Crystals

Some thermal cyclization reactions of allene derivatives in their own crystals are described. Intramolecular cyclization of diallene derivatives within their crystals proceeds stereoselectively according to the Woodward–Hoffmann rule. For example, heating of colorless crystals of *meso*- **19** and *rac*-diallene **23** at 150 °C for 1 h gave the *in,out* **22** and a 1:1 mixture of *in,in* **26** and *out,out* **27** cyclization products, respectively, in quantitative yields (Scheme 2.2.1) [14]. Since no liquid state was observed during the cyclization reactions, the reaction in the crystal is a real crystal-to-crystal reaction. An X-ray analysis showed that **19** and **23** have an *s-trans* configuration in the crystal [15]. In order to cyclize to **22**, and to **26** and **27**, **19** and **23** should first isomerize to their *s-cis*-isomers **20** and **24**, respectively, in the crystal. The conformational change, from the *s-trans* form to the *s-cis* form, requires a rotation of the sterically bulky 1,1-diarylallene moiety around the single bond connecting the two allene groups in the crystalline state, as indicated. Furthermore, thermal conversion of the *s-cis*-diallenes to the corresponding dimethylenecyclobutenes should also be accompanied by a molecular motion of the sterically bulky 1,1-diarylallene groups. These molecular motions in the crystalline state occur stereospecifically, namely, through a [2+2] conrotatory ring closure.

By heating at 180 °C for 1 h, a colorless single crystal of the tetraallene derivative **28** was converted into the crystal of the cyclization product **29** (Scheme 2.2.2). Although the crystal of **29** was not a single crystal any more, it is very clear that the conversion of **28** into **29** occurs by a crystal-to-crystal reaction, as shown in Fig. 2.2.1 [16, 17]. When colorless crystals of the diallene derivative **30** were heated at 140 °C for 12 h, the benzodicyclobutadiene derivative **33** was produced as green crystals [17, 18]. As shown in Scheme 2.2.2, **30** is first isomerized to its *s-cis*-isomer **31**, and an intramolecular thermal reaction of the two allene moieties through a [2+2] conrotatory cyclization gives the intermediate **32**, which upon further thermal

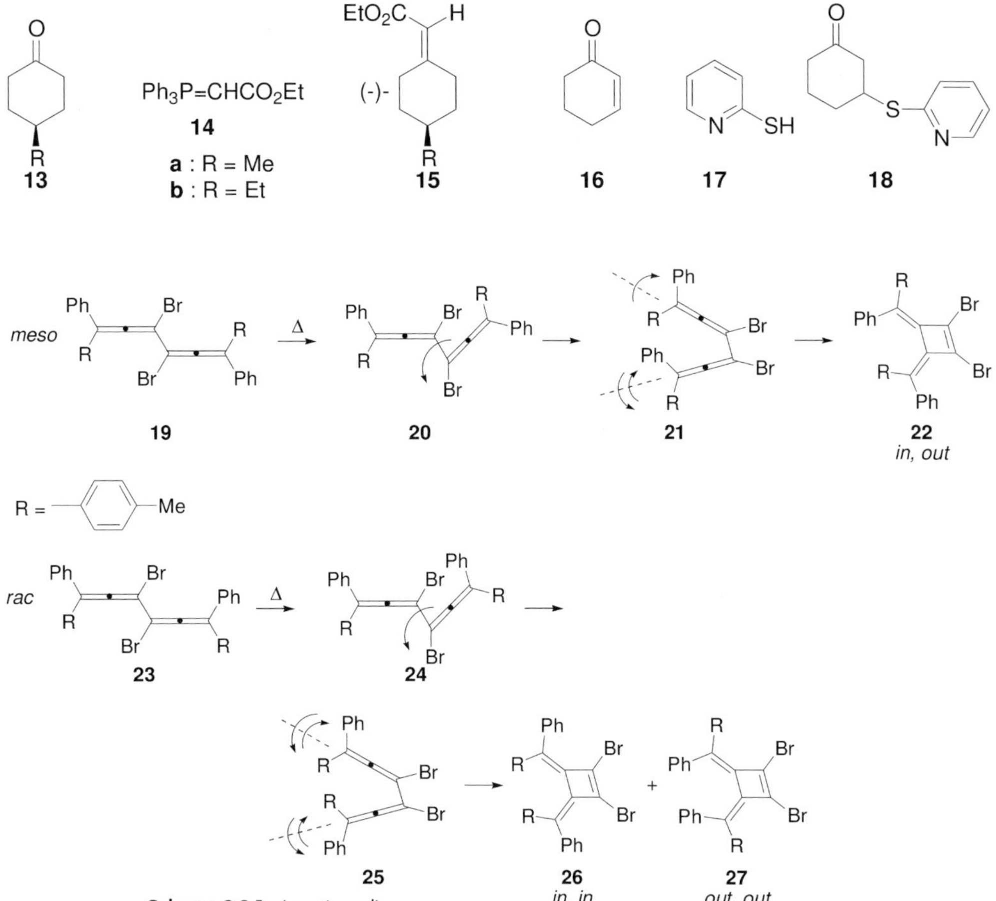

**Scheme 2.2.1** (continued)

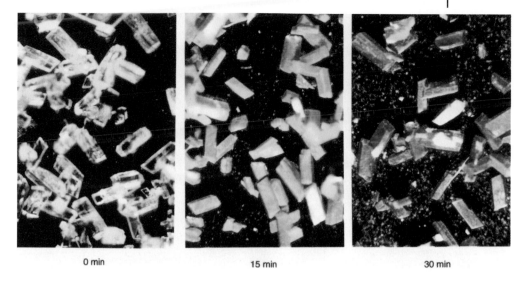

**Figure 2.2.1** Microscopic observation of the thermal conversion of **28** into **29** in the solid state.

reaction between acetylene moieties gives the final product **33**. This thermal cyclization also proceeds by a crystal-to-crystal reaction pathway.

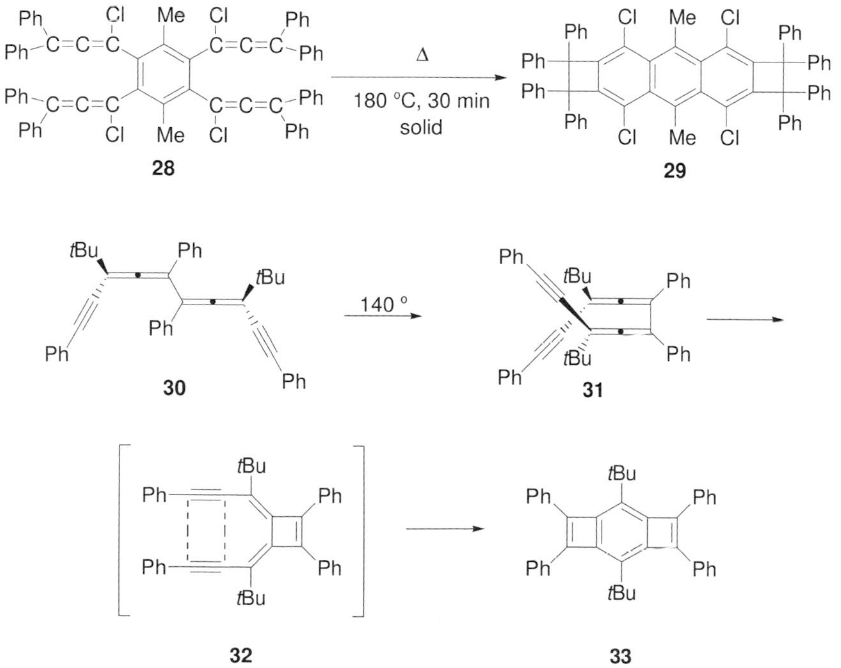

**Scheme 2.2.2**

### 2.2.2.3 Solid–Solid Reactions in the Presence of Solvent Vapor

It was found that some organic solid–solid reactions proceed very efficiently and selectively in the presence of a small amount of solvent vapor. For example, when a solution of **34** and **35** in MeOH was kept at room temperature for 1 h, the Michael addition reaction product **36** was obtained in 71% yield (Table 2.2.2) [19]. In contrast, the reaction in the solid state gave **36** in a poor yield. Heating a mixture of **34** and **35** at 100 °C for 4 h gave **36** in only 21% yield. However, when the reaction was carried out at room temperature in the presence of MeOH vapor for 1 h, the result was **36** in 63% yield. By prolongation of the reaction to 2 and 4 h, the yield increased to 81 and 90%, respectively. The efficiency of these reactions under MeOH vapor is much better than that of the same solid–solid reaction and the solution reaction in MeOH (Table 2.2.2).

**Table 2.2.2** Michael reaction of **34** and **35** in the solid state at room tempreature in the absence of solvent vapor.[a]

| Conditions | | Yield of 36 |
|---|---|---|
| Solvent vapor | Reaction time (h) | (%) |
| – [b] | 4 | 21 |
| MeOH | 1 | 63 |
| MeOH | 2 | 81 |
| MeOH | 4 | 90 |
| MeCN | 2 | 25 |
| MeCN | 4 | 51 |

[a] When the reaction was carried out in MeOH solution for 1 h, **36** was obtained in 71% yield. [b] Reaction was carried out at 100 °C.

The addition reaction of $Br_2$ to **37** was carried out by keeping a mixture of powdered **37** and a pyridine·HBr·$Br_2$ complex **38** at room temperature under various conditions (Table 2.2.3). The solid–solid reaction in the absence of solvent vapor for 4 h gave a 100:0 mixture of *erythro*- **39** and *threo*-addition products **40** in 72% yield. This result was better than that of the solution reaction in $CH_2Cl_2$ which gave a 91:9 mixture of **39** and **40** in 62% yield. The solid–solid reaction was also accelerated by various solvent vapors (Table 2.2.3). All solid–solid reactions under solvent vapor atmosphere were carried out in a sealed flask filled with air and solvent vapor in a roughly 1:1.2:0.25 molar ratio of reactant:reagent:solvent vapor. For example, a mixture of powdered **37** (200 mg, 0.96 mmol), and **38** (369 mg, 1.15 mmol) was kept in a flask (2 ml volume) filled with $CH_2Cl_2$ vapor (20 mg, 0.24 mmol) for 1 h. The reaction product was washed with aqueous $Na_2S_2O_3$ to give the product (336 mg, 95%) as a mixture of **39** and **40** in a 99:1 ratio. All the ratios of isomers were determined by $^1H$ NMR spectra [19].

**Table 2.2.3** Addition reaction of $Br_2$ to **37** in the solid state at room temperature in the absence and presence of solvent vapor.[a]

| Conditions | | Products | |
|---|---|---|---|
| Solvent vapor | Reaction time (h) | Yield (%) | **39:40** ratio |
| – | 4 | 72 | 100:0 |
| $CH_2Cl_2$ | 1 | 95 | 99:1 |
| MeCN | 2 | 82 | 100:1 |
| $CHCl_3$ | 1 | 94 | 98:2 |
| MeOH | 2 | 88 | 100:0 |
| EtOH | 2 | 81 | 99:1 |
| benzene | 4 | 95 | 95:5 |
| $H_2O$ | 4 | 73 | 100:0 |
| $Et_2O$ | 2 | 93 | 100:0 |
| hexane | 4 | 71 | 100:0 |

[a] When the reaction was carried out in $CH_2Cl_2$ and MeCN solution for 1 h, **39** and **40** were obtained in 62 (91:9) and 67% (100:0) yields, respectively, in the **39:40** ratios indicated.

In order to make the contact with solvent vapor efficient, the solid–solid reaction of various chalcone derivatives **41** with **38** was carried out in the presence of $CH_2Cl_2$ vapor by mixing every hour with a spatula (Table 2.2.4). The efficiency and the selectivity of all reactions were much better than those of the corresponding solid–solid reactions in the absence of solvent vapor. Finally, the satisfactory contact of the solvent vapor with **41** and **38** was found to be important. When the reaction of **41** and **38** in $CH_2Cl_2$ was carried out at room temperature for 3 h, a mixture of **42** and **43** (93:3 ratio) was obtained in 69% yield [19].

**Table 2.2.4** Addition reaction of $Br_2$ to **41** in the solid state at room temperature in the absence and presence of solvent vapor.[a]

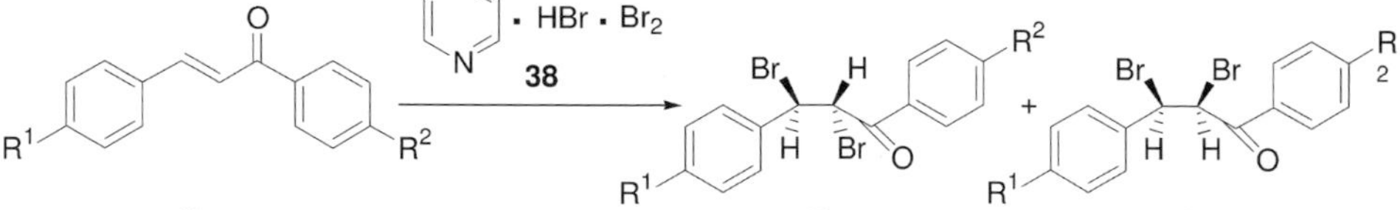

**41**
**a:** $R^1$ = Cl, $R^2$ = H
**b:** $R^1$ = H, $R^2$ = Cl
**c:** $R^1$ = H, $R^2$ = OMe
**d:** $R^1$ = NO$_2$, $R^2$ = H

**42**
erythro

**43**
threo

| | Conditions | | Products | |
|---|---|---|---|---|
| | Solvent vapor | Reaction time (h) | Yield (%) | **42:43** ratio |
| **17a** | – | 3 | 9 | 100:0 |
| | $CH_2Cl_2$ | 3 | 85 | 100:0 |
| **17b** | – | 3 | 4 | 100:1 |
| | $CH_2Cl_2$ | 3 | 93 | 100:0 |
| **17c** | – | 2 | 6 | 100:0 |
| | $CH_2Cl_2$ | 2 | 87 | 100:0 |
| **17d** | – | 2 | 0 | 100:0 |
| | $CH_2Cl_2$ | 6 | 70 | 93:7 |

[a] All reactions were carried out by mixing every 1 h by spatula.

The solid–solid reaction of **44** and **38**, which gives a mixture of *meso-* **45** and *rac*-product **46**, also proceeded efficiently and selectively in the presence of solvent vapor. A solution reaction of **44** and **38** in $CH_2Cl_2$ for 2 h gave a 50:50 mixture of **45** and **46** in 73% yield. Solid–solid reaction for 20 h in the absence of solvent vapor gave a 83:17 mixture of **45** and **46** in 38% yield. In contrast, the yield and the selectivity were improved by carrying out the solid–solid reaction under $CH_2Cl_2$ vapor and an 89:11 mixture of **45** and **46** was obtained in 88% yield (Table 2.2.5). Nevertheless, MeOH and EtOH vapor were not as effective as $CH_2Cl_2$ vapor. Such a difference is also observed sometimes in solution reactions as a solvent effect.

In order to clarify the mechanism of the important role of the solvent vapor in the solid–solid reaction, we studied how the solvent molecule easily comes into the crystal. Powdered **12** is exposed to MeOD vapor at room temperature and deuteration of the OH groups of **12** was monitored by measurement of IR spectra in the solid state (Fig. 2.2.2). At the beginning of the measurement, only the $\nu$OH

**Table 2.2.5** Addition reaction of Br₂ to **44** in the solid state at room temperature in the absence and presence of solvent vapor.[a]

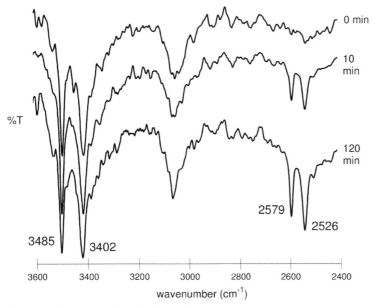

| Conditions | | Products | |
|---|---|---|---|
| Solvent vapor | Reaction time (h) | Yield (%) | **45:46** ratio |
| – | 20 | 38 | 83:17 |
| CH₂Cl₂ | 2 | 88 | 89:11 |
| MeOH | 2 | 34 | 88:12 |
| EtOH | 2 | 36 | 91: 9 |

[a] When the reaction was carried out in CH₂Cl₂ for 2 h, a 50:50 mixture of **45** and **46** was obtained in 73% yield.

**Figure 2.2.2** Monitoring by the ATR method at room temperature of the deuterium exchange reaction between **12** and MeOD vapor.

absorptions of **12** appeared at 3485 and 3402 cm$^{-1}$; these were assigned to inter- and intramolecular hydrogen bonded OH groups, respectively [20]. After 10 min, new inter- and intramolecular hydrogen bonded $v$OD absorptions of deuterated product **12-*d2*** appeared at 2579 and 2526 cm$^{-1}$, respectively. After 120 min, about half of the molecules of **12** were deuterated (Fig. 2.2.2) [19].

### 2.2.3
### Making Inclusion Complex Crystals by Mixing or Grinding Host and Guest Crystals

Host–guest inclusion complex crystals are usually prepared by recrystallization of both components from solvent. In some cases, these complex crystals can also be prepared by simply mixing or grinding host and guest crystals in the solid state. It also happens that the recrystallization from solvent and mixing or grinding of host and guest compounds gives different inclusion crystals. In special cases, inclusion complexes are obtained only by the mixing or grinding method but not by the recrystallization method. When the solubility of host and guest compounds in a solvent is remarkably different, one component crystallizes out preferentially. In special cases in which a new chirality is generated by the inclusion complexation, the chirality changes depending on the complexation method used. When powdered optically active host and racemic guest are mixed and ground together, one enantiomer of the latter is included to give an optically active guest compound in an almost pure state. In all cases described above, the inclusion complexation method by mixing or grinding techniques is important. These are described in the following sections.

### 2.2.3.1  Formation of Host–Guest Inclusion Complexes and Stereoselective Photo-chemical Reactions of the Guest in the Solid State

Making inclusion complex crystals by mixing host and guest crystals in the solid state was first reported in1987 [3]. For example, when a mixture of finely powdered host **47** and an equimolar amount of benzophenone was stirred using a test-tube shaker for 0.2 h at room temperature, a 1:1 complex of these two compounds was formed. Using the same method, chalcone **37**, 2-pyridone **48**, and *p*-dimethylami-nobenzaldehyde formed 1:2 complexes with **47** [3]. In the case of **48**, which is present as a tautomeric mixture with 2-hydroxypyridine even in the solid state [21], inclusion occurs only in the keto form. A similar result has been obtained from the complexation experiment in solution [22]. The formation of quinhydrone by a solid–solid reaction of quinone and hydroquinone has long been known [23]. The mechanism of the solid–solid reaction was found to be sublimation of the quinone whose vapor attacks certain sides of the hydroquinone crystals preferentially [24]. In our cases, however, it is interesting that mixing in the solid state of host and guest compounds of low vapor pressure can form their inclusion complex.

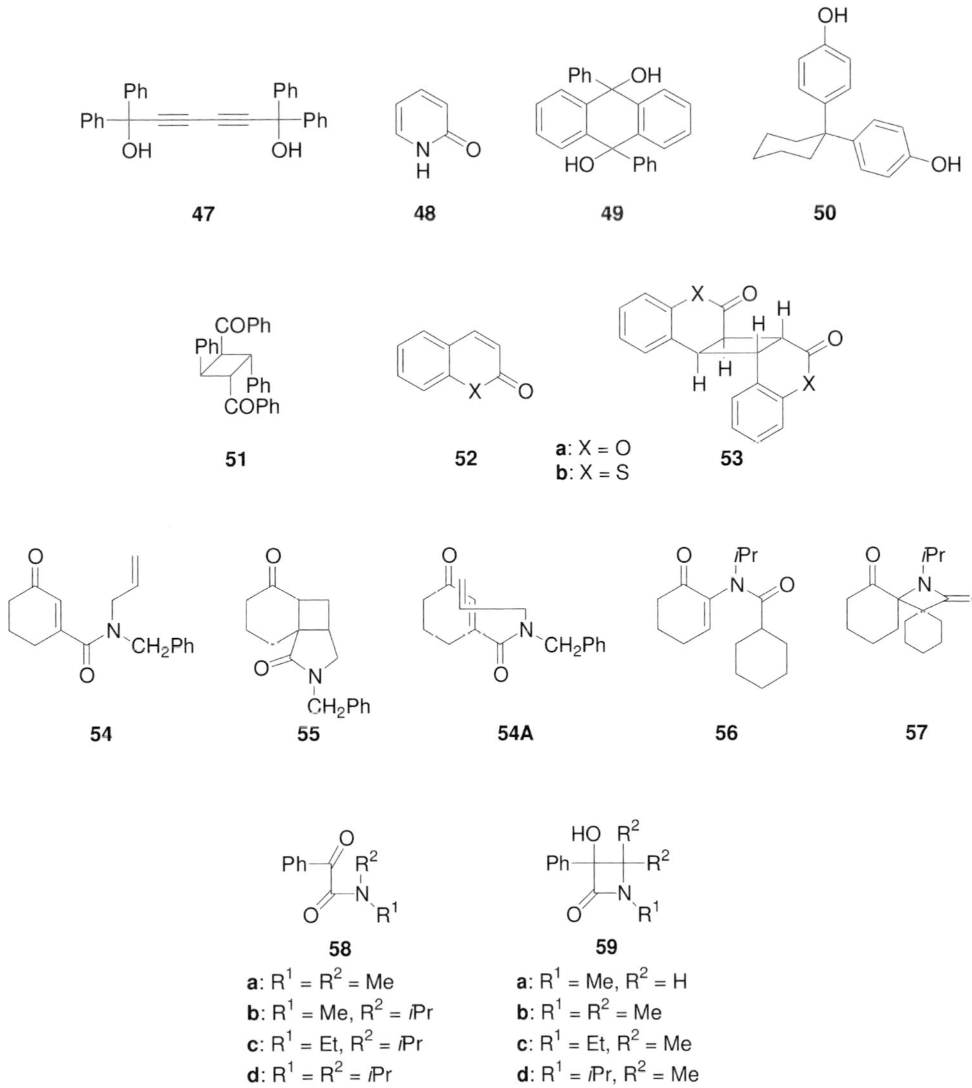

**47**

**48**

**49**

**50**

**51**

**52**

**a:** X = O
**b:** X = S

**53**

**54**

**55**

**54A**

**56**

**57**

**58**

**a:** R$^1$ = R$^2$ = Me
**b:** R$^1$ = Me, R$^2$ = *i*Pr
**c:** R$^1$ = Et, R$^2$ = *i*Pr
**d:** R$^1$ = R$^2$ = *i*Pr

**59**

**a:** R$^1$ = Me, R$^2$ = H
**b:** R$^1$ = R$^2$ = Me
**c:** R$^1$ = Et, R$^2$ = Me
**d:** R$^1$ = *i*Pr, R$^2$ = Me

**Scheme 2.2.3**

The hosts **49** and **50** also form complexes with some guest compounds by intimate mixing in the solid state [3]. However, some differences in complexation behavior were observed between the solid state and solution experiments. In ethyl acetate, **50** formed a mixture of the 3:2 and 2:1 complexes, but only the 2:1 complex was formed in the solid state [3]. Although no complexation of **50** and naphthoquinone occurs in any solvent, a 2:1 complex was obtained from the solid–solid reaction.

It has also been found that a combination of the solid state complexation and irradiation is useful for a continuous stereoselective photoreaction. For example, irradiation of a 1:1 mixture of **47** and **37** during shaking for 10 h gave the *syn-head-to-tail* dimer (**51**) in 80% yield. Since the irradiation of **37** itself in the crystalline state gives a mixture of isomers of **51** in a poor yield [25], and the irradiation of a 1:2 complex of **47** and **37** in the solid state for 6 h gives **51** in 90% yield [22, 26, 27], the 1:2 complex is expected to be produced during the shaking. X-ray analysis of the 1:2 complex of **47** and **37** showed that two **37** molecules are ordered in the complex at close positions so as to give **51** by dimerization [27]. The irradiation is also effective for a 1:2 mixture of **47** and **37**, since after 40 h of reaction under shaking conditions, **51** was obtained in 82% yield. Interestingly, this procedure is also effective for a 1:4 mixture of **47** and **37**, since **51** was obtained in 82% yield after 72 h irradiation under the same conditions. This shows that **47** can be used repeatedly, similar to a catalyst, namely **47**, which is released from the complexation with **37** by its photodimerization, goes on to form a complex with another molecule of **37**.

By inclusion complexation with a chiral host compound, achiral guest molecules can be ordered in a chiral manner, and photoirradiation of the complex gives optically active reaction products. The inclusion complexation can also be accom-

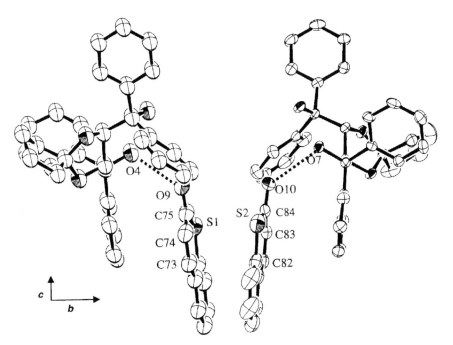

**Figure 2.2.3** ORTEP drawing of molecular structure of a 1:1 complex of **52 b** with **4 b** viewed along the *a*-axis. All hydrogen atoms are omitted for clarity. The hydrogen bonding is shown by dotted lines.

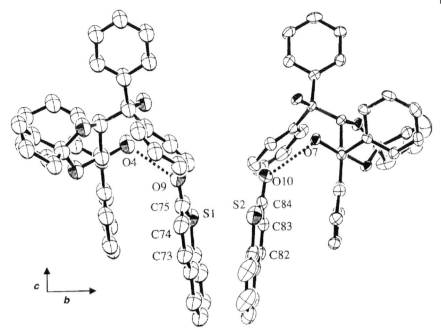

**Figure 2.2.4** ORTEP drawing of molecular structure of a 2:1
complex of (–)-**53 b** with **4 b** viewed along the *a*-axis. All hydrogen
atoms are omitted for clarity. The hydrogen bonding is shown
by dotted lines.

plished by mixing or grinding both crystalline components. For example, grinding
crystalline **4 b** and coumarine (**52 a**) or thiocoumarine (**52 b**) by mortar and pestle
gave the corresponding inclusion complex crystals, and irradiation of these complexes gave (–)-*anti-head-to- head* dimers **53 a** and **53 b**, respectively in optically pure
forms. These reactions were also disclosed to proceed in a single crystal-to-single
crystal manner. X-ray analysis of the inclusion complex was performed for the
single crystal prepared by recrystallization of **4** and **52** [28]. In the 1:1 inclusion
complex of **4 b** and **52 b**, two molecules of **52 b** are ordered at close and chiral
positions so as to give the (–)-dimer ((–)-**53 b**) by dimerization (Fig. 2.2.3). After the
irradiation, the molecule of (–)-**53 b** produced is arranged at almost the same
position as the two **52 b** molecules had been before the irradiation (Fig. 2.2.4).
By the conversion from **52 b** to (–)-**53 b**, no big change of the crystalline cavity
occurred in the complex [28]. This is the reason for the single crystal-to-single
crystal reaction. X-ray structural examination of the **4a**–**52 a** complex before and
after irradiation led to the same conclusion [28].

Photoirradiation of a 2:1 inclusion complex of **4b** and **54**, which had been prepared by recrystallization of these from ether, in a water suspension at room temperature for 10 h gave (–)-**55** of 100% ee in 90% yield [29]. Similar treatment of 2:1 inclusion crystals of **4c** and **54** prepared by recrystallization from benzene gave (–)-**55** of 100% ee in 87% yield [30]. The chirality of **54** in the inclusion complex would result, for example, by locating the allyl group above the cyclohexenone ring (**54A**). Interestingly, however, it was found that the same inclusion complex of **4** and **54** as that prepared by recrystallization can be obtained just by mixing **4** and **54** in the solid state, and that photoirradiation of the inclusion crystals obtained by mixing gives optically active photoreaction product. For example, an occasional grinding of powdered **4b** and a half molar amount of oily **54** using an agate mortar and pestle for 1 h, in order to avoid solidification, gave the 2:1 complex of **4b** and **54**. This complex was identical to that obtained by the recrystallization method. Photoirradiation of the complex in a water suspension for 10 h gave (–)-**55** of 99% ee in 48% yield. Similar mixing of **4c** and **54** in the absence of solvent for 1 h followed by irradiation for 10 h gave (–)-**55** of 87% ee in 40% yield. Such molecular movement in the solid state and arrangement in a chiral form are not special but rather common. Mixing of powdered **4b** and an equimolar amount of powdered **56** followed by irradiation gave (–)-**57** of 81% ee in 37% yield [30]. Chiral arrangement of **56** in its inclusion crystal with **4b** has been studied by X-ray analysis [31].

Very interestingly, irradiation of the 1:1 inclusion complex of **4c** and **58a**, which had been prepared by mixing both components in the absence of solvent, gave (+)-**59a** of 45% ee in 42% yield, although the same irradiation of the complex prepared by recrystallization of **4c** and **58a** from toluene gave the other enantiomer (–)-**59a** of 100% ee in 40% yield [30].

Chiral arrangement of **58a** in an inclusion crystal with **8** also depends on the conditions of the inclusion complexation experiment. Recrystallization of **8** and **58a** from ether gave a 1:1 inclusion complex, which upon irradiation affords (–)-**59a** of 100% ee in quantitative yield. X-ray structural study of the inclusion complex showed that **58a** is arranged in a chiral form [32]. However, irradiation of the inclusion crystal prepared by mixing **8** and **58a** gave *rac*-**59a**. Interestingly, however, both inclusion crystals are thermally interconvertible. After melting (mp 126–127 °C), the former is solidified by cooling to give the latter, but the latter is converted to the former by gradual heating and melts again at 126–127 °C [30].

It has been reported that **58d** forms chiral crystals in which **58d** molecules are ordered in a chiral form produced by a 90° twisting around the single bond between two CO groups [33], and that irradiation of the chiral crystals gives optically active **59d** of 93% ee in 74% yield [34]. Similarly, **58c** also forms chiral crystals, which upon irradiation give optically active **59c**. We tentatively describe the chiral crystals of **58c** and **58d**, which give upon irradiation (+)-**59c** and (+)-**59d**, respectively, as (+)-**58c** and (+)-**58d** crystals, respectively, and vice versa. Very interestingly, by mixing powdered MeOH complex of **4c** (**60**) and powdered (–)-crystal of **58c**, 1:1 inclusion crystals of **4c**, and (+)-**58c** (**61**) were formed (Scheme 2.2.3, continued). This is reasonable, since **4c**, which has (–)-configuration, can include only the (+)-**58c** but not (–)-**58c**. Photoirradiation of the powdered **61** for 7 h gave (+)-**59c** of

**Scheme 2.2.3, continued**

76% ee in 72% yield [30]. Easy conversion of the (–)-configuration in the (–)-**58 c** into the (+)-configuration by complexation with **4 c** in the solid state is surprising (Scheme 2.2.3, continued).

The chiral host **4 b** includes **62** to form two kinds of inclusion compounds, in 1:1 and 2:1 host:guest ratios, which upon photoirradiation in the solid state give (–)-**63** (96% ee, 50% yield) and (+)-**63** (98% ee, 86% yield), respectively, in the optical and chemical yields indicated [35]. Interestingly, irradiation of a 2:1 inclusion complex of **4 b** and **62**, which had been prepared by mixing of both in the solid state, gave (–)-**63** of 97% ee in 16% yield [35]. Structural studies of these inclusion complexes have been accomplished by X-ray analysis [35].

### 2.2.3.2 Enantioselective Inclusion Complexation in the Solid State

Enantioselective complexation of cyanohydrin with brucine in solution has been reported [36]. In this case, racemic cyanohydrin is completely converted into one enantiomer by complexation and racemization through a brucine-catalyzed equilibrium (Scheme 2.2.4). Similar phenomena occurred in the solid state. When brucine and an equimolar amount of *rac*-**64** was shaken for 24 h, a 1:1 complex was formed, which upon distillation gave (+)-**64** of 6.3% ee in almost quantitative yield [3]. This shows that (+)-**64** is included preferentially.

Enantioselective oxidation of **2** to the epoxide **3** in its inclusion complex **5** with **4 c** in the presence of a small amount of water is described in Section 2.2.2.1. However, the same reaction in the absence of water gave *rac*-**3** [5]. For example, when **5** was treated with an equimolar amount of **1** for 2 days at room temperature in the solid state, *rac*-**3** was obtained in 73% yield as an inclusion complex with **4 c**. Nevertheless, the same reaction of **5** with two molar amounts of **1** gave (+)-**3** of 66% ee as an inclusion complex with **4 c** together with the further oxidized product, (–)-**65** of 72% ee, in 43 and 33% yields, respectively. The same reactions of **5** with three and four molar amounts of **1** gave (+)-**3** (88% ee, 29% yield) and (–)-**65** (36% ee, 55% yield) in the optical and chemical yields indicated. Finally, when **5** was reacted with four molar amounts of **1**, *rac*-**55** was obtained in 88% yield. These data can be interpreted by a kinetic resolution of the *rac*-**3** initially formed by complexation

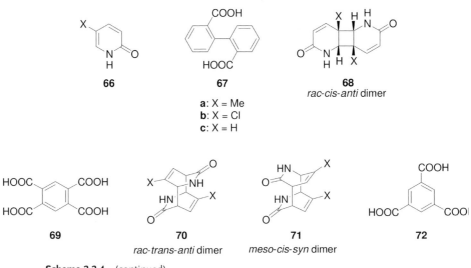

**62**            **63**

(+)- Ph—C—C≡N   ⇌ (brucine)   Ph—C—C≡N + HCN   ⇌   (-)- *t*Bu—C—C≡N
with *t*Bu and OH                with O                        with Ph and OH

(+)-**64**                                                              (-)-**64**

**65**                                **Scheme 2.2.4**

with **4 c**, but not enantioselective oxidation reaction of **2** in **5**. Firstly, reaction of **5** and **1** gives racemic uncomplexed **3** and **4 c**. Secondly, **4 c** includes (+)-**3** selectively in the solid state to form its 1:1 complex with **4 c** and the uncomplexed (–)-**3** is oxidized further to (–)-**65** with **1**. Thirdly, all **3** molecules are oxidized to result in *rac*-**65**. The key step in this kinetic resolution process is the enantioselective

**66**

**67**

**68**
*rac-cis-anti* dimer

a: X = Me
b: X = Cl
c: X = H

**69**

**70**
*rac-trans-anti* dimer

**71**
*meso-cis-syn* dimer

**72**

**Scheme 2.2.4**   (continued)

inclusion complexation between **4 c** and *rac*-**3** in the solid state, which gives finally the 1:1 inclusion complex of **4 c** and (+)-**3**.

This enantioselective complexation in the solid state was proven by the following experiment. A mixture of finely powdered **4 c** (1.2 g, 4.8 mmol) and *rac*-**3** (1.2 g, 2.4 mmol) was kept at room temperature for 1 day and then washed with hexane to give a complex of **4 c** and (+)-**3** (1.3 g) as crystals and hexane solution. From the complex, (+)-**3** of 88 % ee (0.29 g, 24 %) was obtained by distillation *in vacuo*. From the hexane solution, (–)-**3** of 36 % ee (0.6 g, 50 %) was obtained. It is also clear that the Bayer-Villiger oxidation of **3** to **65** proceeds more slowly in the complex **5** than does the oxidation of **3** alone [5].

## 2.2.4
## Making Crystals by Phase Transition

### 2.2.4.1  Phase Transition from Photochemically Nonreactive Inclusion Complexes to Reactive Ones

As a model reaction for DNA damage by photodimerization of its thymine component, which finally causes a skin cancer, photodimerization reactions of 2-pyridone derivatives are interesting. Photoirradiation of a 1:2 inclusion complex of **66 a** and **66 b** with the host **67** gives the corresponding *rac-cis-anti* dimer **68 a** and **68 b**, respectively [37, 38]. Photoirradiations of **66 c** and **66 b** in their inclusion complexes with **67** and **69**, respectively, gives *rac-trans-anti* dimer **70 c** and *meso-cis-syn* dimer **71 b**, respectively [38, 39]. X-ray analysis of these inclusion complexes showed that 2-pyridone molecules are ordered at adequate positions for the corresponding photodimerization reactions in all cases [37–39].

In some cases, however, guest molecules are not ordered at appropriate positions for a photodimerization. A rule for the efficient photodimerization of olefins in the solid state has been established as Schmidt's rule, namely, two olefin molecules should be arranged at parallel and close positions shorter than 4.2 Å in their crystals [40]. Nevertheless, photochemically nonreactive molecular aggregation of 2-pyridone molecules in its inclusion complex with a host can be transferred into a reactive one by phase transition in the solid state. For example, photochemically nonreactive 1:1:1 complex **73** of **66 a**, **72** and MeOH which had been prepared by recrystallization of **66 a** and **72** from MeOH was transformed into the reactive complex **74** by heating or by contact with MeCN vapor (Scheme 2.2.5). Photoirradiation of **74** in the solid state gave **71 a** in 30 % yield (Scheme 2.2.5), which upon heating isomerized to *meso-cis-syn* dimer **75** (Scheme 2.2.6) [41]. The mechanism of the phase transition from **73** to **74** was studied by X-ray analysis. As shown in Fig. 2.2.5, two **66 a** molecules in **73** are ordered at *anti*-positions, and the shortest contacts between the reaction centers of the two **66 a** molecules are, C3–C6′ = 5.04 Å and C6–C3′ = 5.04 Å. These distances are too far for the molecules to react. In **74**, however, two **66 a** molecules are ordered at *syn*-positions and the shortest contact between the reaction center is 3.840 Å (Figs. 2.2.6 and 2.2.7). This distance is short enough for the two molecules to react efficiently. The phase transition from **73** to **74** by molecular movement in the crystal is unusual and

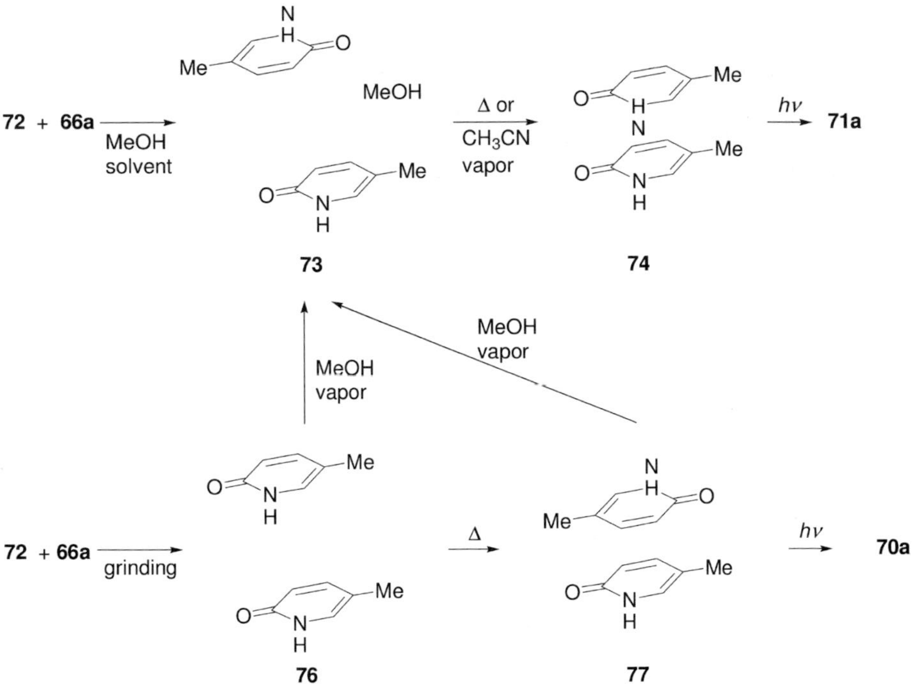


**Scheme 2.2.5** Phase transion from photochemically inert inclusion complexes **73** and **76** into reactive ones, **74** and **77**, respectively, in the solid state. For clarity, host **72** is omitted.

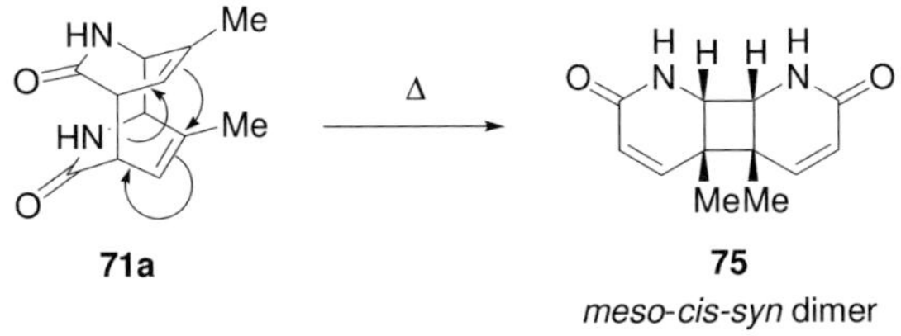


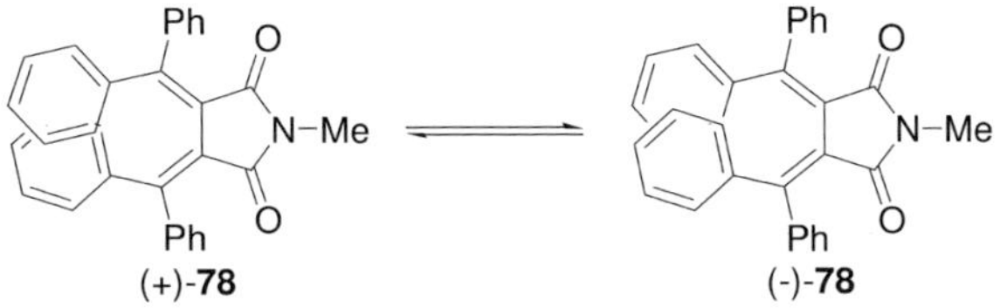


$Me_4NCl$
**80**

**Scheme 2.2.6**

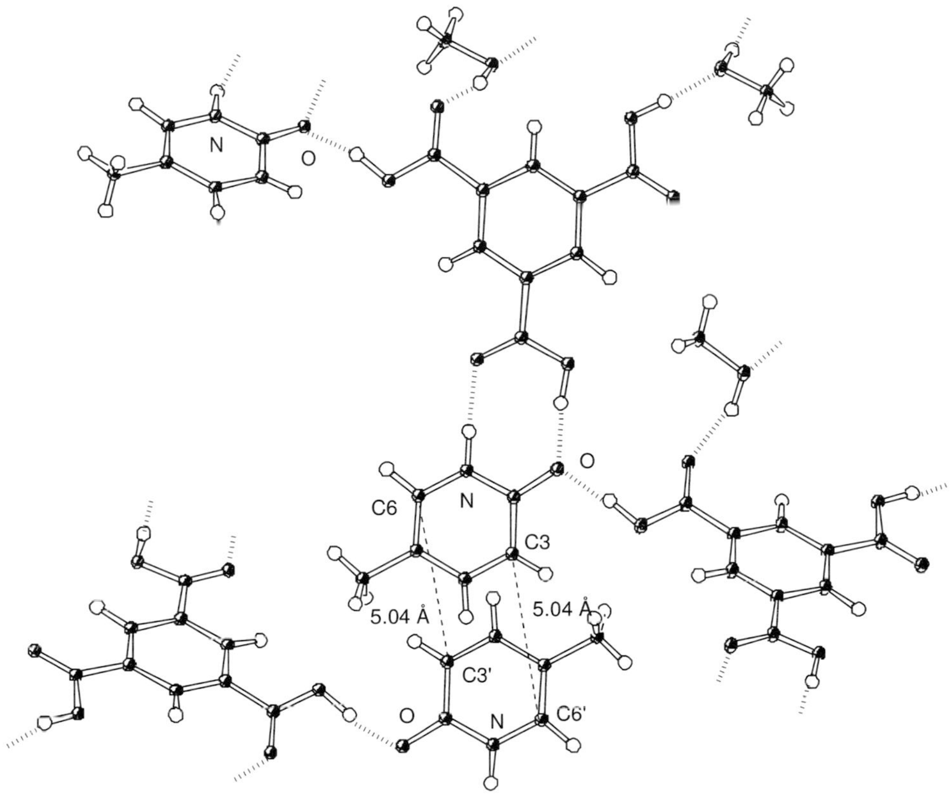

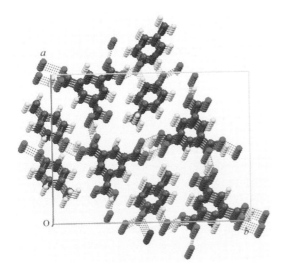

**Figure 2.2.5** Packing diagram of **73**.

**Figure 2.2.6** Packing diagram of **74**.

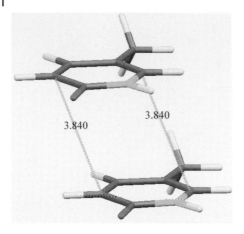

**Figure 2.2.7** Distance between 5-methyl-2-pyridone **66 a** molecules in **74**.

interesting. Especially, the phase transition from **73** to **74** by contact with MeCN vapor is important.

A similar phase transition was observed for a 1:1 complex (**76**) of **66 a** and **72**, which had been prepared by mixing powdered **66 a** and **72** in the solid state (Scheme 2.2.5). Firstly, **76** was prepared as a white powder by mixing **66 a** and an equimolar amount of **72** at room temperature by using an agate mortar and pestle. However, **76** was inert to photoirradiation. Secondly, **76** was transformed, by heating at 120 °C for 10 h, into a different complex **77** as white powder, which upon irradiation gave the *rac-trans-anti* dimer (**70 a**) in 44 % yield. It is also an interesting observation that contact of **76** and **77** with MeOH vapor gives **73** [41].

### 2.2.4.2 Phase Transition from Optically Active Complexes to *Rac*-Ones in Crystals

Since the two inner phenyl groups of **78** are overlapped, **78** becomes chiral in its crystals. Recrystallization of **78** from acetone yielded chiral crystals as orange hexagonal plates (**A**) and two types of *rac*-crystals as orange rectangular plates (**B**, mp 302 °C) and yellow rectangular plates (**C**, mp 297 °C) [42]. X-ray crystal structures of **A**, **B**, and **C** have been analyzed [43]. Crystal structures of **B** and **C** are completely different. When **A** was heated at 260 °C, it was converted to **C** [42]. However, when **A** was put in contact with MeOH vapor, it was converted to **B**.

### 2.2.4.3 Phase Transition from the *Rac*-Inclusion Complex to the Conglomerate One

The *rac*-inclusion complex **79** consisting of **12** and Me₄NCl **80** in a 1:1 ratio was easily converted into conglomerate complex **81** by heating or by contact with MeOH vapors in the solid state [42]. Firstly, authentic samples of **79** and **81** were prepared by the following method. When **12** (1.43 g) and **80** (0.55 g) was recrystallized from 7 and 15 ml of MeOH, *rac*-**79** and conglomerate 1:1 complex **81** were obtained, respectively.

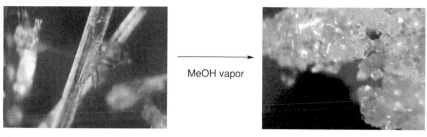

**Figure 2.2.8** Microscopic observation of the phase transition from **79** to **81** by contact with MeOH vapor.

When a mixture of powdered **12** and an equimolar amount of **80** was kept at room temperature for 170 h or contacted with MeOH vapor at room temperature for 1 h, **79** and **81** were obtained, respectively [44]. The easy formation of **81** in solution suggests a possibility of successful optical resolution of **12**. By using preferential crystallization of the chiral complex by adding a seed crystal prepared from optically active **12** and **80** to a solution of **12** and **80**, optical resolution of **12** was accomplished very successfully [44].

Direct formation of **81** from **79** by a phase transition in the solid state was also discovered. Heating of **79** at 160 °C for 5 min or contact of **79** with MeOH vapor at room temperature for 30 min gave **81** [44, 45]. Microscopic observation of the phase transition by MeOH vapor is indicated in Fig. 2.2.8 [44, 45]. Only one other similar phase transition has been reported [46].

## 2.2.5
## Control of Differential Inclusion Complexation by Seed Crystals

Differential inclusion complexation in solution can be controlled by addition of seed crystals during the recrystallization process. It was also found that the control by seed crystals can be accomplished in the solid state (see below). This seeding experiment was carried out for inclusion complexation between **12** and ethyl ether.

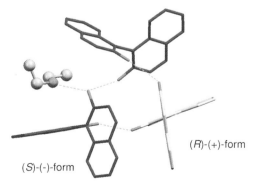

(R)-(+)-form

(S)-(-)-form

**Figure 2.2.9** Crystal structure of the 3:1 complex of **12** and Et₂O (**82**).

When **12** (0.2 g) was recrystallized from 3 ml and 5 ml ethyl ether, the 3:1 **82** and the 1:2 inclusion complex **83** were obtained, respectively. This is the first discovery that differential inclusion complexation is controlled by a concentration of solution. When **12** (0.2 g) was recrystallized from 4 ml of ethyl ether, **82** or **83** was formed, depending on slightly different conditions. In this case, however, control of the differential formation of **82** or **83** can be accomplished completely by addition of their seed crystals. X-ray crystal structures of **82** and **83** are very different, as shown in Figs. 2.2.9 and 2.2.10, respectively. Although both complexes are labile and lose ethyl ether easily, **82** is slightly more stable than **83**, probably because the ether molecule which is surrounded by three host molecules in **82** is only released with difficulty (Fig. 2.2.9) but the ether molecule which is accommodated in the channel of **83** can easily be released through the channel (Fig. 2.2.10) [47].

Ether guest-free hosts **84** and **85** were prepared by evaporation of the guest from **82** and **83**, respectively, by heating at 50 °C under 20 mmHg. Contact of **84** and **85** with ether vapor at room temperature for 3 and 12 h, respectively, gave **82** and **83**, respectively (Scheme 2.2.7). These data strongly suggest that **84** and **85** "remember" the original inclusion pattern of **82** and **83**, respectively. Nevertheless, no structural differences between **84** and **85** could be observed by IR spectroscopy or

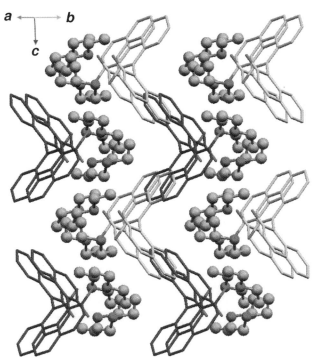

**Figure 2.2.10** Crystal structure of the 1:2 complex of **12** and $Et_2O$ (**83**).

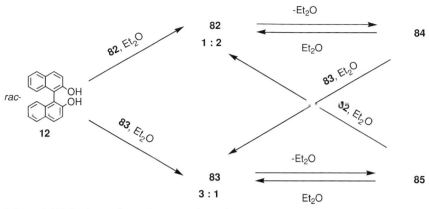

**Scheme 2.2.7** Seed crystal experiments in the inclusion complexation between solid host and gaseous Et₂O guest.

X-ray powder diffraction studies. There is a possibility that **84** and **85** still keep a small amount of ether guest moleclules in their inclusion cavities and that these inclusion complex lattices play a role as the seed crystals during complexation in the solid state. This hypothesis was proven to be correct by the following experiments. When a mixture of powdered **84** and a seed crystal of **83** (2 wt.%) was exposed to ethyl ether vapor for 3 h, **84** was converted to **83** in quantitative yield. By similar contact of a powdered mixture of **85** and a seed crystal of **82** (2 wt.%) for 12 h, **85** was converted to **82**. Furthermore, contact of a powdered mixture of **12** (98 mg, 0.34 mmol) and **83** (2 mg, 2.1 μmol; 2 wt.%, 1.35 mol%) with ethyl ether vapor for 3 h gave **83** in quantitative yield. However, contact of powdered **12** with ethyl ether vapor in the absence of a seed crystal for 72 h did not lead to the formation of any inclusion complexes [47].

Seed crystals were also effective for the inclusion complexation of **12** with some other guests. Contact of powdered **12** with gaseous MeOAc and EtOAc for 3 h in the presence of seed crystals of 3:1 inclusion complexes of **12** and MeOAc and EtOAc, respectively, led to the corresponding inclusion complexes in quantitative yields. A pseudo-seed crystal is also effective. For example, contact of powdered **12** with ethyl ether vapor for 3 h in the presence of a seed crystal of the 3:1 complex of **12** and MeOAc gave the 3:1 complex **82** but not **83**. Similar conctact of powdered **12** with ethyl ether vapor for 12 h in the presence of a seed crystal of the 1:2 complex of **12** and cyclopentyl methyl ether gave the 1:2 complex **83** but not **82** [47].

The efficient role of the seed crystal for inclusion complexation in the solid state is fascinating. Although the mechanism of the role of seed crystals is interesting, it is not easy to clarify. However, as it has been established that molecules in a crystal move quite easily in the solid state [3, 48], easy movement of gaseous molecules into the crystal is understandable. Surprisingly, however, host **12**, which has a helical structure with a twofold symmetry axis [20], is converted efficiently into a sheet structure of **82** and a cyclic structure of **83** in the solid state upon contact with ethyl ether vapor in the presence of a seed crystal.

## 2.2.6
## Conclusions

In this chapter, various molecular dynamics in crystals are described. The author believes that the readers of this chapter are now able to appreciate how various new interesting things can be discovered in the research field of solid-state chemistry. Of these discoveries, crystal-to-crystal reactions which are accelerated by a small amount of water or organic solvent vapor are the most important. When organic reactions can be carried out in the solid state without using solvent, a new green and sustainable synthetic procedure can be established.

## References

1 F. Toda, *Top. Curr. Chem.* **1988**, *149*, 211–238; F. Toda, *Synlett.* **1993**, 303–312; *Acc. Chem. Res.* **1995**, *28*, 480–486; K. Tanaka, F. Toda, *Chem. Rev.* **2000**, *100*, 1025–1074; K. Tanaka, F. Toda, Organic Photoreaction in the Solid State, in *Organic Solid-State Reactions*, F. Toda (Ed.), Kluwer, Dordrecht 2002, pp. 109–158; Z. Urbanczyk-Lipkowska, F. Toda, Selective Reactions in Inclusion Crystals, in *Organic Supramolecular Chemistry*, F. Toda, R. Bishop (Eds.), John Wiley & Sons, New York 2004, pp. 173–184; F. Toda, *Top. Curr. Chem.* **2005**, *254*, 1.

2 F. Toda, M. Yagi, K. Kiyoshige, *J. Chem. Soc., Chem. Commun.* **1988**, 958–959.

3 F. Toda, K. Tanaka, S. Sekikawa, *J. Chem. Soc., Chem. Commun.* **1987**, 279.

4 F. Toda, K. Tanaka, *Tetrahedron Lett.* **1988**, *29*, 551; D. Seebach, A. K. Beck, R. Imwinkelried, S. Roggo, A. Wannacott, *Helv. Chim. Acta* **1987**, *70*, 954.

5 F. Toda, K. Mori, Y. Matsuura, H. Akai, *J. Chem. Soc., Chem. Commun.* **1990**, 1591; F. Toda, K. Tanaka, T. Fujiwara, *Angew. Chem. Int. Ed. Engl.* **1990**, *29*, 662.

6 F. Toda, K. Kiyoshige, M. Yagi, *Angew. Chem. Int. Ed. Engl.* **1989**, *28*, 320.

7 M. Epple, S. Ebbinghaus, A. Reller, U. Gloistein, K. Cammmenga, *Thermochim. Acta* **1995**, 269/270, 433; M. Epple, S. Ebbinghaus, *Therm. Anal.* **1998**, *52*, 165.

8 F. Toda, K. Tanaka, T. Omata, K. Nakamura, T. Oshima, *J. Am. Chem. Soc.* **1983**, *105*, 5151.

9 F. Toda, K. Mori, *J. Chem. Soc., Chem. Commun.* **1989**, 1245.

10 L. R. Nassimbeni, M. L. Niven, K. Tanaka, F. Toda, *J. Cryst. Spectrosc. Res.* **1991**, *21*, 451.

11 F. Toda, K. Tanaka, S. Iwata, *J. Org. Chem.* **1989**, *54*, 3007.

12 F. Toda, H. Akai, *J. Org. Chem.*, **1990**, *55*, 3446.

13 F. Toda, K. Tanaka, J. Sato, *Tetrahedron: Asymm.* **1993**, *4*, 1771.

14 F. Toda, K. Tanaka, T. Tamashima, M. Kato, *Angew. Chem. Int. Ed. Engl.* **1988**, 37, 2724.

15 G. Kaupp, J. Schmeyers, M. Kato, F. Toda, *J. Phys. Org. Chem.* **2002**, *15*, 148.

16 K. Tanaka, N. Takamoto, Y. Tezuka, M. Kato, F. Toda, *Tetrahedron* **2001**, *57*, 3761.

17 F. Toda, *Eur. J. Org. Chem.* **2000**, 1377.

18 F. Toda, M. Ohi, *J. Chem. Soc., Chem. Commun.* **1975**, 506; R. Boese, J. Benet-Buchholz, A. Stanger, K. Tanaka, F. Toda, *Chem. Commun.* **1999**, 319.

19 S. Nakamatsu, S. Toyota, W. Jones, F. Toda, *Chem. Commun.* **2005**, 3808.

20 F. Toda, K. Tanaka, H. Koshima, I. Miyahara, K. Hirotsu, *J. Chem. Soc., Perkin Trans. 2* **1997**, 1877.

21 B. R. Penfold, *Acta Crystallogr.* **1953**, 6, 591.

22 K. Tanaka, F. Toda, *J. Chem. Soc. Jpn.* **1984**, 141.

23 A. R. Ling, J. L. Baker, *J. Chem. Soc.* **1893**, *63*, 1314; A. O. Patil, D. Y. Curtin, I. C. Paul, *J. Am. Chem. Soc.* **1984**, *106*, 348 and references cited therein.

**24** W. T. Pennington, A. O. Patil, I. C. Paul, D. Y. Curtin, J. *Chem. Soc., Perkin Trans.* 2. **1986**, 557.

**25** H. Stobbe, K. Bremer, *J. Prakt. Chem.* **1929**, *123*, 1.

**26** K. Tanaka, F. Toda, *J. Chem. Soc., Chem. Commun.* **1983**, 593.

**27** M. Kaftory, K. Tanaka, F. Toda, *J. Org. Chem.* **1985**, *50*, 2154.

**28** K. Tanaka, F. Toda, E. Mochizuki, N. Yasui, Y. Kai, I. Miyahara, K. Hirotsu, *Angew. Chem. Int. Ed.* **1999**, *38*, 2523; K. Tanaka, E. Mochizuki, Y. Kai, I. Miyahara K. Hirotsu, F. Toda, *Tetrahedron* **2000**, *56*, 6853.

**29** F. Toda, H. Miyamoto, S. Kikuchi, *J. Chem. Soc., Chem. Commun.* **1995**, 621.

**30** F. Toda, H. Miyamoto, *Chem. Lett.* **1995**, 809; F. Toda, H. Miyamoto, K. Takeda, R. Matsunaga, N. Maruyama, *J. Org. Chem.* **1993**, *58*, 6208.

**31** F. Toda, H. Miyamoto, M. Inoue, S. Yasaka, I. Matijasic, *J. Org. Chem.* **2000**, *65*, 2728; F. Toda, H. Miyamoto, K. Takeda, R. Matsunaga, N. Maruyama, *J. Org. Chem.* **1993**, *58*, 6208.

**32** M. Kaftory, M. Yagi, K. Tanaka, F. Toda, *J. Org. Chem.* **1988**, *53*, 4391.

**33** A. Sekine, K. Hori, Y. Ohashi, M. Yagi, F. Toda, *J. Am. Chem. Soc.* **1989**, *111*, 697.

**34** F. Toda, M. Yagi, S. Soda, *J. Chem. Soc., Chem. Commun.* **1987**, 1413.

**35** F. Toda, H. Miyamoto, K. Kanemoto, K. Tanaka, Y. Takahashi, Y. Takenaka, *J. Org. Chem.* **1999**, *64*, 2096.

**36** F. Toda, K. Tanaka, *Chem. Lett.* **1983**, 661.

**37** S. Hirano, S. Toyota, F. Toda, *Chem. Commun.* **2005**, 643.

**38** S. Hirano, S. Toyota, F. Toda, *Heterocycles* **2004**, *64*, 383.

**39** S. Hirano, S. Toyota, F. Toda, *Mendeleev Commun.* **2005**, 247.

**40** G. M. J. Schmidt, *Pure Appl. Chem.* **1971**, *27*, 647.

**41** S. Hirano, S. Toyota, F. Toda, K. Fujii, Y. Ashida, H. Uekusa, unpublished data.

**42** F. Toda, K. Tanaka, *Supramol. Chem.* **1994**, *3*, 87.

**43** F. Toda, K. Tanaka, Z. Stein, I. Goldberg, *Acta Crystallogr., Sect. B* **1995**, *51*, 856, and 2722.

**44** K. Yoshizawa, S. Toyota, F. Toda, *Tetrahedron* **2004**, *60*, 7767.

**45** K. Yoshizawa, S. Toyota, F. Toda, *Chem. Commun.* **2004**, 1844.

**46** P. A. Levkin, E. Schweda, H-J. Kolb, V. Schurig, R. G. Kostyanovsky, *Tetrahedron: Asymm.* **2004**, *15*, 1445.

**47** K. Yoshizawa, S. Toyota, F. Toda, S. Chatziefthimiou, P. Giastas, I. M. Mavridis, M. Kato, *Angew. Chem. Int. Ed.* **2005**, *44*, 5097.

**48** D. Braga, F. Grepioni, *Angew. Chem. Int. Ed.* **2004**, *43*, 4002.

## 2.3
## Making Crystals by Reactions in Crystals.
## Supramolecular Approaches to Crystal-to-Crystal Transformations within Molecular Co-Crystals
*Tomislav Friščić and Leonard R. MacGillivray*

### 2.3.1
### Introduction

Synthetic organic chemistry, which deals with the construction and/or degradation of covalent bonds, is traditionally performed in fluid, or liquid, media (i.e. solutions or melts) [1]. The remaining states of matter (i.e. gas and solid) are not as often considered in the context of performing organic syntheses. That the crystalline solid state is not generally used as an environment to make and/or break covalent bonds is somewhat surprising, owing to the properties that make such materials appealing to organic chemists, as well as materials scientists. The crystalline solid state provides synthetic organic chemists opportunities to conduct regio- and stereoselective transformations by exploiting the topochemical control inherent to the medium [2]. In particular, reactions such as [2+2] photodimerizations [3], [4+4] photodimerizations [4], and Diels-Alder reactions [5] can be controlled in the solid state [6]. From the standpoint of materials science, solid-state reactions can provide an entry to polymers with highly controlled stereochemistry and crystallinity [7, 8]. In addition, solid-state reactions exhibit an environmentally friendly approach to reactivity, owing to an inherent lack of solvent [9]. A major reason why the crystalline solid state is not usually considered by synthetic organic chemists lies in difficulties in predicting, or controlling, solid-state arrangements of molecules [10]. Indeed, the inability to achieve control of molecular arrangements is considered a 'continuing scandal' of solid-state sciences [11]. This means that solids that exhibit reactivity have been obtained, for the most part, by trial and error rather than design.

In this chapter, we will address how reliable methods to control molecular arrangements suitable for reactivity in the solid state are emerging as a result of applications of the principles of supramolecular chemistry [12]. Specifically, we will illustrate how the design of reactive multi-component solids, or co-crystals, can be achieved by exploiting non-covalent forces and the process of self-assembly [13]. The examples that we will provide have been selected to provide insights into the benefits of co-crystals to both synthetic organic chemists and solid-state materials scientists. Specifically, we will show how intermolecular bond-forming reactions that generate crystalline products from crystalline reactants (i.e. crystal-to-crystal reactions) appeal to both areas [14]. Thus, our discussion will be limited to reactions that completely conserve the integrity of a reacting crystal, or single-crystal-to-single-crystal (SCSC) transformations [15]. Reactions of this sort are said to be homogeneous, meaning that the incipient product forms a solid solution with the reactant and does not create a separate phase. Solids that exhibit SCSC reactivity can combine synthetic utilities of topochemically controlled bond-forming processes with possibilities to construct single-crystal polymers, solid-state switches, and/or actuators [16]. The subject of SCSC reactions has recently been treated in the context of topochemical [2+2] photodimerizations [17]. The reactions that will be described here are [2+2] photodimerizations, diacetylene and triacetylene polymerizations, and the Diels-Alder reaction. That our focus is on intermolecular bond-forming processes means that emerging and related processes of crystal solvation/desolvation [18, 19] and intramolecular solid-state reactions [20] will not be covered. We will first provide a brief overview of reactions in single-component solids since such materials have provided the groundwork for studies of more complex supramolecular systems.

## 2.3.2
## Single-component Solids

The ability to conduct chemical reactivity in a SCSC fashion has been realized primarily through investigations of single-component solids. Such investigations have led to the conclusion that SCSC reactivity is the result of interplay between molecular structure, intermolecular forces, and crystal packing. Large, bulky substituents, for example, are thought to favor SCSC reactivity [21]. A similar effect has also been suggested for extended hydrogen-bonded networks [22], although in some cases intermolecular hydrogen bonding has been thought to hinder SCSC reactivity by preventing molecular movement [23]. Furthermore, it is thought that the overall crystal structure needs to provide a flexible reaction cavity, so as to avoid breaking down of the crystal as the result of an accumulation of stress from changes in molecular shape during reaction [24]. Examples of reactions in single-component solids that proceed in a SCSC manner include [2+2] photodimerizations, diacetylene 1,4-polymerizations, and a recently discovered topochemical triene 1,6-polymerization.

### 2.3.2.1 [2+2] Photodimerizations
Olefins that crystallize with double bonds arranged approximately parallel and at a separation less than 4.2 Å generally undergo a topochemical [2+2] photodimerization to produce a cyclobutane product [25]. Among the best known compounds to undergo the photoreaction is cinnamic acid (Scheme 2.3.1). The geometric parameters that describe the orientation of double bonds required for the reaction in the solid state have been established by Schmidt, who also suggested that the reaction may occur in a SCSC fashion in the case of *trans*-cinnamide [26]. However, the first demonstration of a SCSC [2+2] photodimerization was provided by Thomas and co-workers in the case of 2-benzyl-5-benzylidenecyclopentanone [27]. The authors monitored the photodimerization via single-crystal X-ray diffraction. By analyzing structures of a 2-benzyl-5-benzylidenecyclopentanone single crystal at different reaction yields, the course of the photoreaction was followed at the molecular level (Fig. 2.3.1) [28]. Similar reactivity has been reported in several derivatives of the cyclopentanone, as well as cinnamic acid [29, 30]. In this context, the [2+2] photo-

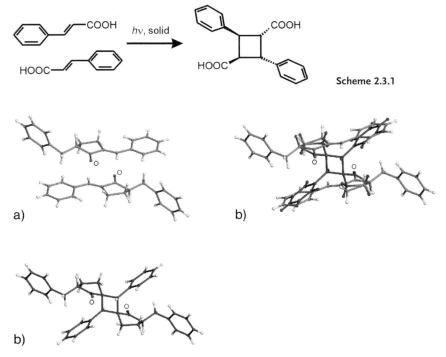

Scheme 2.3.1

a)

b)

b)

**Figure 2.3.1** Ball-and-stick representations of the crystal structure of 2-benzyl-5-benzylidenecyclopentanone at different stages
of reaction: (a) before reaction; (b) at 62% yield; c) after full
conversion. The photoproduct is shown in dark gray.

dimerization of 4-chlorocinnamoyl-*O,O'*-dimethyldopamine, reported by Ito and
co-workers, is especially interesting as the photoreaction occurs on exposure to
X-rays and is further influenced by the wavelength of the X-ray radiation [31].

### 2.3.2.2 [2+2] Photopolymerizations

The use of crystal-to-crystal [2+2] photodimerizations as a means to construct
crystalline polymers has been pioneered by Hasegawa and co-workers [32]. To
construct the polymers, reactants with two double bonds in the form of 1,4-divinyl-
benzenes, such as methyl 4-[2-(4-pyridyl)ethenyl]cinnamate (Scheme 2.3.2) were
employed [33]. One-dimensional chains composed of repeat units of cyclobutane

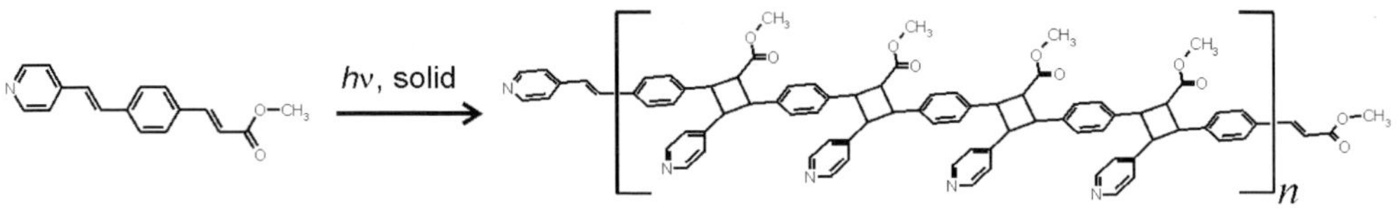

Scheme 2.3.2

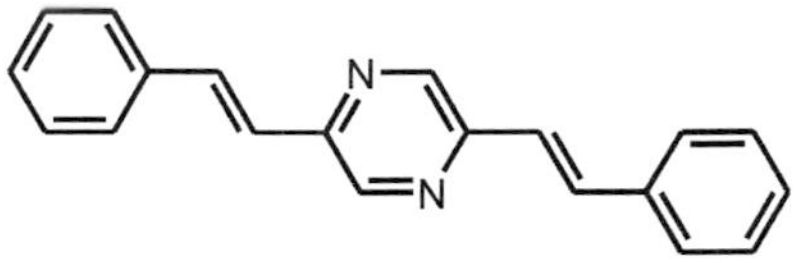

Scheme 2.3.3

rings were obtained through multiple [2+2] photodimerizations. Owing to the top-ochemical nature of the [2+2] photopolymerization, the polymers exhibited well-defined stereochemistries predictable from the arrangements of the reactants in the solids. Whereas a large number of 1,4-divinylbenzene derivatives were found to undergo the polymerization in a crystal-to-crystal fashion, 2,5-distyrylpyrazine (Scheme 2.3.3) presented a unique case in which the reaction was suggested to occur via a SCSC process [34]. Notably, the results of a phonon spectroscopy investigation of the photopolymerization were consistent with a SCSC nature of the reaction. On the other hand, information obtained by way of X-ray diffraction experiments has led Eckhardt and co-workers to suggest that the reaction proceeds heterogeneously within nanosized domains in the solid. Owing to the size of the domains with respect to the wavelength of radiation employed in the measurement, the reaction appears to be homogeneous in nature to phonon spectroscopy, but heterogeneous to X-ray diffraction [35].

### 2.3.2.3 Diacetylene and Triacetylene Polymerizations

In addition to the [2+2] photodimerization of divinylbenzenes, the most significant solid-state reaction that provides a route to highly crystalline polymers is the top-ochemical polymerization of diacetylenes. The reaction, reported and investigated by Baughman and Wegner, produces a polydiacetylene polymer via a 1,4-addition

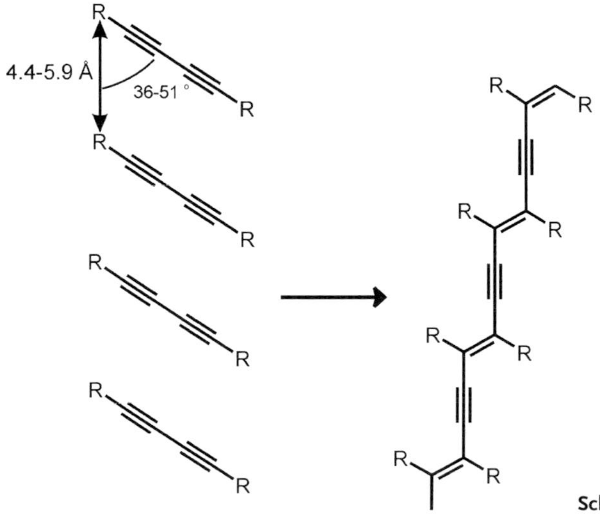

Scheme 2.3.4

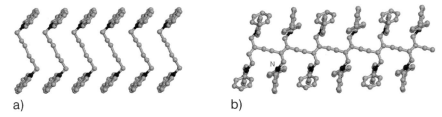

a)                                                                              b)

**Figure 2.3.2** Ball-and-stick views of the crystal structure of
1,6-di-(*N*-carbazolyl)-2,4-hexadiyne: (a) before reaction; (b) after
complete reaction.

reaction [36]. In order for the reaction to occur, diacetylenes in the crystal should be
juxtaposed with a repeat distance of 4.4 to 5.9 Å. In addition, the polydiacetylene
moiety needs to be tilted at an angle of approximately 36–51° with respect to the
direction of the repeat (Scheme 2.3.4) [37]. Indeed, the reaction has been found to
proceed via a SCSC process in a large number of cases. The ability to conserve the
single-crystal nature of the sample during the reaction has been explained by way of
two different mechanisms that involve either a continuous deformation of the
monomer lattice or a phase transformation of the solid during the course of the
reaction. The latter has been reported by Enkelmann and co-workers in the case of
1,6-di-(*N*-carbazolyl)-2,4-hexadiyne (Fig. 2.3.2) [38].

The first report of a topochemical 1,6-polymerization of a conjugated triene was
recently described by Lauher and co-workers [39]. In particular, the double ester of
1,8-dihydroxy-2,4,6-octatriene and nicotinic acid (Scheme 2.3.5) was found to adopt
a solid-state packing arrangement conducive for the 1,6-polymerization. Specifically, the molecules were oriented at a repeat distance of 7.2 Å (Fig. 2.3.3(a)).
Although the separation between terminal carbon atoms of adjacent triene moieties was somewhat large (4.09 Å), the solid underwent a SCSC topochemical
1,6-polymerization upon heating at 110 °C for 8 h (Fig. 2.3.3(b)).

**Scheme 2.3.5**

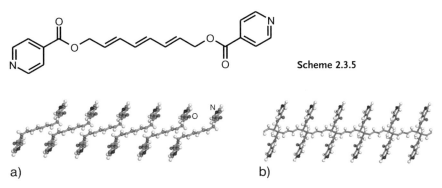

a)                                                                              b)

**Figure 2.3.3** Ball-and-stick representations of the crystal structure of the diester of 1,8-dihydroxy-2,4,6-octatriene and nicotinic
acid: (a) before reaction; (b) after complete reaction.

## 2.3.3
## Co-crystals

Recent years have witnessed rapid growth in the use of non-covalent interactions, the most prominent being hydrogen bonds [40], to construct crystalline solids composed of more than one molecular component [41, 42]. The construction of such multi-component solids, or co-crystals, has been recognized as a way to achieve solids that exhibit diverse properties and applications [43]. A significant advantage of multi-component over single-component solids lies in modularity. Modularity is achieved through an ability to combine and interchange different molecules. Such modularity has made co-crystals especially attractive, for example, in the context of pharmaceutical solids, by allowing the construction of materials that contain pharmaceutically active ingredients that avoid difficulties of polymorphism and/or bioavailability [44, 45].

That the formation of a multi-component solid depends on an association between two or more molecules means that the principles of supramolecular chemistry may be exploited to construct reactive co-crystals [46]. Such solid-state reactivity would be achieved by utilizing one component – the co-crystal former [47] – to dictate suitable positioning of the other molecule – the reactant. In this way, a solid may be constructed in which the reactivity is independent of crystal packing. In principle, the co-crystal former may direct chemical reactivity by either forming a host crystal lattice or acting as a linear template that juxtaposes the reactants for reaction [42]. That a SCSC reaction may be achieved in a multi-component solid was first established by Desiraju and co-workers using mixed crystals of 2-benzylidene-5-(4-chlorobenzyl)cyclopentanone and 2-benzylidene-5-(4-methylbenzyl)cyclopentanone (Scheme 2.3.6) [48]. Although the system involved a flexible ratio of the two components, rather than a stoichiometric composition of a co-crystal, the study nevertheless demonstrated that SCSC reactivity could be achieved in a solid with more than one molecule.

### 2.3.3.1    [2+2] Photodimerization
The [2+2] photodimerization of olefins is the best studied reaction in the context of SCSC reactions in both co-crystals and single-component solids. Co-crystals that exhibit SCSC [2+2] photoreactivity have been achieved using both the host–guest and linear template approach. The reactive solids have involved neutral and ionic [49] species, and have been constructed primarily using hydrogen bonds.

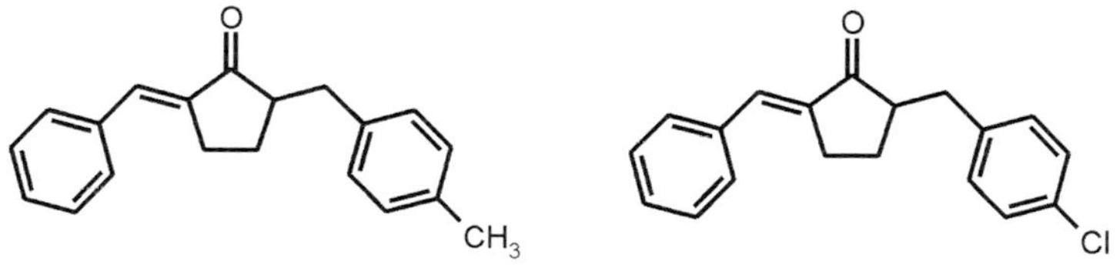

**Scheme 2.3.6**

### 2.3.3.1.1 Host–Guest Systems

Examples of SCSC reactions achieved through host–guest systems have been reported by Toda and co-workers. Olefins such as cyclohex-2-enone, coumarin, and thiocoumarin (Fig. 2.3.4(a)), for example, have been discovered to undergo topochemical [2+2] photodimerizations within hydrogen-bonded host frameworks constructed by optically active diols (Fig. 2.3.4(b)). In each case, the reactants interacted with the host via O–H · · · O hydrogen bonds that contributed to positioning the coumarin and thiocoumarin molecules for an enantiospecific [2+2] photodimerization (Fig. 2.3.5(a)) [50]. In the case of the cyclohex-2-enone, the enantioselectivity of the reaction was diminished. The change in the stereochemical properties of the reaction on going from the coumarins to the cyclohex-2-enone was attributed to increased mobility of the smaller reactant where rotational free-

a)
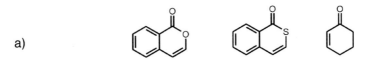

**Figure 2.3.4** Schematic representations of: (a) coumarin, thiocoumarin and cyclohex-2-enone molecules; (b) optically active diols employed as hosts.

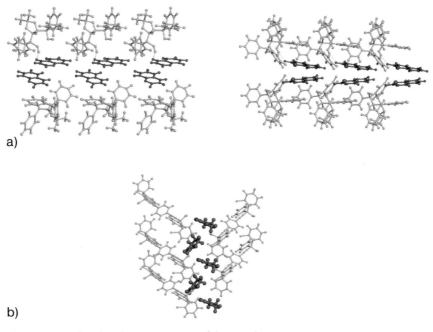

**Figure 2.3.5** Ball-and-stick representations of the crystal structure of the host-guest co-crystals involving: (a) coumarin (left) and thiocoumarin (right); (b) cyclohex-2-enone. Guest molecules shown in dark gray.

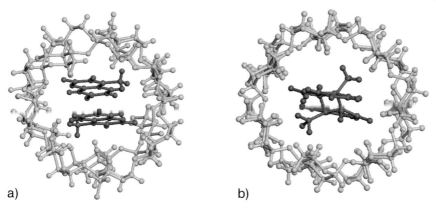

**Figure 2.3.6** Ball-and-stick representations of the host-guest complex of 7-hydroxy-4-methylcoumarin and β-cyclodextrin in the solid state. Coumarin molecules shown in dark gray.

dom of cyclohex-2-enone within the co-crystal allowed for two competing reaction pathways (Fig. 2.3.5(b)) [52].

A SCSC [2+2] photodimerziation involving 7-hydroxy-4-methylcoumarin has been shown to occur in the cavity of β-cyclodextrin, as reported by Stezowski and co-workers (Fig. 2.3.6) [53]. The SCSC reaction was facilitated by the inclusion of both reactants within a single cyclodextrin cavity. In that way, the photodimerization produced very small changes in the overall shape of the inclusion complex. In addition, water molecules of crystallization were shown to afford an environment favorable for the SCSC transformation by providing a way to relieve the strain induced by the reaction.

### 2.3.3.1.2 Linear Templates and Self-assembly

The application of linear templates to form reactive co-crystals relies on the ability of template molecules to juxtapose reactant olefins within finite molecular assemblies for reaction (Scheme 2.3.7) [42]. By virtue of the finite nature of the molecular assembly, the reaction is expected to occur largely without disturbing the overall crystal structure. In that way, co-crystals involving linear templates can be expected to provide an environment suitable to achieve SCSC reactivity. The linear template

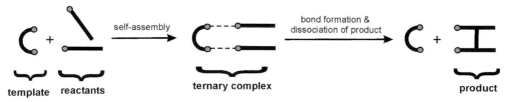

**Scheme 2.3.7**

approach also provides an opportunity to utilize multiple templates to assemble reactants and thereby take advantage of the process of self-assembly.

### Hydrogen-bond donor templates

The first example of a SCSC [2+2] photodimerization within a molecular assembly involving a co-crystal based on a linear template was reported by Ito and co-workers in the case of benzene-1,2-dicarboxylic acid and *trans*-cinnamide [54]. The two compounds formed a solid composed of *trans*-cinnamide olefins juxtaposed for a [2+2] photodimerization, via O–H···O hydrogen bonds, involving a single benzene-1,2-dicarboxylic acid molecule. The hydrogen-bonded assemblies of the diacid and cinnamide were also connected via N–H···O forces involving neighboring cinnamide molecules to give an infinite structure (Fig.2.3.7). The reaction proceeded in a SCSC fashion to a maximum of 13 % yield.

The ability of a SCSC reaction to occur within a finite molecular assembly based on a linear template was first demonstrated by MacGillivray and co-workers who utilized 1,8-naphthalenedicarboxylic acid as a template to assemble *trans*-1-(3-pyridyl)-2-(4-pyridyl)ethylene, via O–H···N hydrogen bonds, for a regioconotrlled [2+2] photodimerization [55]. The SCSC reaction was found to proceed, upon UV-irradiation with 420 nm light, quantitatively to give the head-to-head photoproduct *rctt*-1,2-bis(3-pyridyl)-3,4-bis(4-pyridyl)cyclobutane. A comparison of the single-

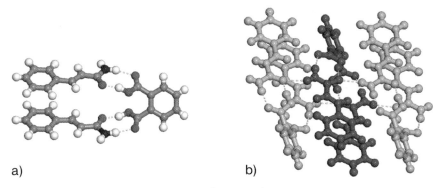

a)                                         b)

**Figure 2.3.7** Ball-and-stick representations of: (a) a single assembly and (b) three hydrogen-bonded assemblies of *trans*-cinnamide with 1,2-benzenedicarboxylic acid in the solid state. For clarity, the central assembly is shown in dark gray.

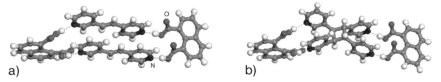

a)                                         b)

**Figure 2.3.8** Ball-and-stick representation of a single hydrogen-bonded assembly in the co-crystal involving 1,8-naphthalenedicarboxylic acid: (a) before reaction; (b) after complete reaction.

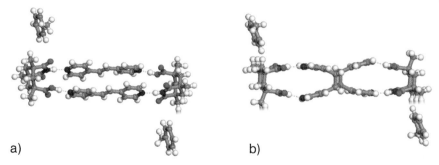

**Figure 2.3.9** Ball-and-stick representation of a single hydrogen-bonded assembly involving Rebek's imide, surrounded by two toluene molecules: (a) before reaction; (b) after complete reaction.

crystal structures of the solid before and after the photoreaction revealed that structural changes in the dicarboxylic acid template, as well as the hydrogen-bonded assembly, occurred (Fig. 2.3.8). In particular, the carboxylic acid moieties of each template underwent an approximate 4° rotation upon photoreaction while the templates also experienced sliding. Presumably, such structural flexibility was achieved by isolating the reaction within the finite assembly, which permitted stress to be reduced homogeneously within the solid.

MacGillivray and co-workers have also shown that a SCSC [2+2] photodimerization can be achieved with a linear template based on two different hydrogen-bond donor groups. In particular, Rebek's imide was shown to assemble *trans*-1,2-bis (4-pyridyl)ethylene via a combination of O–H···O and N–H···N forces involving the carboxyl and imide groups, respectively [56]. The components formed solvated hydrogen-bonded assemblies involving guest toluene molecules (Fig. 2.3.9(a)). The [2+2] photodimerization proceeded, upon irradiating the solid with 420 nm UV light, in a SCSC manner to a maximum of 27% yield to give, stereospecifically, *rctt-*

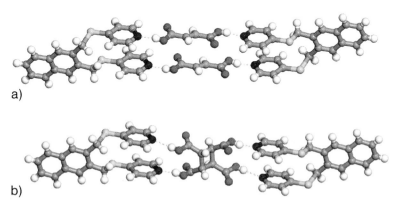

**Figure 2.3.10** Ball-and-stick representation of a single hydrogen-bonded assembly involving fumaric acid: (a) before reaction and (b) after complete reaction.

tetrakis(4-pyridyl)cyclobutane (Fig. 2.3.9(b)). That the solvent was present in the co-crystal throughout the photoreaction makes the material a unique example of SCSC reactivity achieved in a three-component solid.

#### Hydrogen-bond acceptor templates

MacGillivray and co-workers have recently demonstrated that the linear template approach can also be expanded to hydrogen-bond acceptor templates. In particular, co-crystallization of 2,3-bis(4-thiopyridylmethyl)naphthalene as a template with fumaric acid as the reactant produced a finite, four-component molecular assembly held together by O–H···N hydrogen bonds (Fig. 2.3.10(a)) [57]. The double bonds of the stacked fumaric acid molecules were aligned suitably for a photodimerization that proceeded, upon UV-irradiation with 300 nm light, in a SCSC fashion to a maximum of 36% yield. The photoreaction provided, stereospecifically, the *rctt* isomer of 1,2,3,4-cyclobutane-tetracarboxylic acid (Fig. 2.3.10(b)).

#### 2.3.3.1.3  Diacetylene and Triacetylene Polymerizations

The well-defined geometric parameters that dictate polymerizations of diacetylenes in the solid state have led Lauher and Fowler to employ principles of supramolecular chemistry to control the reaction within co-crystals. In particular, a urea derivative has been shown to self-assemble to form a hydrogen-bonded β-array that preorganizes complementary diacetylenes for a 1,4-photopolymerization. Through the use of a derivative with lateral carboxylic acid groups (Scheme 2.3.8(a)) and a bis(4-pyridyl)acetylene as the reactant (Scheme 2.3.8(b)), a solid-state arrangement suitable for the reaction was achieved (Fig. 2.3.11(a)) [58]. Indeed, general applic-

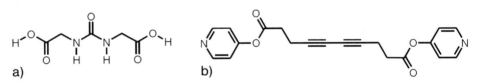

a)  b)

**Scheme 2.3.8**

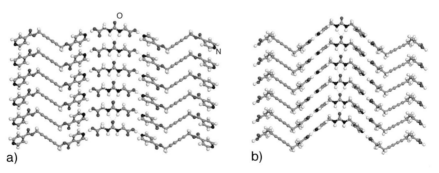

a)  b)

**Figure 2.3.11**  Ball-and-stick representations of the crystal structures of co-crystals of diacetylene derivatives involving: (a) a dicarboxylic acid host; (b) a bis(pyridine) host.

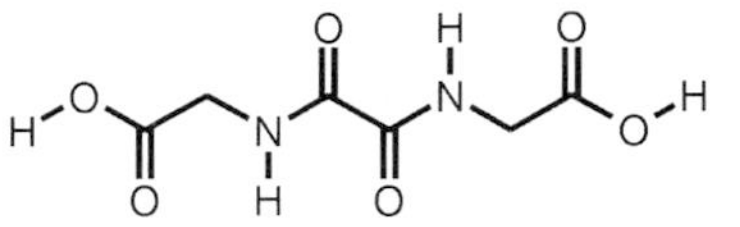

**Scheme 2.3.9**

ability of the urea β-stacking motif to align diacetylenes for reaction was established by obtaining an analogous solid involving bis(pyridine) and dicarboxylic acid derivatives of the urea host and diacetylene guest molecules, respectively (Fig. 2.3.11(b)). In both cases the co-crystals underwent a slow polymerization. The low rate of the reaction was explained in terms of an imperfect stacking motif dictated by the urea host. In order to achieve a more suitable stacked architecture of diacetylenes, the authors subsequently applied an alternative host based on a double oxalamide of glycine (Scheme 2.3.9). Co-crystals of the oxalamide with the bis(pyridine) derivative of diacetylene provided a solid-state architecture that exhibited SCSC reactivity so as to yield the corresponding polydiacetylene [59]. Notably, the same oxalamide host has also been shown to facilitate a previously unobserved topochemical 1,4-polymerization of a terminal diacetylene. This unique reaction occurred via a SCSC transformation within helical hydrogen-bonded assemblies of the oxalamide and terminal diacetylene to give the corresponding polydiacetylene (Fig. 2.3.12) [60].

The elegant approach of Lauher and Fowler has also been used to align triacetylenes for a topochemical 1,6-polymerization in the solid state [61]. The reaction was achieved using a vinylogous amide as a host that self-assembled via N–H···O hydrogen bonds to enforce stacking of a triacetylene functionalized with pyridyl groups. The reaction provided the corresponding polytriacetylene in a SCSC fashion (Fig. 2.3.13) [62].

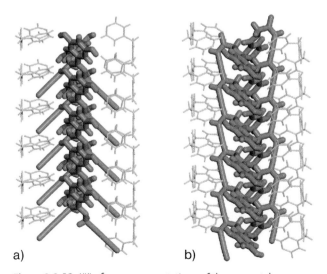

a)   b)

**Figure 2.3.12** Wireframe representations of the co-crystals involving the terminal diacetylene reactant: (a) before reaction; (b) after complete reaction.

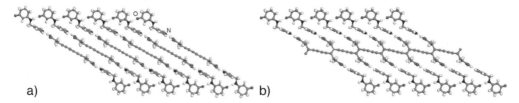

**Figure 2.3.13** Ball-and-stick representations of the co-crystals involving the triacetylene reactant: (a) before reaction; (b) after complete reaction.

### 2.3.3.1.4 Diels-Alder Reaction

Through the use of $\pi \cdots \pi$ donor $\cdots$ acceptor interactions, Kochi and co-workers have constructed two-component co-crystals involving bis($N$-methylimino)dithiin and anthracene that have been shown to undergo a SCSC Diels-Alder cycloaddition reaction (Fig. 2.3.14(a)) [63, 64]. The compounds provided co-crystals that consist of infinite stacks of anthracene and dithiin molecules (Fig. 2.3.14(b)). In each stack, neighboring molecules of dithiin and anthracene were positioned suitably for a Diels-Alder reaction, which occurred at temperatures between 50 and 80 °C, involving the 9 and 10 carbon atoms of the anthracene moiety as the diene and a double bond of the dithiin moiety as the dienophile.

### 2.3.4
### SCSC Reactivity by Modifying Experimental Conditions

In addition to the synthetic methods described above, two physical methods directed towards facilitating SCSC reactivity involving the [2+2] photodimerization

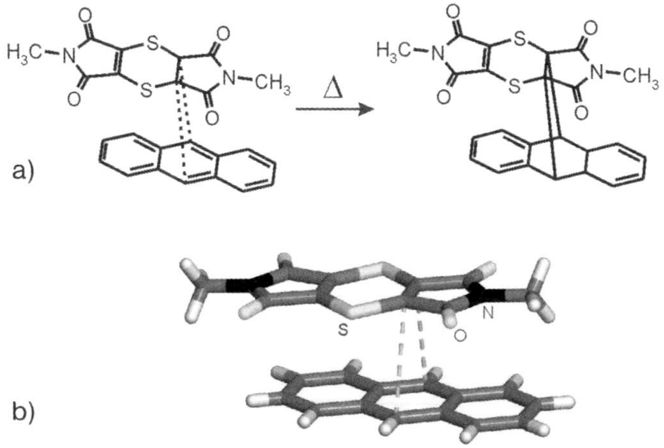

**Figure 2.3.14** (a) Schematic representation of the SCSC Diels-Alder reaction; (b) a wireframe representation of a thiine and an anthracene molecule in the reactant co-crystal.

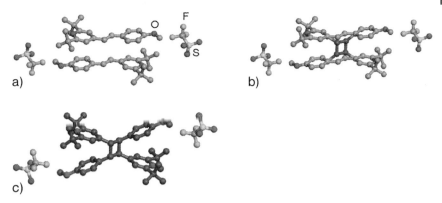

**Figure 2.3.15** Ball-and-stick representations of the crystal structure of the pyrillium triflate salt at different stages of reaction: (a) before reaction; (b) at 12 % yield; (c) after complete reaction.

have been reported: (i) adjustment of the UV source and (ii) the use of crystals of nanometer-scale dimensions.

The first method, introduced by Enkelmann and co-workers, relies on the use of radiation with a wavelength that corresponds to the absorption tail-end of the reactant crystal [65]. By employing such a wavelength, rapid build-up of photo-dimers and resulting strain at or near the crystal surface is avoided and, at the same time, a greater penetration depth of the radiation into the solid is achieved [66]. As a result, the photoreaction is expected to occur uniformly through the solid. The application of this absorption-edge irradiation method allowed Enkelmann and co-workers to monitor, for the first time, the course of the [2+2] photodimerization of α-cinnamic acid via dynamic X-ray crystallography [67]. Moreover, the method displayed somewhat general applicability to reactant diversity. In this context, an application of the method to solids based on styrylpyrillium cations [68] has been shown to produce photochromic systems that serve as holographic grating devices (Fig. 2.3.15). However, attempts to employ the approach for an SCSC construction of a polymer using styrylpyrillium cations with multiple double bonds (Scheme 2.3.10) were met with partial success. The limited success has been attributed to changes in the absorption edge energy throughout the [2+2] photodimerization process. In particular, the SCSC polymerization has been shown to stop at the level of pentamers owing to a shift of the absorption edge to shorter wavelengths [69].

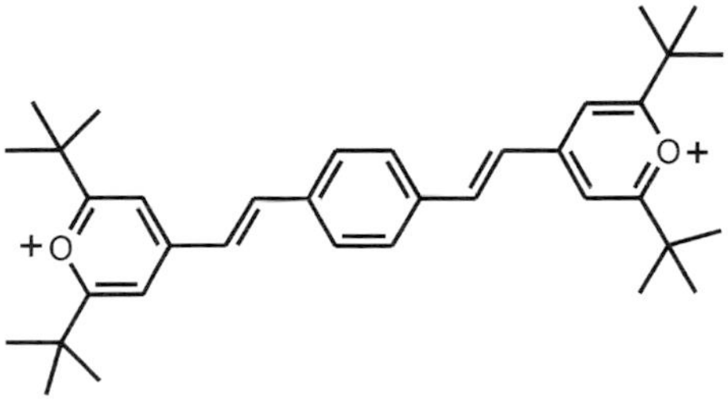

**Scheme 2.3.10**

A more recent method of inducing SCSC [2+2] photoreactivity involves the use of crystals of nanometer-scale dimensions [70]. The method has been successful for the construction of polymers of 2,5-distyrylpyrazine, as well as methyl *p*-phenylene-diacrylate. Presumably, the nanoscopic crystal size provides the reactant crystals with enough flexibility to withstand strain induced by the photopolymerization process.

2.3.5
**Conclusion**

Supramolecular approaches to direct reactivity in the solid state are emerging as reliable means to control chemical reactivity. The general approach has involved directing reactivity within molecular co-crystals. An increase in research on reactive co-crystals has also led to an increase in the number of solid-state reactions that proceed via SCSC transformations. This increase has also been accompanied by an increase in the scope of reactions that proceed in the SCSC manner. The rapid discovery of SCSC reactions that occur in co-crystals suggests that such multi-component solids provide an environment that is conducive to SCSC transformations. Such an environment has enabled significant amounts of atomic movement to occur in the solid state without disrupting overall crystal structures. Indeed, we expect that explorations and advances in the design of reactive co-crystals will lead to discoveries of new reactive solid-state systems that display crystal-to-crystal reactivity in the form of SCSC behavior. Such expansion is expected to facilitate advancement of knowledge on the mechanistic detail of chemical reactions (e.g. atomic movement), as well as applications of SCSC reactions for the construction of technologically useful materials.

**References**

1 Tanaka, K.; Toda, F. *Chem. Rev.* **2000**, *100*, 1025–1074.

2 Schmidt, G. M. J. *Pure Appl. Chem.* **1971**, *27*, 647–678.

3 Ramamurthy, V; Venkatesan, K. *Chem. Rev.* **1987**, *87*, 433–481.

4 Hirano, S.; Toyota, S.; Toda, F. *Mendeleev Commun.* **2004**, *6*, 247–249; Shizuka, H.; Ishii, Y.; Hoshino, M.; Morita, T. *J. Phys. Chem.* **1976**, *80*, 30–32.

5 (a) Radha Kishan K. V.; Desiraju, G. R. *J. Org. Chem.* **1987**, *52*, 4640–4641; (b) Desiraju, G. R.; Radha Kishan, K. V. *J. Am. Chem. Soc.* **1989**, *111*, 4838–4843.

6 Ramamurthy, V. *Tetrahedron* **1986**, *42*, 5753–5839

7 Hasegawa, M. *Pure Appl. Chem.* **1986**, *58*, 1179–1188.

8 Baughman, R. H. *J. Polym. Sci.* **1974**, *12*, 1511–1535.

9 Trask, A. V. T.; Jones, W. *Top. Curr. Chem.* **2005**, *254*, 41–70.

10 (a) Gavezzotti, A. *CrystEngComm* **2002**, *4*, 343–347; (b) Ashwini, N. *CrystEng-Comm* **2002**, *4*, 93–101.

11 Maddox, J. *Nature,* **1988**, *335*, 201.

12 Hosseini, M.-W. *Chem. Commun.* **2005**, 5825–5829.

13 Etter, M. C.; Panunto, T. W. *J. Am. Chem. Soc.* **1988**, *110*, 5896–5897.

14 Roth, H. D. *Angew. Chem. Int. Ed.* **1989**, *28*, 1193–1207.

15 Thomas, J. M. *Philos. Trans. R. Soc. London, Ser. A* **1974**, *277*, 251–286.

16 (a) Keating, A. E.; Garcia-Garibay, M. A. in *Molecular and Supramolecular Photo-*

*chemistry Vol. 2*, V. Ramamurthy, K. Schanze (Eds.), 195–248, Marcel Dekker, New York 1998, pp. 195–248; (b) Irie, M.; Kobatake, S.; Horichi, M. *Science* 2001, *291*, 1769–1772.

17 Friščić, T.; MacGillivray, L. R. *Z. Kristallogr.* 2005, *220*, 351–363.

18 (a) Atwood, J. L.; Barbour, L. J.; Jerga, A. *Science* 2002, *296*, 2367–2369; (b) Atwood, J. L.; Barbour, L. J.; Jerga, A. *Angew. Chem. Int. Ed.* 2004, *43*, 2948–2950; (c) Thallapally, P. K.; Lloyd, G. O.; Atwood, J. L. *Angew. Chem. Int. Ed.* 2005, *44*, 3848–3851; (d) Cote, A. P.; Benin, Ai. I.; Ockwig, N. W.; O'Keefe, M.; Matzger, A. J.; Yaghi, O. M. *Science* 2005, *310*, 1166–1170; (e) Rosswell, J. L. C.; Spencer, E. C.; Eckerten, J.; Howard, J. A. K.; Yaghi, O. M. *Science* 2005, *309*, 1350–1354.

19 (a) Brunet, P.; Demers, E.; Maris, T.; Enright, G. D.; Wuest, J. D. *Angew. Chem. Int. Ed.* 2003, *42*, 5303–5306; (b) Saied, O.; Maris, T.; Wuest, J. D. *J. Am. Chem. Soc.* 2003, *125*, 14956–14957; (c) Subramanian, S.; Zaworotko, M. J. *Angew. Chem. Int. Ed.* 1995, *34*, 2127–2129.

20 (a) Mortko, C. J.; Garcia-Garibay, M. A. *J. Am. Chem. Soc.* 2005, *127*, 7994–7995; (b) Yamada, T.; Muto, K.; Kobatake, S.; Irie, M. *J. Org. Chem.* 2001, *66*, 6164–6168.

21 Marubayashi, N.; Ogawa, T.; Hamasaki, T.; Hirayama, N. *J. Chem. Soc., Perkin Trans. 2* 1997, 1309–1314.

22 Hosomi, H.; Ito, Y.; Ohba, H. *Acta Crystallogr., Sect. B* 2000, *56*, 682–689.

23 Turowska-Tyrk, I.; Grz'esniak, K.; Trzop, E.; Zych, T. *J. Solid State Chem.* 2003, *174*, 459–465.

24 Hollingsworth, M. D.; McBride, J. M. *Adv. Photochem.* 1990, *15*, 279–379.

25 Cohen, M. D.; Schmidt, G. M. J.; Sonntag, F. I.; *J. Chem. Soc.* 1964, 2000–2013.

26 Osaki, K.; Schmidt, G. M. J. *Isr. J. Chem.* 1972, *10*, 189–193.

27 Nakanishi, H.; Jones, W.; Thomas, J. M. *Chem. Phys. Lett.* 1980, *71*, 44–48.

28 Honda, K.; Nakanishi, F.; Feeder, N. *J. Am. Chem. Soc.* 1999, *121*, 8246–8250; Turowska-Tyrk, I. *Acta Crystallogr., Sect. B* 2003, *59*, 670–675.

29 Jones, W.; Nakanishi, H.; Theocharis, C. R.; Thomas, J. M. *J. Chem. Soc., Chem. Commun.* 1980, 610–611.

30 (a) Chakrabarti, S.; Gantait, M.; Misra, T. N. *Proc. Indian Acad. Sci. Chem. Sci.* 1990, *102*, 165–172; (b) Atkinson, S. D. M.; Almond, M. J.; Hibble, S. J.; Hollins, P.; Jenkins, S. J.; Tobin, M. J.; Wiltshire, K. S. *Phys. Chem. Chem. Phys.* 2004, *6*, 4–6.

31 Ohba, S.; Ito, Y. *Acta Crystallogr., Sect. B* 2003, *59*, 149–155.

32 Hasegawa, M. *Chem. Rev.* 1983, *83*, 507–518.

33 Chung, C.-M.; Hasegawa, M. *J. Am. Chem. Soc.* 1991, *113*, 7311–7316.

34 Hasegawa, M.; Suzuki, Y.; Suzuki, F.; Nakanishi, H. *J. Polym. Sci. A-1* 1969, *7*, 743–752.

35 Stezowski, J. J.; Peachey, N. M.; Goebel, P.; Eckhardt, C. J. *J. Am. Chem. Soc.* 1993, *115*, 6499–6505.

36 Wegner, G. *Z. Naturforsch.b* 1969, *24*, 824–832.

37 Baughman, R. H. *J. Polym. Sci.* 1974, *12*, 1511–1535.

38 (a) Enkelmann, V.; Leyrer, R. J.; Shleier, G.; Wegner, G. *J. Mater. Sci.* 1980, *15*, 168–176; (b) Enkelmann, V.; Schleier, G.; Wegner, G.; Eichele, H.; Schwoerer, M. *Chem. Phys. Lett.* 1977, *52*, 314–319.

39 Hoang, T.; Lauher, J. W.; Fowler, F. W. *J. Am. Chem. Soc.* 2002, *124*, 10656–10657.

40 Wuest, J. D. *Chem. Commun.* 2005, 5830–5837.

41 Zerkowski, J. A.; MacDonald, J. C.; Seto, C. T.; Wierda, D. A.; Whitesides, G. M. *J. Am. Chem. Soc.* 1994, *116*, 2382–2391.

42 MacGillivray, L. R. *CrystEngComm* 2002, 4.

43 Trask, A. V.; Motherwell, W. D. S.; Jones, W. *Cryst. Growth Des.* 2005, *5*, 1013–1021.

44 Bailey Walsh, R. D.; Bradner, M. W.; Fleischman, S.; Morales, L. A.; Moulton, B.; Rodríguez-Hornedo, N.; Zaworotko, M. J. *Chem. Commun.* 2003, 186–187.

45 Remenar, J. F.; Morisette, S. L.; Peterson, M. L.; Moulton, B.; MacPhee, J. M.; Guzman, H. R.; Almarsson, Ö. *J. Am. Chem. Soc.* 2003, *125*, 8456–8457.

46 (a) MacGillivray, L. R.; Papaefstathiou, G. S.; Friščić, T.; Varshney, D. B.; Ham-

ilton, T. D. *Top. Curr. Chem.* **2005**, *248*, 201–221; (b) Friščić, T.; MacGillivray, L. R. *Supramol. Chem.* **2005**, *17*, 47–51.

47 McMahon, J. A.; Bis, J. A.; Vishweshwar, P.; Shattock, T. R.; McLaughlin, O. L.; Zaworotko, M. J. *Z. Kristallogr.* **2005**, *220*, 340–350.

48 Theocharis, C. R.; Desiraju, G. R.; Jones, W. *J. Am. Chem. Soc.* **1984**, *106*, 3606–3609.

49 (a) Papaefstathiou, G. S.; Zhong, Z.; Geng, L.; MacGillivray, L. R. *J. Am. Chem. Soc.* **2004**, *126*, 9158–9159; (b) Amirsakis, D. G.; Garcia-Garibay, M. A.; Rowan, S. J.; Stoddart, J. F.; White, A. J. P.; Williams, D. J. *Angew. Chem. Int. Ed.* **2001**, *40*, 4256–4261; (c) Metrangolo, P.; Neukirch, H.; Pilati, T.; Resnati, G. *Acc. Chem. Res.* **2005**, *38*, 386–395.

50 Tanaka, K.; Toda, F.; Mochizuki, E.; Yasui, N.; Kai, Y.; Miyahara, I.; Hirotsu, K. *Angew. Chem. Int. Ed.* **1999**, *38*, 3523–3525.

51 Tanaka, K.; Mochizuki, E.; Yasui, N.; Kai, Y.; Miyahara, I.; Hirotsu, K.; Toda, F. *Tetrahedron*, **2000**, *56*, 6853–6865.

52 Tanaka, K.; Mizutani, H.; Miyahara, I.; Hirotsku, K. Toda, F. *CrystEngComm* **1999**, 3.

53 Brett, T. J.; Alexander, J. M.; Stezowski, J. J. *J. Chem. Soc., Perkin Trans. 2* **2000**, 1105–1111.

54 Ohba, S.; Hosomi, H.; Ito, Y. *J. Am. Chem. Soc.* **2001**, *123*, 6349–6352.

55 Varshney, D. B.; MacGillivray, L. R. *Chem. Commun.* **2002**, 1964–1965.

56 Varshney, D. B.; Gao, X.; Friščić, T.; MacGillivray, L. R. *Angew. Chem. Int. Ed.* **2006**, *45*, 646–650.

57 Friščić, T.; MacGillivray, L. R. *Chem. Commun.* **2005**, 5748–5750.

58 Kane, J. J.; Liao, R.-F.; Lauher, J. W.; Fowler, F. W. *J. Am. Chem. Soc.* **1995**, *117*, 12003–12004.

59 Fowler, F. W.; Lauher, J. W. *J. Phys. Org. Chem.* **2000**, *13*, 850–857.

60 Quyang, X.; Fowler, F. W.; Lauher, J. W. *J. Am. Chem. Soc.* **2003**, *125*, 12400–12401.

61 Xiao, J.; Yang, M.; Lauher, J. W.; Fowler, F. W. *Angew. Chem. Int. Ed.* **2000**, *39*, 2132–2135.

62 Enkelmann, V. *Chem. Mater.* **1994**, *6*, 1337–1340.

63 Kim, J. H.; Hubig, S. M.; Lindemann, S. V.; Kochi, J. K. *J. Am. Chem. Soc.* **2001**, *123*, 87–95.

64 Kim, J. H.; Lindeman, S. V.; Kochi, J. K. *J. Am. Chem. Soc.* **2001**, *123*, 4951–4959.

65 Enkelmann, V. *Mol. Cryst. Liq. Cryst.* **1998**, *313*, 15–23.

66 Buchholz, V.; Enkelmann, V. *Cryst. Liq. Cryst.* **1998**, *313*, 309–314.

67 Enkelmann, V.; Wegner, G.; Novak, K.; Wagener, K. B. *J. Am. Chem. Soc.* **1993**, *115*, 10390–10391.

68 Köhler, W.; Novak, K. Enkelmann, V. *J. Chem. Phys.* **1994**, *101*, 10474–10480.

69 Buchholz, V.; Enkelmann, V. *Mol. Cryst. Liq. Cryst.* **2001**, *356*, 315–325.

70 Takahashi, S.; Miura, H.; Kasai, H.; Okada, S.; Hidetoshi, O.; Nakanishi, H. *J. Am. Chem. Soc.* **2002**, *124*, 10944–10945.

## 2.4
## Making Coordination Frameworks
*Neil R. Champness*

### 2.4.1
### Introduction

The study of coordination frameworks is an area of research that has risen rapidly to the forefront of modern chemistry [1]. The area has grown from the initial, and now seminal, reports of Richard Robson [2] to the highly advanced design strategies [3–5], structural appreciation [6, 7] and fascinating materials properties [8–13] now regularly reported in the highest profile scientific journals. The area of coordination frameworks now brings together researchers from across scientific disciplines, not only drawing upon synthetic chemists but increasingly upon structural scientists, materials scientists (from a variety of different disciplines), physicists and mathematicians. Indeed, the diversity of research issues that arise from studying coordination frameworks has brought the research area from its earliest origins amongst coordination and supramolecular chemists to such a wide audience that leading researchers in the area will no longer be readily identified as coming from a particular subsection of chemistry, or even be identified as chemists at all.

Underpinning all of this research is the primary requisite for the synthesis of these materials in an appropriate form for further study, whether investigating structure or properties. This chapter will attempt to address some of the issues that need to be considered when preparing coordination frameworks. Hopefully the chapter will provide not only a guide for those coming into the area but also food for thought to those who are established practitioners in the field. It aims to guide the reader through the process of coordination framework synthesis, from initial design, through choice of the components of the desired framework, choice of experimental conditions, considerations with respect to structural analysis and the ultimate understanding of framework structure and description.

### 2.4.2
### Coordination Framework Design Criteria

When preparing a coordination framework a variety of factors need to be taken into account. Firstly, and most importantly, the strategy to be employed for the synthesis of the coordination framework material depends entirely upon the desired outcome. Two main features come into play here, is the coordination framework being prepared purely for a structural arrangement or is a particular property/ function required from the resultant material? In the first instance, preparing a particular structural type, the design criteria relate almost entirely to the components of the coordination framework and the issues surrounding this are detailed below. If material properties are required then many other factors need to be anticipated, depending on the targeted property. For example, for magnetic materials the relative placement and orientation of spins within the components of the coordination framework array are critical to the resultant properties of the array [14]. This reliance on component orientation is similarly true in the design of materials with electrical/conducting properties [15] and those with optical properties [16]. It is not the intention of this chapter to address the properties of coordination frameworks but rather to address how one might target particular orientations and this will be discussed further below. The other major targets for material properties are porous materials for guest molecular entrapment, whether gases or guest solvent species. In this context there is a clear need for preparing coordination frameworks with channels, or pores, of defined dimensions [17]. Targeting such arrays brings the potential for further complications in the form of interpenetration [18] and more complex arrangements such as self-entangled structures [19]. The criteria that need to be considered when designing any of these structures will be considered below.

## 2.4.3
### Coordination Polymer Design Approaches

#### 2.4.3.1 The Building-block Methodology
Structural design of coordination frameworks, whether considering materials' properties or not, can be achieved using a very simple approach known as the building-block methodology. The basic principle of this methodology, or design strategy as proposed by Robson [2], relies upon the assembly of molecular, and/or ionic, components into extended arrays by using complementary interactions between the components. The strategy relies upon the use of divergent building blocks, i.e. those that encourage the formation of extended arrays rather than discrete, cyclic species. Thus, ligands such as 4,4'-bipyridine, that possesses two N-donors oriented at 180° to each other such that coordination to two distinct metals is ensured by the donor arrangement, will always encourage the formation of extended arrays with suitably compatible building-blocks [20]. In contrast a ligand such as 2,2'-bipyridine favours coordination of a single metal in a chelating fashion, discouraging polymer formation [21].

A similar approach can be considered for metal centers. A transition metal cation that bears no strongly coordinating ligands, such as $[Cu(MeCN)_4]^+$ upon which the MeCN ligands are highly labile [22], can act as a linking unit between divergent ligands such as 4,4'-bipyridine to afford an extended array, $\{[Cu(4,4'\text{-bipyridine})_2]^+\}_\infty$ [23]. However, in contrast to divergent ligands, metal cations with potentially divergent sites do not always generate extended arrays. For example, *cis*-protected metal centers such as $[M(\text{ethylenediamine})_2(\text{solv})_2]^{2+}$ (M = Pd, Pt), that react with divergent ligands to form cyclic species as their thermodynamic products, are capable of forming polymeric arrays as kinetic products [24].

In its most basic form the building-block methodology provides an excellent starting point for any coordination framework design. In essence, by linking building blocks through simple linear rods (connecting units) the geometric properties of the building block are expressed throughout three-dimensional space.

For example, one may consider the properties of a transition metal ion. Thus, if one links metal cations with a linear coordination geometry, e.g. Ag(I), through a simple connecting unit, a one-dimensional chain will be constructed. Similarly, a square-planar metal cation can be linked to generate a two-dimensional sheet structure. Three-dimensional architectures can be achieved by linking nonlinear or nonplanar metal centers, tetrahedral, octahedral or five-coordinate metals. For example, perhaps the most commonly studied example of a three-dimensional coordination framework is that constructed from tetrahedral metal centers, such as Cu(I), and linear bridging ligands, including 4,4'-bipyridine [23], or trans-1,2-bis (4-pyridyl)ethane [25]. The tetrahedral metal nodes will be linked such that the tetrahedral geometry of the cation is expressed in three dimensions to generate a structure with a diamond-like structure. Considering only the metal geometry, a wide variety of networks is possible. Whereas linear, tetrahedral and octahedral building-blocks lead to a limited number of framework arrangements, other building-block geometries lead to a greater variety of network types. This simple observation from the geometrical patterns is also observed in real chemical systems [3].

In addition, the geometry of the ligand can also be readily adapted and, in principle, combinations of different building-block geometric properties may be combined to generate increasingly complex multidimensional architectures. This concept of using building-block geometries to develop extended arrays is highly

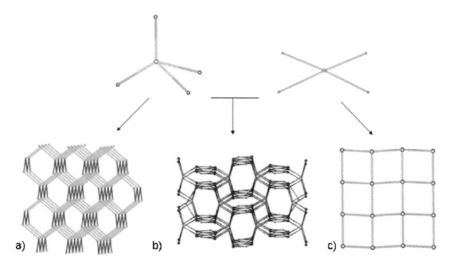

**Figure 2.4.1** The building-block modular approach to the synthesis of coordination frameworks developed an appreciation of not only the range of materials that might be prepared but also the material properties that could be targeted. This concept is illustrated by the combination of tetrahedral and square-planar building-blocks. Connection of tetrahe dral or square-planar building-blocks by simple linear connectors leads to the formation of (a) diamondoid ($6^4$ topology) or (c) square-grid ($4^4$ topology) structures, respectively, whereas combination of tetrahedral and square-planar building-blocks affords (b) PtS –type structures ($4^2.8^4$ topology).

adaptable. It is demonstrated by considering the frameworks generated from tetrahedral and square-planar building-blocks. Thus by connection of tetrahedral centers, or nodes, with linear bridges, diamondoid networks will be the most favourable configuration of the product coordination frameworks [23, 25]. Similarly, connecting square-planar nodes by linear bridges would most favorably form two-dimensional nets [26], although three-dimensional nets are also possible, such as structures with NbO or $Cd(SO_4)$ topologies [27]. However, combination of tetrahedral and square-planar nodes in a single framework would be expected to produce a three-dimensional net of PtS topology [28]. See Fig. 2.4.1 for a simple representation of this principle.

The building-block methodology is, therefore, a highly adaptable approach that can be used to generate a highly diverse range of coordination frameworks. Indeed, even with uninodal networks, i.e. those which possess a single type of node, examples of two- [29], three- [30], four- [23, 25], five- [31], six- [32], seven- [33], eight- [33, 34] and twelve-connected [35] frameworks have been generated by utilising the geometry of either single d- or f-block metal center cations, or by using larger metal-based clusters. This latter approach, where a larger node, based on multiple metal centers, is utilised has become extremely popular and successful in generating framework structures. Indeed, this approach where the metal-based node is sometimes called the secondary building unit (SBU) [36] has led to some of the most structurally robust frameworks that exhibit extraordinary material properties, such as Yaghi's highly significant compound MOF-5 [37]. An example of such an approach is where carboxylate-based ligands are reacted with Cu(II) salts to generate the highly familiar paddle-wheel structures, based on dinuclear $[Cu_2(O_2R)_4]$ units or SBUs [38]. By judicious choice of the carboxylate ligand, extended framework structures can be produced. Such SBUs build upon the knowledge and traditions of the well-established chemistry of discrete, molecular, coordination compounds. However, it is equally possible to serendipitously encounter SBUs during the preparation of a coordination framework where new, previously discovered metal-based cluster arrangements are stabilised by the framework environment [35].

Thus, we see that the metal-based building-block cannot be treated simply as a reliable cation of predefined geometry, but rather that the reactions of this building-block must be taken into account, and even utilised for the generation of specific frameworks. Similarly, the diverse possibilities of ligand design can also be utilised by the researcher wishing to make a coordination polymer. The majority of coordination frameworks reported to date have used relatively simple ligand systems exploiting bridging ligands that provide two–four monodentate donors in a predefined geometry [20, 23, 25, 31–34, 39, 40]. Such ligands give rise to a vast range of structural types but are limited in their influence of network structure in two ways. Firstly, as the ligands use only monodentate donors they offer little control over the interaction between the ligand and the metal center. The ligand is able to occupy a single metal coordination site, offering influence over the rest of the metal center only through steric crowding or through electronic effects (*trans*-effect etc.). Secondly, the ligand can bridge metal centers in a single manner. This means that the ligand is controlling the network structure only by means of the ligand length or potentially via steric bulk or weaker supramolecular interactions. However successful this strategy has been thus far, it should be clear from the wealth of highly advanced molecular coordination chemistry that ligand design can have a significant influence on the metal geometry and ultimately on metal properties. In contrast to the much more established molecular coordination chemistry, ligand design has only just begun to be probed in the context of coordination polymer structures, and has much to offer in providing control over coordination polymer structure and function [41].

### 2.4.3.2 Complications: An Inclusive Building-block Methodology

The metal cation and bridging ligands represent perhaps the most significant building-blocks in terms of their influence over coordination polymer structure. However, as with all aspects of crystal engineering, one cannot just consider the major energetic contributions, metal–ligand coordination bonds in the case of coordination frameworks, rather one must consider the crystal as a whole including weak interactions, which may have a significant effect upon the stability of a particular structural arrangement over that of a different assembly. Indeed, the individual components of a network system cannot be treated separately, but the network must be thought of as a combined entity drawing upon the unified properties of the building-blocks used in its construction. Thus, in the context of coordination polymers, other building-blocks or components can become highly significant, notably the anion (required to balance the charge of the metal cation) and the reaction solvent.

The very nature of coordination polymers requires the use of a metal cation and accordingly the charge that this cation brings, normally as $M^+$, $M^{2+}$ or $M^{3+}$, must be balanced by anions for the system to achieve electro-neutrality. This charge balancing can be achieved in two main ways. Firstly, by utilising anionic ligands and, secondly, by including a distinct anion. In many ways the use of anionic ligands precludes other possible complications. For, example the use of di-, or even tri-, carboxylates as bridging ligands can result in the formation of electrically neutral polymeric structures. In suitable cases porous frameworks can be formed, such as MOF-5 [37], such that the open cavities and channels in the structure are inherently not occupied by anions, leaving the space open for guest molecule inclusion.

In contrast, if neutrally charged ligands, such as bipyridyl ligands, are used then the cationic charge must be compensated by anions. The nature of these anions can have a significant influence on the overall structure. In particular, three factors need to be considered in the choice of anion. The first and perhaps most crucial property of any anion is its relative coordinating ability. Thus, anions such as cyanide, halides, thiocyanate, nitrate and sulfates have a strong coordinating ability for many metals. Metal coordination by auxiliary anions limits the number of available coordination sites of other, perhaps desired, bridging ligands. The coordinating anions may also control the arrangement of other ligands by adopting a preferred orientation. Lastly, coordinating anions may even act as bridging ligands themselves, thus significantly affecting the overall coordination polymer structure. The effect of different modes of anion coordination on coordination polymer structure has been reviewed for the nitrate anion [42].

An example of the effect of the change in anion involving nitrate is demonstrated by the following example of the Ag(I) complexes of 4-pytz (4-pytz = 1,4-bis-(4-pyridyl)-2,3,4,5-tetrazine). In contrast to the linear chain complexes {[Ag(4-pytz)]PF$_6$}∞ and {[Ag(4-pytz)]BF$_4$}∞, the complex formed between AgNO$_3$ and 4-pytz forms what has been termed a helical staircase structure [29], Fig. 2.4.2. The structure of the complex comprises Ag(I) ions each coordinated to two 4-pytz ligands in a linear arrangement to give linear chains of alternating Ag(I) ions and bipyridyl ligands. Each Ag(I) ion also exhibits weak interactions with two NO$_{3-}$ ions, Ag···O 2.787(2) Å, which adopt

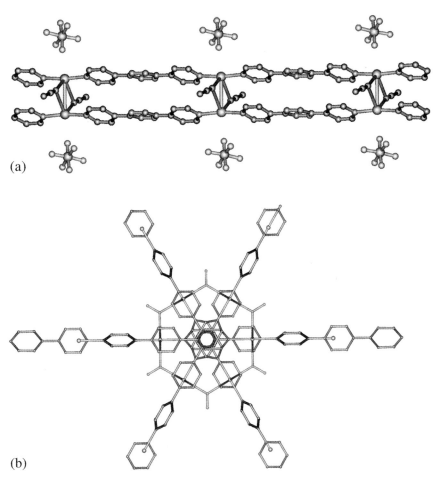

(a)

(b)

**Figure 2.4.2** Views of (a) a simple linear chain complex {[Ag(4-pytz)]PF$_6$}∞ that contrasts with the "helical staircase" structure formed by {[Ag(4-pytz)]NO$_3$}∞ [29].

positions perpendicular to the pyridyl–Ag–pyridyl axis. The $NO_{3-}$ ions bridge adjacent Ag–bipyridyl chains through two of their oxygen atoms so that each Ag(I) chain is related to the next by a 60° rotation and a step of 5.18 Å, generating a helical staircase. Thus, it can be seen that the three-dimensional structure is controlled via the $Ag–ONO_2$ interactions.

Although it is clear that the coordinating ability of an anion can have the most significant effect upon a coordination polymer structure, it is also possible for non-coordinated, or non-coordinating anions to affect coordination polymer structure in more subtle ways. Many anions have significant hydrogen bonding ability such that the anion may interact with hydrogen bond donors, either as part of the ligands used or in the form of other species connected to the coordination framework structure such as coordinated water molecules. Detailed studies of anion effects are scarce, although examples are known for some Ag(I)-pyridyl-based systems, amongst others [43]. Similarly, other supramolecular interactions such as π–π interactions may influence the structure [44, 45], although it is not clear that this potential factor has been studied in any detail for coordination polymer systems.

Lastly, it is conceivable that the size of the anion may have an influence over coordination polymer structure. Such an effect can be difficult to assess, and in particular, de-convolute from other factors, most noticeably from weak interactions with the polymer structure such as CH···acceptor interactions. If the size of an anion is influencing a structure, it is likely that it is in close contact with the polymer and therefore the formation of hydrogen bonds, or other supramolecular interactions, could be the main factor in affecting the structure.

The reaction/crystallisation solvent may also have a significant impact on coordination polymer structure, in a similar manner to that observed for anions. The coordinating ability of a solvent can have a particular influence on polymer structure, occupying coordination sites and influencing coordination sphere and geometry. This is demonstrated by the structures formed from the reaction of $Zn(NO_3)_2$ with 3-pytz [46] (3-pytz = 1,4-bis-(3-pyridyl)-2,3,4,5-tetrazine). When 3-pytz is reacted with $Zn(NO_3)_2$ in an EtOH (or $^i$PrOH) – $CH_2Cl_2$ mixture each Zn(II) center adopts a T-shaped motif which is utilised in the formation of a ladder coordination polymer $[Zn_2(NO_3)_4(3\text{-pytz})_3]_\infty$, Fig. 2.4.3(a). However, the T-shaped geometry at the Zn(II) center is disrupted when MeOH is used as the alcoholic solvent, with the MeOH molecule coordinating the metal center, allowing only two residual sites for pyridyl coordination. This results in the formation of a hydrogen-bonded 1D polymer, Fig. 2.4.3(b).

Reaction solvent can also have a more subtle effect upon coordination polymer reactions, and this is discussed in more detail below. However, in terms of the building-block methodology, if a solvent is particularly effective at binding and solvating a cation, it has been noticed that the metal cation can be kept back from participating in coordination polymer formation; the choice of solvent can thus influence the polymer product stoichiometry [47].

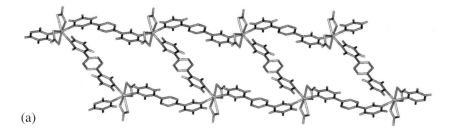

(a)

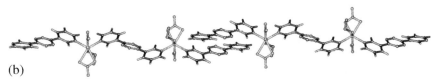

(b)

**Figure 2.4.3** View of (a) a ladder coordination polymer, [Zn$_2$(NO$_3$)$_4$(3-pytz)$_3$]$_\infty$, prepared when using an EtOH or $^i$PrOH as crystallisation solvent. However, the T-shaped geometry at the Zn(II) center is disrupted when MeOH is used as the crystallisation solvent, with the MeOH molecule coordinating the metal center resulting in the formation of a hydrogen-bonded 1D polymer (b).

### 2.4.3.3  Porosity and Interpenetration

One of the major goals currently facing researchers interested in coordination polymers is that of creating robust porous materials capable of absorbing guest molecules [10, 11, 13, 17], notably dihydrogen [8, 9, 48, 49]. Such a target requires the rational, designed, synthesis of porous coordination polymer structures (Fig. 2.4.4). In principle, porous structures are an easy target – simply by increasing the separation between nodes of the network by increasing ligand length (donor separation) should lead to the formation of porous frameworks. However, this is a simplification of the issues that are encountered. Firstly, any three-dimensional coordination framework will contain guest molecules filling the pores upon initial synthesis. The nature of this guest varies, depending upon the coordination framework, and can vary from anions, in the case of cationic frameworks, to guest solvent molecules and even uncoordinated ligands. Of these potential guests it is clear that anions should be avoided as they cannot be removed entirely, only exchanged for other anions, thus not aiding the design of a porous framework. Guest solvent molecules or "free" ligands can potentially be removed [50] although the ease of removal will depend on the nature of the guest, for example molecules which interact only through loose interactions (C–H$\cdots$O interactions for example) can be readily removed. In contrast, molecules that are strongly hydrogen bonded to the framework or have strong intermolecular interactions with other guest molecules are potentially difficult to remove from the structure without the framework undergoing a significant structural change or even collapse. The other factor to consider is that if the guest molecule has a boiling-point that is higher than the

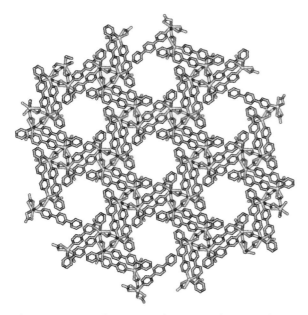

**Figure 2.4.4** View of a porous cadmium coordination polymer, {Cd(NO₃)₂[2,4'-(1,4-phenylene)bispyridine]}∞ [50]. Such materials are a particular target for coordination polymer chemists and require rational, designed, synthetic approaches.

decomposition temperature of the framework, then removal of the guest molecule would have to be achieved by non-thermal methods (possibly exchange with a more volatile guest followed by thermal removal).

The other key issue that requires addressing when attempting the preparation of porous coordination framework structures is interpenetration [18]. Interpenetration is the observation of two or more polycatenated coordination polymer arrays within a single compound, see Fig. 2.4.5(a). Although the majority of interpenetrated coordination polymer structures are formed by polycatenanes other systems involving polyrotaxanes and self-entangled structures are also known [19] (Fig. 2.4.5(b)). Many beautiful structures exhibiting interpenetration have been observed and the control of interpenetration, whether in polycatenanes or self-entangled structures, has been a focus of research over recent years. However, from the perspective of porous structures, interpenetration may be perceived as unfortunate, as it leads to the blocking of available space, limiting the porosity of the material product. However, it has also been noted that interpenetration may act to reinforce the stability of the coordination framework, with respect to structural collapse and potentially enhances absorption of certain molecules such as $H_2$ where large pore volume is not necessarily beneficial [49].

Strategies to avoid interpenetration do exist [3, 52] and have been successfully employed to generate porous framework structures. In particular Yaghi et al. have successfully prepared a vast array of framework structures utilising the defined

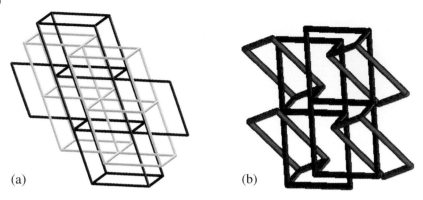

**Figure 2.4.5** Unusual structural arrangements can be observed for coordination polymers including (a) interpenetration, or polycatenation, and (b) self-entanglement.

geometry of the $Zn_4O$ SBU (see above) with carboxylate-based bridging ligands [51]. Indeed some of these materials represent the most porous materials yet reported, with free space within each framework far in excess of the required 50% at which interpenetration may occur. It is apparent, therefore that interpenetration does not have to present a problem, rather it is an issue that the researcher designing a coordination polymer needs to be aware of.

### 2.4.4
### Synthetic Considerations and Approaches

The most important factor to be considered by a researcher wishing to prepare a coordination polymer structure is the requirement that the product must be characterised. In contrast to many of areas of chemistry, those working with coordination polymers are restricted to the use of predominantly one technique, single-crystal X-ray diffraction. Although solution of powder X-ray diffraction patterns has become significantly more advanced over recent years [53], it is still not in a position to rival single-crystal approaches. This significant factor leads to a requirement for single crystals from reactions to prepare coordination polymers, a significant challenge to those working in the area.

The requirement for single crystals is perhaps more of a challenge considering the highly insoluble nature of coordination polymers. It is also notable that recrystallisation, often used for molecular systems, is not as relevant for coordination polymer systems, as not only are coordination polymers typically highly insoluble but even if they are dissolved in aggressive solvents, such as DMSO or DMF, the polymer structure will break up and any coordination polymer that re-forms may bear no resemblance to the product from the original reaction.

Why is making single crystals of a coordination polymer a challenge? Many coordination polymer reactions precipitate a product rapidly from solution. Although it is possible to encounter systems that give single crystals of sufficient quality for X-ray diffraction studies, many systems afford either microcrystalline or amorphous products following rapid precipitation.

Approaches to growing crystals of coordination polymers must be found and these may take two major directions. The first involves the judicious choice of building-blocks. Coordination polymer systems that involve the formation of kinetically inert coordinate bonds rarely, if ever, form highly crystalline material. In simplistic terms the reason behind this is that once a bond has been formed in building a coordination polymer, if that bond is inert, it is difficult for that bond to be broken under conventional crystallisation conditions. Thus, "mistakes" in the coordination polymer formation cannot be readily corrected, typically leading to small crystallites or microcrystalline material.

It is not insignificant that a large proportion of reported coordination polymers contains highly labile metal centers, in particular $d^{10}$ centers Cu(I) [54], Ag(I) [55], Zn(II) [51] and Cd(II) [42], but also $d^9$ Cu(II) [38] and $d^7$ Co(II) [56] cations and increasingly with lanthanide cations, Ln(III) [7]. Examples of second or third row transition metal based coordination polymers, with the exception of Ag(I) and Cd(II), are extremely rare amongst reported examples.

It is also not insignificant that many coordination polymers have been synthesised using bridging ligands with monodentate donors, 4,4'-bipyridine etc. [20], and that multidentate chelating ligands have a much lower profile in the field to date [4]. Carboxylate-based systems [51] provide a slight exception in that they can adopt $\eta^2$ modes of coordination, although these can be intermediate between $\eta^1$ and $\eta^2$, but such systems are often prepared under hydrothermal/solvothermal conditions.

The second consideration when attempting to prepare crystals of coordination polymers is the crystallisation condition used. As stated above, recrystallisation approaches are not usually appropriate for the growth of coordination polymer systems, due to solubility reasons. Another method of crystallisation often used for molecular systems, that of vapor diffusion of an anti-solvent into solutions of the product, is also rarely possible for coordination polymer systems. As with traditional recrystallisation this is due to the insolubility of the product material and consequently to the difficulty of obtaining solutions of either the coordination polymer or its constituent parts.

Thus, crystallisation of coordination polymers is often achieved by using less common approaches (than for molecular systems) that rely upon the slow diffusion of separate solutions of the building-blocks, thus ensuring slow introduction of the components of the target material. Thus, layering methods, use of "U-tubes", "H-tubes", introduction of buffers (whether solvent buffers, silica gels or sintered glass) between separate solutions of the building-blocks have all been used to prepare coordination polymers. It is not the intention in this chapter to discuss the specifics of crystal growth, but rather I would direct the reader to a valuable resource [57] that describes in detail many crystallisation methods.

These approaches to crystal growth rely upon slow introduction of the component building-blocks, however, another approach is possible. The key issue in coordination polymer crystal growth is to slow the crystal formation process. This can be achieved by providing greater solubility of the building-blocks, or possibly small oligomers of the coordination polymer. This can be achieved by judicious choice of solvent. For example, Ag(I) and Cu(I) coordination polymers are commonly grown from MeCN solutions, as this solvent exhibits a high affinity for either Ag(I) or Cu(I), solubilising these cations as $[M(MeCN)_4]^+$ and providing chemical competition for ligands that are being used to form the coordination polymer structure [54, 55]. Similar approaches are used when strongly coordinating solvents such as *N,N*-dimethylformamide (DMF), are used with dicationic metal cations, notably Zn(II) [51].

Hydrothermal or solvothermal methods of crystallisation are increasingly common in preparing coordination polymers and this approach utilises the increased solubility of building-blocks and coordination polymer oligomers under the conditions of the reaction [58]. It is also notable that the elevated temperatures of the solvothermal processes are also likely to increase the lability of the coordinate bonds in the reaction process, thus slowing coordination polymer formation. The slow cooling of the solutions, following heating at high temperatures under pressure for prolongued periods, also favors the formation of crystalline product.

Recently it has also become possible to prepare coordination polymers via grinding of the components in the solid state, to afford microcrystalline material that can be analysed using powder X-ray diffraction techniques [59]. Although it is difficult to envisage the preparation of single crystals large enough for single crystal X-ray diffraction studies using this approach (the grinding disfavors the formation of large crystals!) it seems likely that this simple and highly versatile approach will become increasingly important for the preparation of coordination polymers. At present, structural elucidation for microcrystalline materials remains a significant challenge (see above) but solid-state grinding approaches do allow facile, large scale preparation of coordination polymers of known materials, originally characterised by single-crystal diffraction studies.

## 2.4.5
### Structural Evaluation and Analysis

Structural evaluation of coordination polymer structure is clearly extremely important if the synthesis and material properties of these species are to be understood. As stated above, by far the most efficient and conclusive method of assessing such structures is by single-crystal X-ray diffraction methods and this technique is so crucial, at present, that it strongly influences the choice of building-blocks and synthetic methodology used. *Ab initio* powder X-ray diffraction structure solution has made significant progress over recent years and has led to the evaluation of coordination polymer structures. Thus the major tools open to the coordination polymer chemist are these two diffraction techniques, with powder X-ray diffraction structure solution likely to become increasingly important. Neutron diffraction studies are particularly important in studies of weakly diffracting guest molecules, including $H_2$ [60] and $O_2$ [10]. Other tools are clearly important in evaluating material properties, but are less important in direct structural determination.

Two problems are regularly encountered by those investigating coordination polymer structures – evaluation of diffuse areas of highly disordered guest molecules and structural description. Although the former problem may be addressed by skilful structural modelling, it is increasingly common for guest molecules to be so disordered and/or diffuse that traditional methods of structural modelling become impossible. Structural evaluation of the material then becomes reliant upon the supporting evidence of combined techniques, potentially including elemental analysis, thermal analysis (TGA, DSC), spectroscopy (IR, Raman) and solid-state NMR. It is worth noting an approach which is becoming increasingly commonly reported, the SQUEEZE procedure as used with the PLATON suite of crystallographic tools [61]. SQUEEZE can be used to evaluate residual electron density, giving an estimated number of residual, unassigned electrons within a unit cell. This information can be used to aid the identification of any unidentified guest species, but needs to be supported by experimental evidence from other techniques.

## 2.4.6
## Structural Description

The description of coordination polymer structures can be a challenging task and many approaches have been proposed. The most common approach that has been used is to describe the structure in terms of the connectivity of nodes and in terms of the sequencing of shortest circuits within the two- or three-dimensional structure. This Wellsian approach [62] has received considerable attention and has been utilised in various forms to describe increasingly complex structures. The advantage of the approach, which leads to the use of topological terms that describe the extended structure, is that any structure can be unequivocally assigned to a topological term symbol as long as sufficient depth of description is given. This chapter does not aim to describe topological nomenclature but rather directs the reader to excellent works on the subject, notably a recent book by Öhrström [63].

The objective of topological nomenclature is not only to accurately describe a structural arrangement but also to inform the reader of the overall structure in terms of its components. It is often noted that the casual reader of a paper that uses topological descriptions of coordination frameworks does not appreciate the subtleties of topological nomenclature. Thus alternative approaches have been described. Structure with high connectivity can be particularly difficult to appreciate and thus an alternative approach based on the interconnectivity of two-dimensional nets, or subnet tectons, has been described [7]. This approach relies on the basis that many simple two-dimensional nets are readily visualised (such as simple square-based $4^4$ nets, or honeycomb arrangements, $6^3$ nets) and many highly connected nets are built from the intersection of these relatively simple nets. This approach works for all structures reported to date and may prove valuable for complex and intricate structures, notably those with connectivity over six.

In addition a recent approach to structural description has been reported in which common structural types have been given a three letter designation relating the structure to a well-known solid-state, typically inorganic, compound [6]. For example the diamond-like net constructed from the connection of tetrahedral nodes has $6^6$ topology but is simply described as **dia**, relating the structure directly to the familiar diamond arrangement. Similarly the quartz net is given the designation **qtz** and a primitive cubic net is called a **pcu** net. The approach is simple and yet can allow for more complex nets by simple augmentation of the three letter code by the addition of a fourth letter describing modifications of the existing net. Although the approach is still relatively recent, it does have the advantage of being brief and, in its simplest form (just three letters), accounts for the majority of structures that are reported.

The other common difficulty that can arise in structural description is the assignment of nodes in the net described by a structure. In a simple coordination polymer structure a single metal cation may act as a node, exhibiting four- or six-fold connectivity for example. However, as more complex systems are developed larger units can be considered as nodes, providing a much more useful and visibly accessible description of the structure. For example SBUs (see above) are widely regarded as topological nodes, such as the $Zn_4O$ unit that acts as a six-connected node in MOF-5 [37]. It is also worth noting that ligands should be considered as nodes when exhibiting greater than two-fold connectivity (i.e. simple bridging). For example, three-connected ligands should be considered as distinct nodes, often leading to frameworks with more than one topological node in their structure [41].

## 2.4.7
### Conclusions

This chapter has endeavored to address the issues that need to be considered when preparing a coordination polymer. It has raised a number of issues that need to be considered when designing a coordination polymer, from the initial building-block approach to other factors that need to be addressed such as influence of solvent, anion etc. Once having designed the components to be used in constructing the desired system, synthetic aspects need to be taken into account, dealing with the choice of reaction conditions and the requirement for single crystals at the end of the experiment. Lastly, the analysis of coordination polymer structures requires not only careful experimental consideration but also, perhaps more than in any other area of current research, an appreciation of structural description to allow effective communication of the results. Although coordination polymers offer many exciting possibilities, both in terms of structural science and materials properties they also represent a significant challenge to chemists.

## References

1 N. R. Champness, *Dalton Trans.*, **2006**, 877.
2 B. F. Hoskins, R. Robson, *J. Am. Chem. Soc.*, **1990**, *112*, 1546; R. Robson, B. F. Abrahams, S. R. Batten, R. W. Gable, B. F. Hoskins, J. P. Liu, *ACS Symp. Ser.*, **1992**, *499*, 256.
3 M. Eddaoudi, D. B. Moler, H. L. Li, B. L. Chen, T. M. Reineke, M. O'Keeffe, O. M. Yaghi, *Acc. Chem. Res.*, **2001**, *34*, 319.
4 N. R. Champness, N. S. Oxtoby, *Encyclopaedia of Nanoscience and Nanotechnology*, Ed. J. A. Schwartz, C. I. Contescu, K. Putyera, Dekker **2004**, p. 845.

5 M.W. Hosseini, *Acc. Chem. Res.*, 2005, *38*, 313.

6 N. W. Ockwig, O. Delgado-Friedrichs, M. O'Keeffe, O. M. Yaghi, *Acc. Chem. Res.*, 2005, *38*, 176.

7 R. J. Hill, D.-L. Long, N. R. Champness, P. Hubberstey, M. Schröder, *Acc. Chem. Res.*, 2005, *38*, 337.

8 N. L. Rosi, J. Eckert, M. Eddaoudi, D. T. Vodak, J. Kim, M. O'Keeffe, O. M. Yaghi, *Science*, 2003, *300*, 1127.

9 X. B. Zhao, B. Xiao, A. J. Fletcher, K. M. Thomas, D. Bradshaw, M. J. Rosseinsky, *Science*, 2004, *306*, 1012.

10 R. Kitaura, S. Kitagawa, Y. Kubota, T. C. Kobayashi, K. Kindo, Y. Mita, A. Matsuo, M. Kobayashi, H. C. Chang, T. C. Ozawa, M. Suzuki, M. Sakata, M. Takata, *Science*, 2002, *298*, 2358.

11 D. Maspoch, D. Ruiz-Molina, K. Wurst, N. Domingo, M. Cavallini, F. Biscarini, J. Tejada, C. Rovira, J. Veciana, *Nat. Mater.*, 2003, *2*, 190.

12 E. Coronado, J. R. Galan-Mascaros, C. J. Gomez-Garcia, V. Laukhin, *Nature*, 2000, *408*, 447.

13 A. R. Millward, O. M. Yaghi, *J. Am. Chem. Soc.*, 2005, *127*, 17998.

14 O. Kahn, *Acc. Chem. Res.*, 2000, *33*, 647.

15 H. Miyasaka, C. S. Campos-Fernandez, R. Clerac, K. R. Dunbar, *Angew. Chem. Int. Ed.*, 2000, *29*, 3831.

16 O. R. Evans, W. B. Lin, *Acc. Chem. Res.*, 2002, *35*, 511.

17 D. Bradshaw, J. B. Claridge, E. J. Cussen, T. J. Prior, M. J. Rosseinsky, *Acc. Chem. Res.*, 2005, *38*, 273.

18 S.R. Batten, *CrystEngComm*, 2001, *3*, 67.

19 L. Carlucci, G. Ciani, D. M. Proserpio, *Coord. Chem Rev.*, 2003, *246*, 247; L. Carlucci, G. Ciani, D. M. Proserpio, *CrystEngComm*, 2003, *5*, 269.

20 H. W. Roesky, M. Andruh, *Coord. Chem. Rev.* 2003, *236*, 91.

21 C. Kaes, A. Katz, M. W. Hosseini, *Chem. Rev.*, 2000, *100*, 3553.

22 J. M. Knaust, D. A. Knight, S. W. Keller, *J. Chem. Crystallogr.*, 2003, *33*, 813.

23 L. R. Mac Gillivray, S. Subramanian, M. J. Zaworotko, *J. Chem. Soc. Chem. Commun.*, 1994, 1325.

24 M. Fujita, M. Tominaga, A. Hori, B. Therrien, *Acc. Chem. Res.*, 2005, *38*, 369.

25 A. J. Blake, N. R. Champness, S. S. M. Chung, W.-S. Li, M. Schröder, *Chem. Commun.* 1997, 1005.

26 S. Noro, R. Kitaura, M. Kondo, S. Kitagawa, T. Ishii, H. Matsuzaka, M. Yamashita, *J. Am. Chem. Soc.*, 2002, *124*, 2568; L. Carlucci, G. Ciani, D. M. Proserpio, S. Rizzato, *CrystEngComm*, 2002, *4*, 413; L. Carlucci, G. Ciani, D. M. Proserpio, S. Rizzato, *CrystEngComm*, 2002, *4*, 413; K. A. Hirsch, S. R. Wilson, J. S. Moore, *Inorg. Chem.*, 1997, *36*, 2960.

27 L. Carlucci, G. Ciani, P. Macchi, D. M. Proserpio, *Chem. Commun.*, 1998, 1837; K. N. Power, T. L. Hennigar, M. J. Zaworotko, *Chem. Commun.*, 1998, 595; S. A. Barnett, A. J. Blake, N. R. Champness, C. Wilson, *Chem. Commun.*, 2002, 1640; M. Eddaoudi, J. Kim, M. O'Keeffe, O. M. Yaghi, *J. Am. Chem. Soc.*, 2002, *124*, 376.

28 O. V. Dolomanov, D. B. Cordes, N. R. Champness, A. J. Blake, L. R. Hanton, G. B. Jameson, M. Schröder, C. Wilson, *Chem. Commun.*, 2004, 642.

29 M. A. Withersby, A. J. Blake, N. R. Champness, P. Hubberstey, W.-S. Li, M. Schröder, *Angew. Chem., Int. Ed. Engl.*, 1997, *36*, 2327.

30 S. R. Batten, B. F. Hoskins, R. Robson, B. Moubaraki, K. S. Murray, *Chem. Commun.*, 2000, 1095.

31 D-L. Long, A. J. Blake, N. R. Champness, C. Wilson, M. Schröder, *J. Am. Chem. Soc.*, 2001, *123*, 3401.

32 D.-L. Long, R. J. Hill, A. J. Blake, N. R. Champness, P. Hubberstey, C. Wilson, M. Schröder, *Chem. Eur. J.*, 2005, *11*, 1384 and references therein.

33 D-L. Long, A. J. Blake, N. R. Champness, C. Wilson, M. Schröder, *Angew. Chem., Int. Ed.*, 2001, *40*, 2444.

34 D-L. Long, R. J. Hill, A. J. Blake, N. R. Champness, P. Hubberstey, D. M. Proserpio, C. Wilson, M. Schröder, *Angew. Chem., Int. Ed.*, 2004, *43*, 1851.

35 X-M. Zhang, R-Q. Fang, H-S. Wu, *J. Am. Chem. Soc.*, 2005, *127*, 7670; D. Li, T. Wu, X.-P. Zhou, R. Zhou, X.-C. Huang, *Angew. Chem. Int. Ed.*, 2005, *44*, 4175.

36 O. M. Yaghi, M. O'Keeffe, N. W. Ockwig, H. K. Chae, M. Eddaoudi, J. Kim, *Nature*, 2003, *423*, 705.

37 M. Eddaoudi, H. L. Li, O. M. Yaghi, *J. Am. Chem. Soc.*, **2000**, *122*, 1391.

38 B. L. Chen, M. Eddaoudi, S. T. Hyde, M. O'Keeffe, O. M. Yaghi, *Science*, **2001**, *291*, 1021.

39 A. J. Blake, N. R. Brooks, N. R. Champness, P. A. Cooke, A. M. Deveson, D. Fenske, P. Hubberstey, W.-S. Li, M. Schröder. *J. Chem. Soc., Dalton Trans.*, **1999**, 2103.

40 C. Klein, E. Graf, M. W. Hosseini, A. De Cian, *New J. Chem.* **2001**, *25*, 207.

41 N. S. Oxtoby, A. J. Blake, N. R. Champness, C. Wilson, *Proc. Natl. Acad. Sci. (USA)*, **2002**, *99*, 4905; J. J. M. Amoore, L. R. Hanton, M. D. Spicer, *Supramol. Chem.*, **2005**, *17*, 557.

42 S. A. Barnett, N. R. Champness, *Coord. Chem. Rev.*, **2003**, *246*, 145.

43 A. J. Blake, G. Baum, N. R. Champness, S. S. M. Chung, P. A. Cooke, D. Fenske, A. N. Khlobystov, D. A. Lemenovskii, W.-S. Li, M. Schröder, *J. Chem. Soc., Dalton Trans.*, **2000**, 4285.

44 A. J. Blake, N. R. Champness, A. N. Khlobystov, D. A. Lemenovskii, W.-S. Li, M. Schröder, *Chem. Commun.*, **1997**, 1339.

45 I. Dance, *Mol. Cryst. Liq. Cryst.*, **2005**, *440*, 265.

46 M. A. Withersby, A. J. Blake, N. R. Champness, P. A. Cooke, P. Hubberstey, W.-S. Li, M. Schröder. *Inorg. Chem.*, **1999**, 38, 2259.

47 A. J. Blake, N. R. Champness, P. A. Cooke, J. E. B. Nicolson, C. Wilson, *J. Chem. Soc., Dalton Trans.*, **2000**, 3811.

48 G. Ferey, C. Mellot-Draznieks, C. Serre, F. Millange, *Acc. Chem. Res.*, **2005**, *38*, 217; J. L. C. Rowsell, O. M. Yaghi, *Angew. Chem. Int. Ed.*, **2005**, *44*, 4670; S. S. Kaye, J. R. Long, *J. Am. Chem. Soc.*, **2005**, *127*, 6506.

49 B. Chen, M. Eddaoudi, S. T. Hyde, M. O'Keeffe, O. M. Yaghi, *Science*, **2001**, *291*, 1021; J. L. C. Rowsell, O. M. Yaghi, *J. Am. Chem. Soc.* **2006**, *128*, 1304.

50 A. J. Blake, N. R. Champness, A. N. Khlobystov, S. Parsons, M. Schröder. *Angew. Chem., Int. Ed.*, **2000**, *39*, 2317.

51 N. L. Rosi, M. Eddaoudi, J. Kim, M. O'Keeffe, O. M. Yaghi, *CrystEngComm*, **2002**, *4*, 401.

52 X. L. Xu, M. Nieuwenhuyzen, S. L. James, *Angew. Chem. Int. Ed.*, **2002**, *41*, 764.

53 N. Masciocchi, S. Galli, A. Sironi, *Comments Inorg. Chem.*, **2005**, *26*, 1.

54 A. J. Blake, N. R. Champness, P. Hubberstey, W.-S. Li, M. A. Withersby, M. Schröder, *Coord. Chem. Rev.*, **1999**, *183*, 117.

55 A. N. Khlobystov, A. J. Blake, N. R. Champness, D. A. Lemenovskii, A. G. Majouga, N. V. Zyk, M. Schröder, *Coord. Chem. Rev.*, **2001**, *222*, 137.

56 T. L. Hennigar, D. C. MacQuarrie, P. Losier, R. D. Rogers, M. J. Zaworotko, *Angew. Chem. Int. Ed. Engl.*, **1997**, *36*, 972.

57 See http://www.nottingham.ac.uk/~pczajb2/growcrys.htm and references therein.

58 P. D. C. Dietzel, R. Blom, H. Fjellvag, *Dalton Trans.*, **2006**, 586; Y. J. Kim, Y. J. Park, D. Y. Jung, *Dalton Trans.*, **2005**, 2603; B. Zhao, L. Yi, Y. Dai, X. Y. Chen, P. Cheng, D. Z. Liao, S. P. Yan, Z. H. Jiang, *Inorg. Chem.*, **2005**, *44*, 911.

59 D. Braga, M. Curzi, A. Johansson, M. Polito, K. Rubini, F. Grepioni, *Angew. Chem. Int. Ed.*, **2006**, *45*, 142. A. Pichon, A. Lazuen-Garay, S. L. James, *CrystEngComm*, **2006**, *8*, DOI: 10.1039/b513750k.

60 J. L. C. Rowsell, J. Eckert, O. M. Yaghi, *J. Am. Chem. Soc.*, **2005**, *127*, 14904; E. C. Spencer, J. A. K. Howard, G. J. McIntyre, J. L. C. Rowsell, O. M. Yaghi, *Chem. Commun.*, **2006**, 278.

61 A. L. Spek, *Acta Crystallogr., Sect. A*, **1990**, *46*, 194.

62 A. F. Wells, *Structural Inorganic Chemistry*, 5th edn., Clarendon Press, Oxford, 1984; A. F. Wells, *Three-Dimensional Nets and Polyhedra*, John Wiley and Sons, New York, 1977.

63 L. Öhrström, K. Larsson, *Molecule-Based Materials: The Structural Network Approach*, Elsevier, Amsterdam, 2005.

## 2.5
## Assembly of Molecular Solids via Non-covalent Interactions
*Christer B. Aakeröy and Nate Schultheiss*

## 2.5.1
## Introduction

### 2.5.1.1 How Do We Design, Engineer and Build a Molecular Crystal?

A central aspect of crystal engineering, which in itself can be viewed as a sub-discipline of supramolecular chemistry, concerns the construction of crystalline materials from discrete molecular building blocks using non-covalent interactions [1, 2]. An overriding goal for such efforts is to acquire and develop reliable and practical means for the synthesis of molecular materials with specific and tunable properties [3]. For example, we may want to design materials that can perform chemical separations, or that have nonlinear optical [4], magnetic [5], or catalytic properties [6]. In order to obtain nonlinear optical materials, individual molecules must be organized in a noncentrosymmetric arrangement with appropriately aligned dipole moments, for magnetic materials they must be positioned so that communication between spins is facilitated and optimized, and for chemical separations the host-material must be able to selectively entrap molecules or ions. In other words, if we wish to build functional materials, we must be able to control *how* molecular building blocks can be assembled into architectures with desirable connectivity and precisely defined metrics. Supramolecular synthesis is a proactive process – molecular building blocks are typically designed such that directional intermolecular interactions are expressed in terms of specific structural consequences and it may therefore be more appropriate to describe targeted supramolecular synthesis as "directed assembly" instead of "self-assembly".

### 2.5.1.2 The Power of Covalent Synthesis

For more than a century, synthetic chemists have devised a vast number of reactions allowing more and more complicated compounds to be made using elaborate processes [7]. Today, we are capable of making extraordinary molecules that rival some of Nature's best efforts when it comes to structural complexity [8, 9]. Covalent synthesis has become such a powerful discipline [8, 9] partly because organic chemists have been able to establish reproducible links between molecular structure, reactivity, and reaction pathways through systematic studies of innumerable organic reactions. The explicit correspondence between molecular structure and function has provided a plethora of "named reactions" that collectively create an invaluable toolbox for covalent synthesis. Individual reactions can often be described in a brief, yet fully comprehensive manner – "...this is the transformation of A into B, by treatment with X and Y". These recipes, often involving specific catalysts and reagents that facilitate coupling reactions between two different molecular fragments through the making and breaking of covalent bonds, play an essential role in every aspect of synthetic chemistry. In contrast, supramolecular synthesis [8] has yet to reach the same level of sophistication [2, 10, 11], and despite much progress we do not yet have access to a 'dictionary' that allows us to translate from molecular structure to supramolecular assembly.

Chemistry is the science of communication and change, and these interrelated processes are primarily initiated and controlled by reversible interactions *between* molecules. Consequently, the behavior of an assembly of molecules, be it in solution or in the solid phase, is determined by cooperative intermolecular communication. Unfortunately, our understanding of many fundamental physical properties as displayed by relatively simple crystalline materials, e.g. color, solubility, and thermal stability, is still rudimentary. This lack of comprehension and control stems from an incomplete insight into the forces that determine molecular recognition, binding and multi-molecular association [12, 13, 14, 15, 16]. Crystal engineering is still at the stage where much experimentation is concerned with *how* to control assembly [10, 17]. Thus, predicting and controlling the competition between intermolecular forces such as weak and strong hydrogen bonds, halogen–halogen interactions, $\pi \cdots \pi$ contacts, van der Waals forces, etc. are part of every crystal engineering effort. The challenge comes down to the following: how do we unravel the bundle of often interrelated interactions and identify strong, directional, reliable non-covalent interactions? How do we develop a reliable synthetic protocol for supramolecular synthesis?

### 2.5.1.3 The Case for Supramolecular Chemistry

*"...the emerging science of supramolecular chemistry represents a natural extension of the science of molecular chemistry"* [18].

Despite the inherent difficulties with using weak forces as primary synthetic tools, considerable progress has been made, and the preponderance of strategies for supramolecular synthesis prompted Desiraju [19] to introduce the term "supramolecular synthon". Synthons describe the precise recognition events that take place when molecules assemble into supermolecules and provide an important illustration of the conceptual similarities between retrosynthetic organic synthesis and supramolecular assembly [20].

Identifying supramolecular synthons is clearly as important to crystal engineering as is an understanding of reaction mechanisms and reagents in conventional covalent synthesis. It is essential to find the limits and conditions under which intermolecular recognition can be used to reliably assemble structures and materials with specific topologies. Such insight is most effectively obtained through systematic structural studies, and recent developments in crystallography/diffraction hardware and software have provided unprecedented access to structural information, both in terms of sheer numbers of new crystal structures and the type (size/quality) of samples that can be examined.

This chapter outlines some advances that have been made specifically through the use of hydrogen-bond interactions in directed assembly of organic molecular solids with desirable and well-defined supramolecular connectivity [21]. However, the scope of this chapter only allows us to mention but a fraction of the many efforts that have been put forth, and the main focus will be on extended networks and infinite architectures; directed assembly of discrete entities has been reviewed extensively [22].

This chapter is broadly divided into two subject areas. First, an overview of strategies for the directed assembly of homomeric 0-D, 1-D, 2-D and 3-D molecular architectures with specific, pre-determined, and desirable connectivities and metrics is offered. Second, a description of some design principles that have been devised and tested for the assembly of co-crystals, heteromeric molecular solids, is given.

## 2.5.2
## Directed Assembly of Homomeric Molecular Solids

### 2.5.2.1 Hydrogen-bond Interactions – Essential Tools for Constructing Molecular Solids

The intermolecular interaction that lends itself most readily to chemical or geometric fine-tuning is, arguably, the hydrogen bond, and the strength and directionality of this interaction, as compared to most other intermolecular forces, account for its significance in supramolecular synthesis of molecular solids [23, 24]. Many design strategies have relied upon hydrogen-bond complementarity [2, 7, 10, 11, 13–16, 25], which can involve both geometric factors and a suitable balance between the number of hydrogen-bond donors and hydrogen-bond acceptors. Self-complementary homomeric interactions have been studied extensively and they include the well-known carboxylic acid dimer, although other homomeric synthons have also been employed in crystal engineering, Scheme 2.5.1.

Not only do these complementary supramolecular synthons bring molecules together, they also constrain the relative orientation of those components in much the same way as carbon–carbon double bonds impart specific stereochemistry to individual molecules.

An increased degree of complementarity (triple [26], quadruple [27], or quintuple [28]) will increase selectivity but may also present considerable covalent synthetic challenges; a balance has to be struck in order to develop cost-effective and versatile supramolecular synthesis.

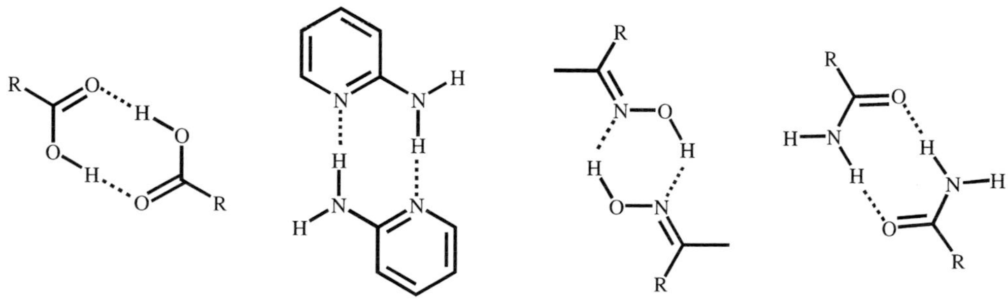

**Scheme 2.5.1** Examples of self-complementary (homomeric) hydrogen-bond interactions.

The importance of complementary hydrogen bonds for providing selectivity and specificity in biological processes is also well documented. A crucial consequence of the commonly occurring eight-membered hydrogen-bonded rings in such systems is that the participating molecules only fit together in one orientation, thereby providing an essential organizational handle that provides a mechanism for coding and encoding information upon the resulting assembly (cf. base pairing in DNA).

### 2.5.2.2 Hydrogen-bond-based Assemblies

#### 2.5.2.2.1 Homomeric 0-D Assemblies Constructed from Hydrogen Bonds

Carboxylic acids, oximes, diones and amino-pyridines can all form homomeric 0-D motifs through self-complementary and, frequently, symmetry-related hydrogen-bond interactions,

4-Chlorophenylcarboxylic acid produces a dimer through self-complementary O–H···O hydrogen bonds [29], while 4-chlorophenyloxime dimerizes through O–H···N hydrogen bonds [30]. Similarly, 5-chloro-2-pyridone forms a dimer through N–H···O hydrogen bonds [31], while 2-amino-5-chloropyridine forms a dimer through N–H···N hydrogen bonds [32]. In the absence of other strong hydrogen-bond donors or acceptors such as a pyridine or an –OH moiety, these interactions can be relied upon to appear, regardless of the overall shape and size of the molecule to which they are covalently attached. All four functionalities could, in principle, also lead to infinite 1-D assemblies (see Section 2.5.2.3.2) or cyclic

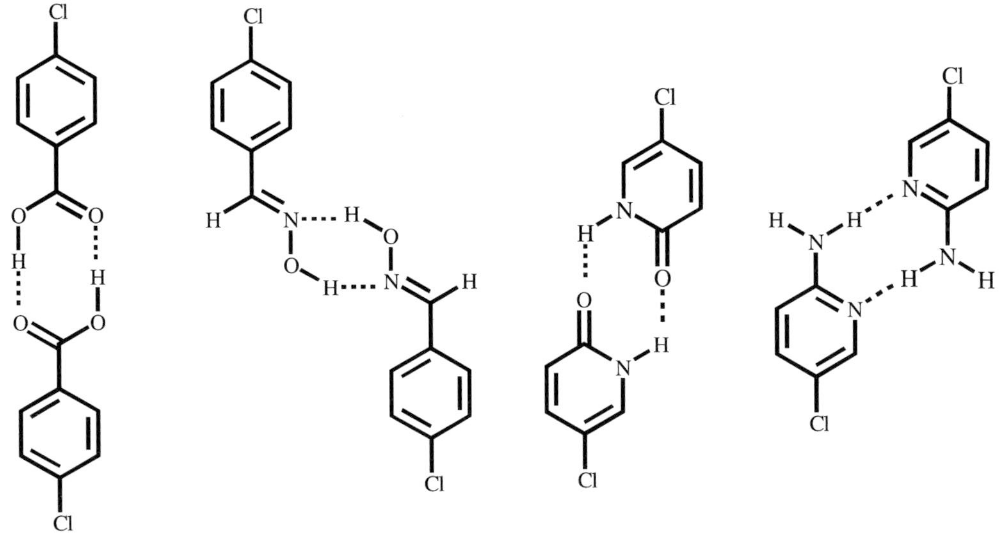

**Scheme 2.5.2** Homomeric 0-D motifs formed from acids, oximes, pyridones and amino-pyridines dimers.

structures without altering hydrogen-bond connectivities, but such aggregates are generally far less common than their dimeric counterparts. One exemption is noteworthy however. If the oxime is equipped with a strongly electron-withdrawing group on the imine moiety (such as –C≡N) the most likely motif is a 1-D chain (O–H···N≡C–) instead of a dimeric motif involving the imine nitrogen atom as the hydrogen-bond acceptor.

### 2.5.2.2.2    Hydrogen-bonded Homomeric 1-D Architectures

Amides [33], pyrimidines [34], dicarboxylic acids [35], dioximes [36, 37] and dipyridones [38] can all form homomeric 1-D motifs through self-complementary interactions, Scheme 2.5.3.

Amides frequently combine two unique structural motifs, dimers and catemers, into a well-known 1-D ribbon-like structure. The dimer is constructed from self-complementary N–H···O hydrogen bonds, while the catemeric N–H···O hydrogen bonds result from the anti-proton of the amine and the bifurcated carbonyl oxygen atom.

Symmetric dipyridones form 1-D arrays through pairs of N–H···O interactions. Similarly, diacids, both aliphatic and aromatic, commonly produce infinite 1-D chains in the absence of strong competitive hydrogen-bond acceptors.

Oxime functionalities have the capability of forming either dimers or catemers both of which result from O–H···N hydrogen bonds. The balance between the two types of motifs is very subtle, which is illustrated by the fact that 4-chloropheny-

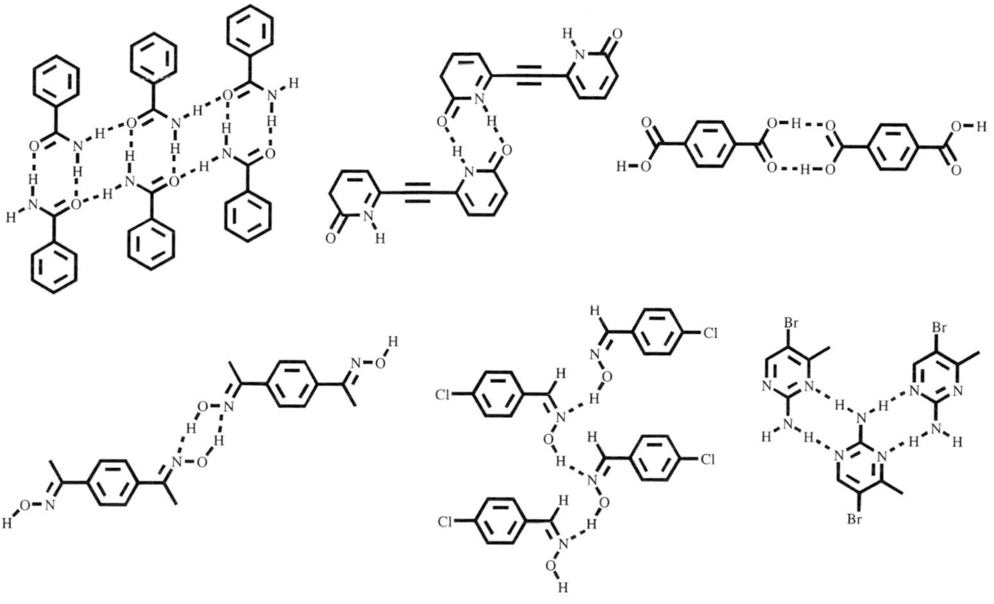

**Scheme 2.5.3** Homomeric 1-D motifs formed from a variety of reliable hydrogen-bond synthons.

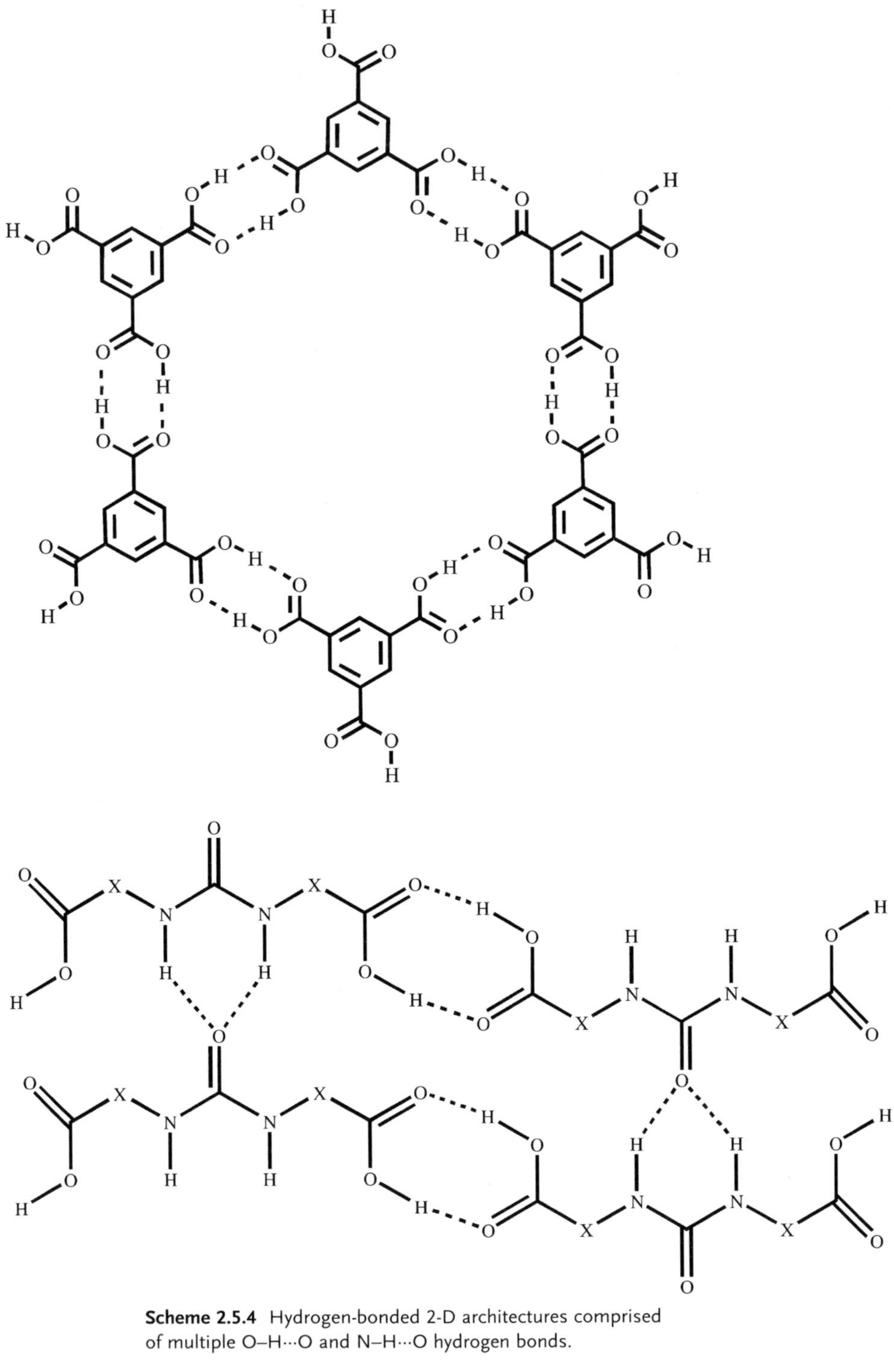

**Scheme 2.5.4** Hydrogen-bonded 2-D architectures comprised of multiple O–H⋯O and N–H⋯O hydrogen bonds.

laldoxime, which is known to be dimorphic, appears both in a dimeric as well as in a catemeric version in the solid state [30, 37]. A search through the CSD on aromatic molecules containing the oxime fragment, in the absence of competing moieties, revealed approximately 20 1-D chain motifs, 2 cyclic motifs and 45 dimeric motifs.

### 2.5.2.2.3 Hydrogen-bonded Homomeric 2-D Architectures

Any reasonably reliable supramolecular synthon can, in principle, provide the basis for molecular building blocks capable of creating extended 2-D architectures. For example, trimesic acid (benzene-1,3,5-tricarboxylic acid) forms hexagonal networks with the aid of self-complementary O–H$\cdots$O hydrogen bonds [39], Scheme 2.5.4. These hexagonal sheets are catenated, leading to an interpenetrated 3-D framework. Extended 2-D arrays are produced from a variety of symmetrical and unsymmetrical substituted urea and oxalamidecarboxylic acids utilizing acid–acid dimers to form the 1-D chains while further propagation is achieved in two dimensions through multiple N–H$\cdots$O hydrogen bonds [24, 40], Scheme 2.5.4. Not surprisingly, when the acid moiety of the symmetrical substituted ureas and oxalamides is changed to a phenol group, a 2-D sheet is once again formed through a series of O–H$\cdots$O and N–H$\cdots$O hydrogen bonds [41]. It has also been shown that pyrazinecarboxamide forms 2-D architectures through multiple N–H$\cdots$O and C–H$\cdots$N hydrogen bonds [42].

### 2.5.2.2.4 Hydrogen-bonded Homomeric 3-D Architectures

Carboxylic acids, such as adamantane-1,3,5,7-tetracarboxylic acid [43] and tetrapyridones (silane and carbon based) [44, 45], have been employed in the construction of 3-D diamondoid networks through self-complementary interactions, Scheme 2.5.5.

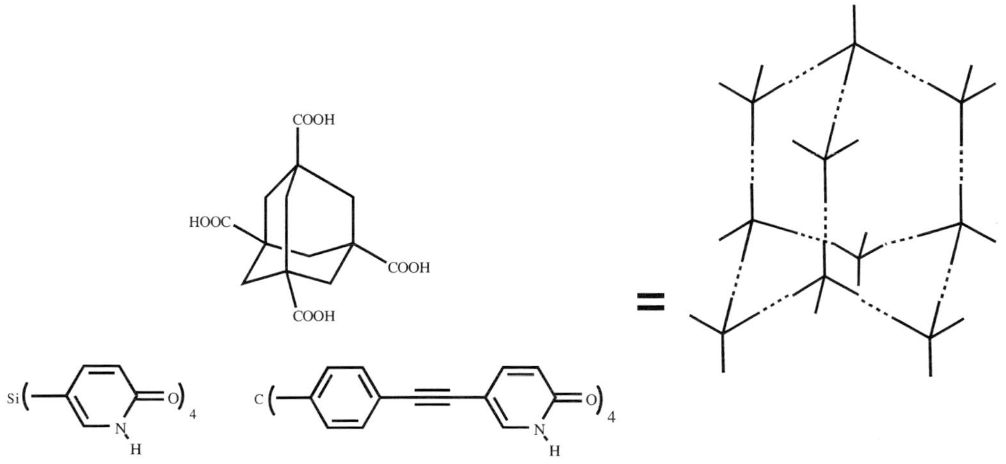

**Scheme 2.5.5** Examples of molecules that form 3-dimensional self-complementary hydrogen-bonded networks.

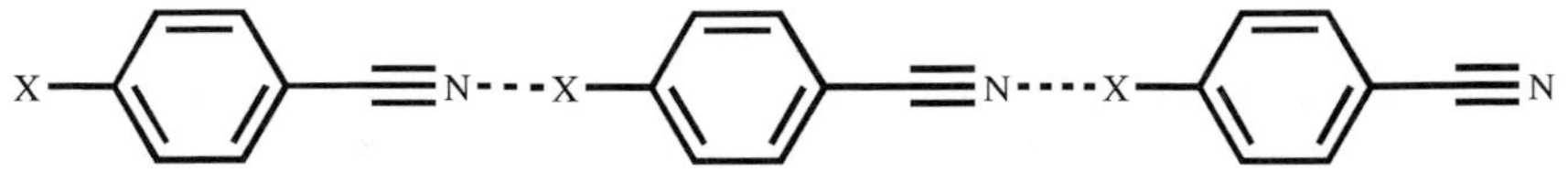

$$X=I, Br$$

**Scheme 2.5.6** Homomeric 1-D chains formed from nitrile halogen interactions.

### 2.5.2.3 Halogen-bond-based Assembly Strategies

This section covers structures and assemblies that have been constructed principally through $X \cdots X$, $N \cdots X$, and $C \equiv N \cdots X$ (where X is either Cl, Br, or I). For information on the nature of such interactions, see some recent review articles [46].

#### 2.5.2.3.1 Halogen Bonds through $N \cdots X$ (X = Cl, Br, I)

4-Iodobenzonitrile and 4-bromobenzonitrile form one-dimensional head-to-tail chains through $C \equiv N \cdots X$ interactions, Scheme 2.5.6 [47, 48].

Fluorination of 4-iodobenzonitrile or 4-bromobenzonitrile molecules does not disrupt the primary nitrile $\cdots$ halogen interactions, in fact, the $C \equiv N \cdots X$ distances become somewhat shorter. This can be expected, since the Lewis acidity of the halogen atoms increases due to the electron-withdrawing effects of the four fluorine substituents on the phenyl ring. In changing the halogen to a chloro group, the nitrile $\cdots$ chloro interaction does not form, even with the ring being fluorinated.

4-iodopyridine forms 1-D head-to-tail chains through $N \cdots I$ interactions, Scheme 2.5.7 [49]. Similarly, 4'-bromo-2',3',5',6'-tetrafluorostilbazole forms 1-D head-to-tail through $N \cdots Br$ interactions, Scheme 2.5.7 [50]. The $N \cdots I$ and $N \cdots Br$ interactions are fundamentally similar, but the former are generally regarded as being significantly stronger [51].

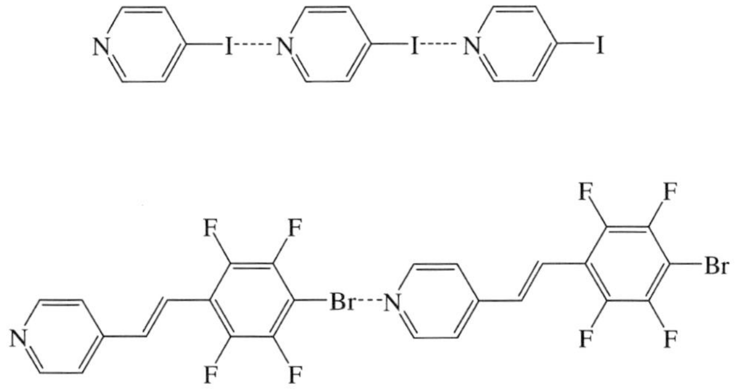

**Scheme 2.5.7** Homomeric 1-D chains formed from nitrogen iodo or bromo interactions.

### 2.5.2.3.2    Halogen···Halogen Interactions

A variety of halogen···halogen interactions have been used to guide the orientation of molecular building blocks in homomeric solids into predictable architectures. For example, 1,4-dihalogenated aryls (X = Cl, Br, I) consistently give rise to two-dimensional sheets through a series of bifurcated halogen···halogen interactions, Scheme 2.5.8 [52, 53, 54].

Fluorinated compounds do behave much differently compared to their heavier congeners. Desiraju has proposed that F···F interactions are unlikely to play a structure-directing role and, instead, fluorine atoms prefer to interact weakly with neighboring C–H protons forming C–H···F interactions [55]. Interestingly, Dunitz has reported, based on molecular modeling and specific CSD searches, that organic fluorine atoms hardly ever accept a hydrogen atom to form C–H···F interactions and therefore, should not be regarded as a weak hydrogen bond [56].

Halogen···halogen interactions have also been used to design even more complex architectures, and the crystal structure determination of 1,3,5,7-tetraiododoadamantane reveals the presence of a 3-D diamondoid-like network established through multiple iodo···iodo interactions [57, 58].

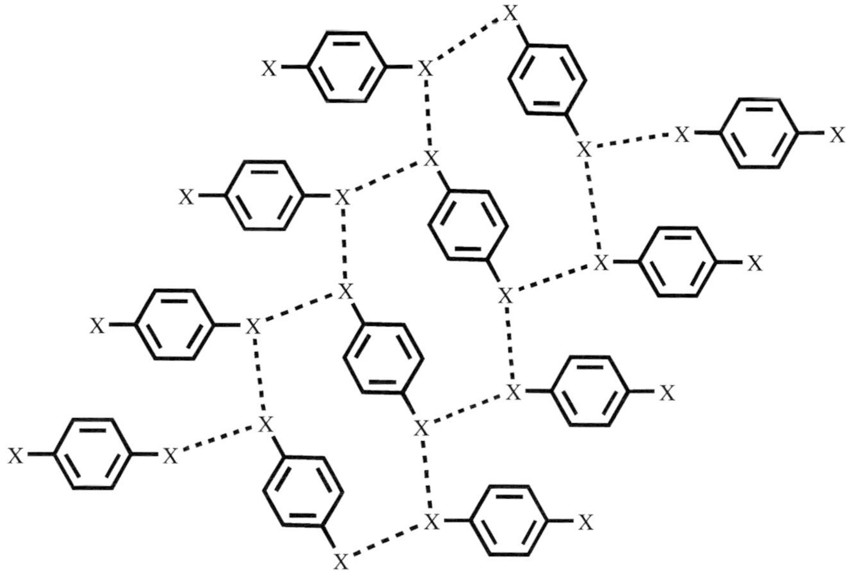

**Scheme 2.5.8** Extended 2-D sheets produced from bifurcated halogen halogen interactions.

## 2.5.3
## Design and Synthesis of Co-crystals

### 2.5.3.1 Background

#### 2.5.3.1.1 Supramolecular Synthesis
The term "synthesis", when used in chemical sciences, has come to mean the construction of new species, most commonly by bringing together two different discrete entities, followed by the breaking and making of covalent bonds. Since conventional synthesis relies on relatively stable covalent bonds, it is possible to devise assembly processes composed of several independent steps performed in a sequential manner, e.g. protection of active groups, followed by coupling reaction, then deprotection, etc. In supramolecular synthesis, on the other hand, the desired product is typically held together by reversible intermolecular interactions [2, 10, 11] and solution-phase supramolecular chemistry is therefore normally restricted to one-pot processes. A supramolecular "intermediate" cannot ordinarily be prepared, isolated and then added to another solution in order to perform sequential synthesis and herein lies a challenge at the core of supramolecular synthesis: How to devise reliable synthetic pathways towards sophisticated supramolecular structures if we are frequently confined to one-pot reactions? Is it somehow possible to develop a 'dictionary' that allows us to translate from molecular structure to supramolecular assembly [59]?

A possible solution to the problem of making one-pot synthesis "sequential" may be to devise *modular* assembly processes based upon a hierarchy of intermolecular interactions, Scheme 2.5.9.

#### 2.5.3.1.2 Co-crystals: Solid-state Targets for Non-covalent Synthesis
What is the most likely outcome when a solution containing different molecular solutes is allowed to evaporate to dryness? Unless a chemical reaction driven by the formation of covalent bonds takes place between the solutes, one would normally expect the appearance of separate molecular solids. This is a demonstration of the innate structural selfishness of molecules [60], and it is utilized every time recrystallization is employed as a method of purification. Recrystallizations are per-

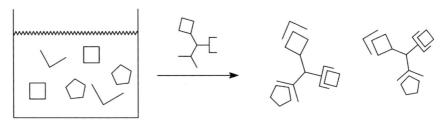

**Scheme 2.5.9** A modular one-pot supramolecular reaction driven by three different selective intermolecular interactions resulting in quaternary supermolecules.

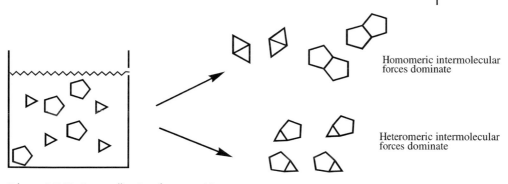

**Scheme 2.5.10** Recrystallization (homomeric) or co-crystallization (heteromeric)?

formed on a daily basis in every synthetic laboratory around the world but in the supramolecular laboratory, however, the very same process provides the supramolecular chemist with an opportunity to move in a completely different direction. A co-crystallization is an attempt at bringing together different molecular species within one periodic crystalline lattice without making or breaking covalent bonds. Recrystallization and co-crystallization processes are essentially only distinguishable by their intents. The goal of the former is a homomeric product, whereas the latter procedure strives for a heteromeric product. In general, the odds are stacked firmly in favor of the former, Scheme 2.5.10, so how do we go about developing reliable, effective, and versatile synthetic methods for the directed assembly of molecular co-crystals?

Since a crystalline material can be constructed in an enormous number of ways, these recognition mechanisms are extraordinarily selective and, consequently, each crystal structure contains important information about the way in which intermolecular forces compete and collaborate and eventually create an energetically balanced system. The use of co-crystallizations for studying packing patterns, hydrogen-bond motifs, and intermolecular forces, was brought to the forefront by Etter and coworkers [61–64]. By introducing several molecular species to the same solution, followed by systematic structural studies, one can begin to identify patterns of behavior, structural selectivity, and trends in recognition and binding that will furnish an improved understanding of the balance between solvent–solute and solute–solute interactions.

This section will provide some practical approaches to the design of binary and ternary supermolecules and co-crystals by outlining supramolecular synthetic strategies based upon modular hydrogen-bond driven approaches.

## 2.5.3.1.3 Nomenclature

Crystal engineering is a highly interdisciplinary area, which to some extent explains why unambiguous definitions and systematic nomenclatures have yet to be fully developed. The term co-crystal is not well-defined, and the existing literature contains terms such as molecular complexes, co-crystals, molecular adducts, molecular salts, clathrates, inclusion compounds, etc., that are often meant to describe one and the same family of chemical compounds. The purpose of this chapter is not to suggest definitions or to comment on the recent debate on nomenclature [65], but it is necessary to define the scientific scope of this overview. Thus, the primary subject matter of this next section will be delineated with the aid of the following provisions:

1. Only compounds constructed from discrete *neutral* molecular species will be considered as co-crystals, and all solids containing ions, including complex transition-metal ions, are excluded from this overview.
2. Co-crystals are made from reactants that are solids at ambient conditions [66]. Therefore all hydrates and other solvates are excluded which, in principle, eliminates compounds that are typically classified as clathrates or inclusion compounds (where the guest is a solvent or a gas molecule).
3. A co-crystal is a structurally homogeneous crystalline material that contains two or more neutral building blocks that are present in definite stoichiometric amounts.

At this point we are essentially left with two families of compounds: binary charge-transfer complexes and hydrogen-bonded or halogen-bonded co-crystals. This section will focus almost exclusively upon representatives from the latter categories [67]. The terms "binary/ternary supermolecule" indicates a discrete species with predictable and desirable connectivity, constructed from two/three different molecular species and assembled via directional non-covalent forces [68–78].

   Despite the abundance of papers dealing with all aspects of design and assembly of organic extended networks with desirable connectivities and shapes, it remains exceedingly difficult to bring more than two different molecular species into one crystalline lattice, without making or breaking covalent bonds.

### 2.5.3.2  Developing Hydrogen-bond Design Strategies for Synthesis of Co-crystals

The reliability of and competition between synthons are most effectively determined through systematic database studies [42, 79, 80], not through detailed analyses of individual structures. Etter and co-workers [63, 81] were pioneers in this area, and after extensive studies of preferential hydrogen-bond patterns in organic crystals they presented the following empirical 'rules' as a guide to the deliberate design of hydrogen-bonded solids;

1. All good proton donors and acceptors are involved in hydrogen bonding [82].
2. Six-membered ring intramolecular hydrogen bonds form in preference to intermolecular hydrogen bonds.
3. The best proton donor and acceptor remaining after intramolecular hydrogen bond formation will form intermolecular hydrogen bonds.

These guidelines are based upon observed correlations between hydrogen-bond patterns and functional groups and generally, as a proton increases in acidity, its

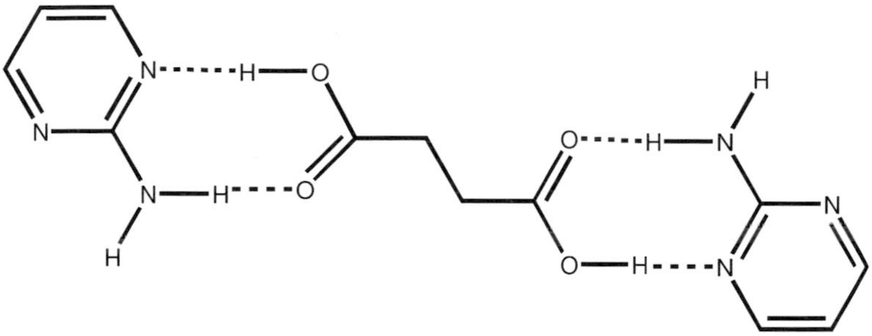

**Scheme 2.5.11**  The primary hydrogen bonds in co-crystals of 2-aminopyrimidine and dicarboxylic acids.

donor properties will improve. The second rule derives from competition studies showing that intramolecular hydrogen bonds are more difficult to break than comparable intermolecular bonds formed from similar donors. The third rule can be illustrated by examining the structure of 2-aminopyrimidine/carboxylic acid co-crystals, where the best donors (acidic proton) are paired with the best acceptors (ring nitrogen atom) [61], Scheme 2.5.11.

### 2.5.3.2.1  Heteromeric Halogen-bonded Co-crystals

Intermolecular interactions of the type $D\cdots X–Y$, where D – electron donor/halogen-bond acceptor, X = halogen atom/halogen-bond donor, Y = carbon moiety, have shown to be reliable and robust supermolecular synthons. The strength of the halogen bond increases with the increase in the Lewis acidity of the acceptor atom, thus I > Br > Cl > F. Likewise, an increase in the electron density on the electron donor, will strengthen its donor ability, as a result N > O > S. Additionally, many halogen-bond acceptor molecules will contain fluorine atoms; this enhances the acceptor strength of the halogen nuclei, thus strengthening the overall halogen bond. As shown previously, nitrile moieties can also act as halogen-bond donors. An example of this is shown in the 1:1 binary co-crystal of 1,4-dicyanobutane and diiodoperfluorohexane, Scheme 2.5.12 [83].

4,4'-Bipyridine and 1,4-diiodobenzene crystallize in a 1:1 ratio forming 1-D chains through nitrogen–iodine interactions with $N\cdots I$ distances of 3.023 Å [51]. Similarly, the crystallization of 4,4'-bipyridine with 1,4-diiodotetafluorobenzene produces a 1-D chain with $N\cdots I$ distances of 2.851 Å, which is considerably

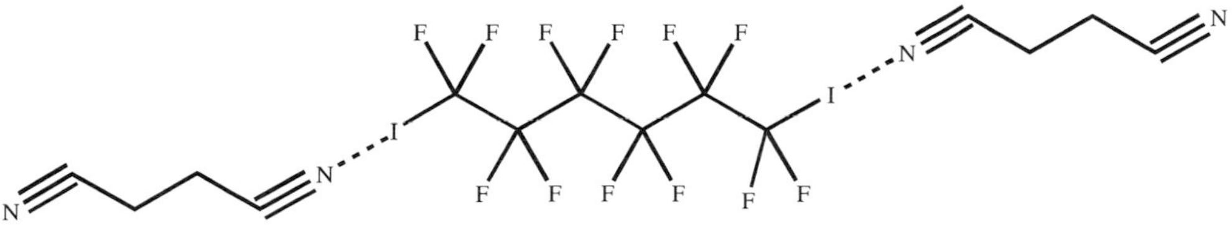

**Scheme 2.5.12**  The primary halogen bonds in co-crystals of dicyanoalkyls and diiodoperfluoroalkanes.

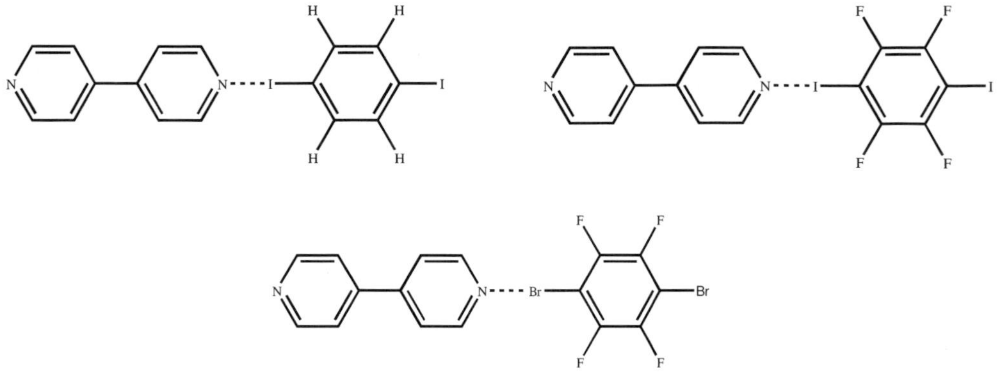

**Scheme 2.5.13** 1-D motifs in co-crystals of 4,4'-bipyridine and 1,4-dihalogenated aryls.

shorter than for the non-fluorinated derivative. Not surprisingly, 4,4'-bipyridine and 1,4-dibromotetrafluorobenzene also form a 1:1 co-crystal containing 1-D chains constructed from nitrogen–bromine interactions with N⋯Br distances of 2.878 and 2.979 Å respectively, Scheme 2.5.13 [84].

The number of co-crystal structures assembled primarily through nitrile–halogen or nitrogen–halogen interactions is still quite small. A search through the CSD showed a total of 10 co-crystals containing C≡N⋯I interactions, 13 co-crystals of Br⋯N interactions and 53 co-crystals containing N⋯I contacts.

Finally, some examples of 3-D motifs in halogen-based binary co-crystals include tetrabromoadamantane:hexamethylenetetramine [85], 1,4-di-iodotetrafluorobenzene: hexamethylenetetramine [51] and iodoform:hexamethylenetetramine [86].

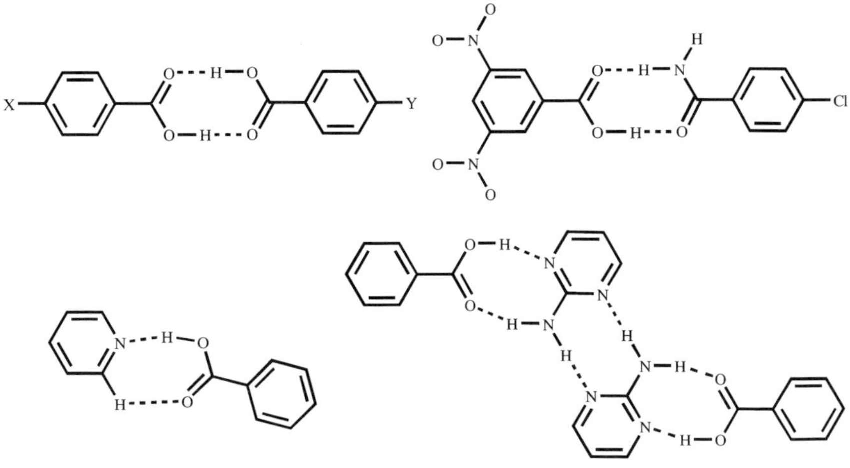

**Scheme 2.5.14** Four 0-D motifs found in co-crystals created by heteromeric hydrogen-bond interactions.

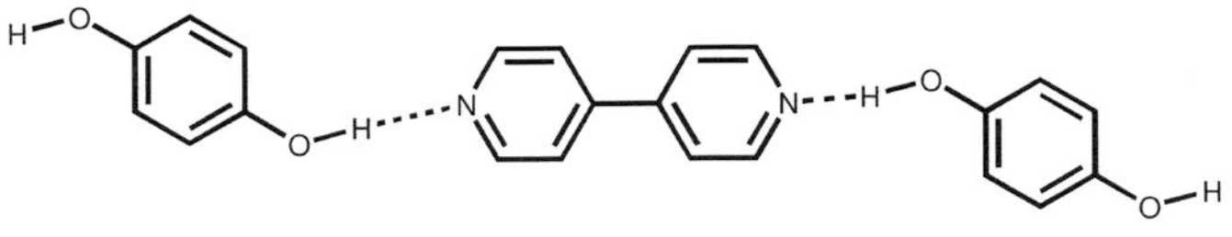

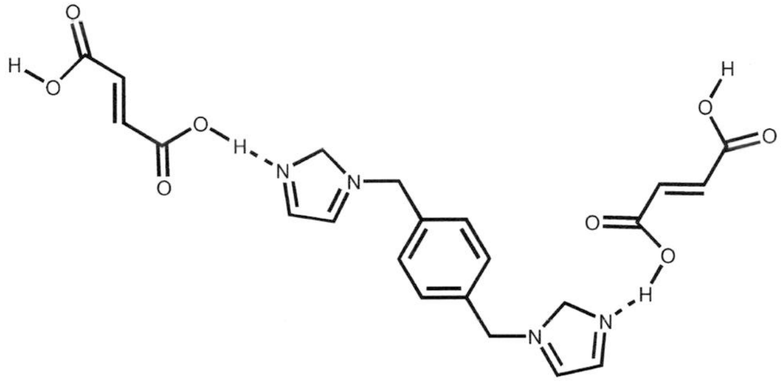

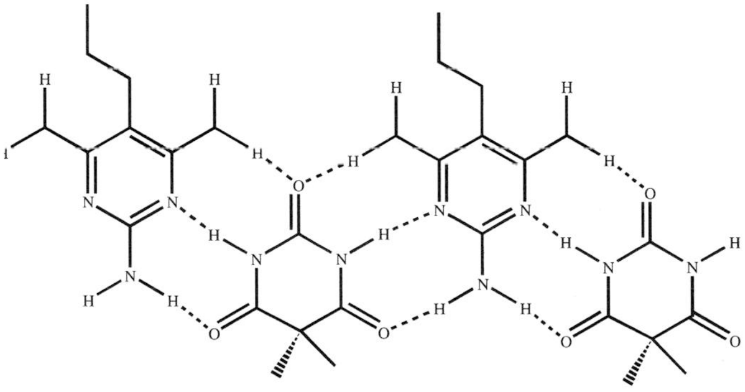

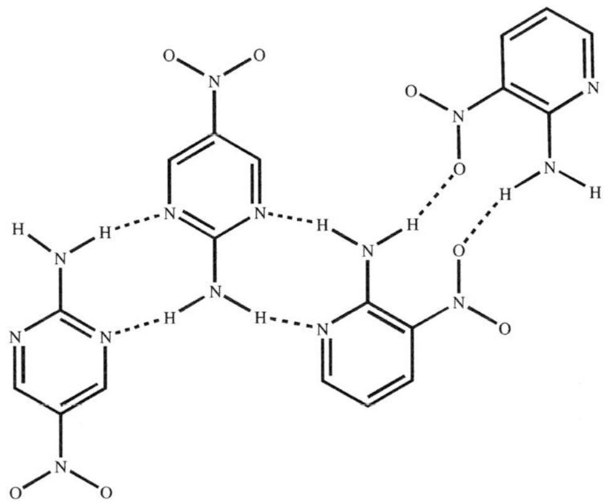

**Scheme 2.5.15** Examples of 1-D motifs generated by heteromeric hydrogen-bond interactions.

### 2.5.3.3 Examples of Binary Hydrogen-bonded Co-crystals

There are, strictly speaking, no "discrete" aggregates within a solid-state framework but it is nevertheless convenient to classify an assembly as being 0-D, 1-D, 2-D or 3-D, depending upon the type of intermolecular interactions that are present between and within an array of molecules.

Some examples of 0-D assemblies, Scheme 2.5.14, in co-crystals include heteromeric carboxylic acid:carboxylic acid dimers (1:1) [87], pyridine:carboxylic acid dimers (1:1) [88], 2-aminopyrimidine:carboxylic acid trimers (1:2) [89], pyridine:dihydroxybenzene trimers (2:1) [90], bipyridine:carboxylic acid trimers (1:2) [73], 2-aminopyrimidine:carboxylic acid tetramers (2:2) [89, 91], bipyridine:resorcinol tetramers (2:2) [74], 3,5-dinitrobenzoic acid:nicotinic acid tetramers (2:2) [92], *iso*nicotinamide:carboxylic acid tetramers (2:2) [93], 2-pyridone:carboxylic acids pentamers (4:1) [94], melamine:thymine tetramers (1:3) [95], melamine:barbital hexamers (3:3) [96] and tripenylphosphine oxide with a variety of hydrogen-bond donors (1:1) [62, 97, 98].

Examples of co-crystals containing chains, ribbons, and other infinite 1-D motifs, Scheme 2.5.15, include: bipyridine:dihydroxybenzene [99], melamine:cyanuric acid [100], bipyridine:(fluorinated)dibromobenzene [101], bypyridine:(diiodobenzene, tetraiodoethylene or diiodine) [102], 2-aminopyridine:dicarboxylic acids

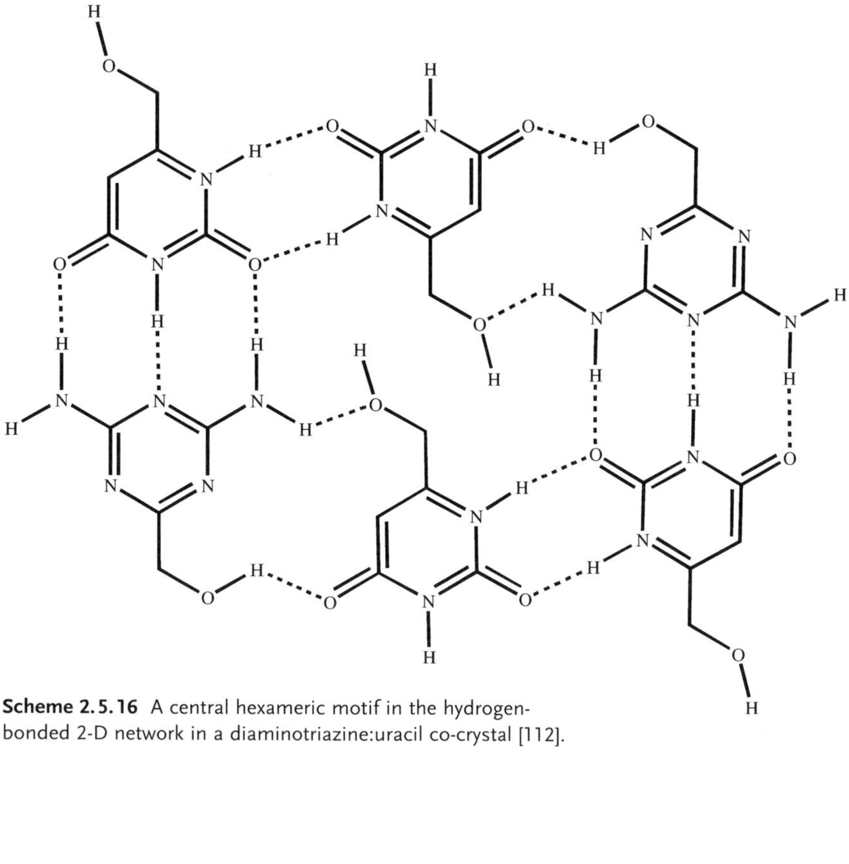

**Scheme 2.5.16** A central hexameric motif in the hydrogen-bonded 2-D network in a diaminotriazine:uracil co-crystal [112].

[103], triaminopyrimidine:barbituric acid [72], 2-aminopyrimidine:dicarboxylic acid [61], 1,2,3-trihydroxybenzene:hexamethylenetetramine [104], diols:diamines [105], bis-benzimidazole:dicarboxylic acids [106] and 2-amino-5-nitropyrimidine:2-amino-3-nitropyridine [107].

Many infinite 2-D assemblies have also been constructed including (but not limited to): piperazine:carboxylic acid [108], trithiocyanuric:bipyridine [109], pyridyloxamide:dicarboxylic acid [110], picolylaminocyclohexenone:dicarboxylic acid [111], triazine:uracil [112], tris-(4-pyridyl)triazine:trimesic acid [113], bipyridine:ureylene dicarboxylic acid [76] and *iso*nicotinamide:dicarboxylic acid [93, 114], Scheme 2.5.16.

Finally, carbamazepine:tetracarboxylic acid-adamantane [115] represents one of the few examples of clearly defined and predictable 3-D motif observed in a hydrogen-bond-based binary co-crystal.

### 2.5.3.4 From Pattern Recognition to Practical Crystal Engineering

The idea of structural motifs and patterns within a solid framework is closely linked to the notion that some intermolecular interactions are more important than others [34]. The challenge of recognizing and classifying motifs generated by multiple intermolecular interactions was addressed by Etter et al. [116], using a graph-set notation that allows structural motifs to be described in a consistent and 'user-friendly' way. The graph-set approach employs four principal motifs, chains ($C$), dimers ($D$), rings ($R$), and intramolecular hydrogen bonds ($S$), as descriptors of hydrogen-bonded molecular solids. Although this notation does not offer an unambiguous assignment of every structural arrangement, it is flexible enough to facilitate a systematic description of a wide range of structures [117]. The simplicity of this approach is undoubtedly an important reason why it has become so successful, and descriptors such as $R_4^2(8)$ (an eight-membered ring with four hydrogen-bond donors and two acceptors) and $R_2^2(8)$ (e.g. the carboxylic head-to-head dimer), have become widely recognizable.

One of the attractions of supramolecular chemistry is the extraordinary potential for synthesis of new materials that can be achieved much more rapidly and more effectively than with conventional covalent means. For supramolecular synthesis to advance, it is obviously important to characterize, classify, and analyze structural patterns, space group frequencies, and symmetry operators [118]. However, at the same time we also need to bring together this information with the explicit aim of improving and developing supramolecular synthesis – the deliberate combination of *different* discrete molecular building blocks within periodic crystalline materials.

### 2.5.3.5 Synthesis of Co-crystals and the Supramolecular Yield

The fact that in the crystal structures of 4-bromo-4'-cyanobiphenyl [119] and 4-bromobenzonitrile [120] the molecular components are aligned in a head-to-tail fashion with relatively short Br $\cdots$ N contacts indicates that there are stabilizing intermolecular interactions between cyano and bromo moieties [121], Scheme

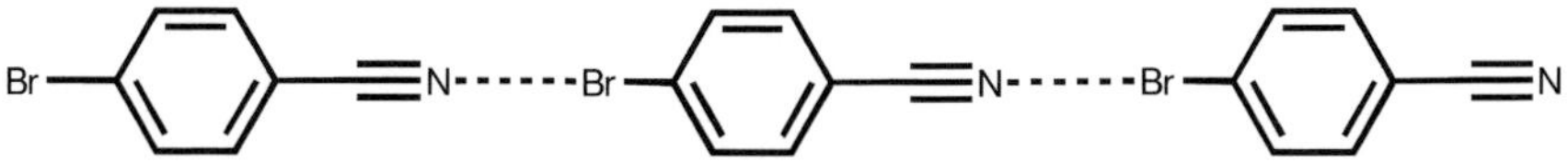

**Scheme 2.5.17** A chain of molecules organized in a head-to-tail manner in 4-bromobenzonitrile.

2.5.17. However, there is still no example of successful synthesis of binary co-crystals driven by CN$\cdots$Br interactions. Such interactions can organize molecules within a lattice but have yet to bring about the assembly of heteromeric co-crystals. There is clearly a difference between observing a large number of short-contacts in molecular crystal structures composed of only one type of building block and translating such interactions into useful synthetic tools for constructing *heteromeric* architectures.

Much current work in organic crystal engineering is now geared towards synthesizing co-crystals using supramolecular reactions based upon reliable synthons and, so far, the vast majority of organic molecular co-crystals have been assembled via conventional, strong hydrogen bonds [122, 123]. Weaker hydrogen bonds and many other intermolecular interactions such as nitro$\cdots$iodo, cyano$\cdots$nitro, halogen$\cdots$halogen, etc., have not yet been found to be useful tools for construction of co-crystals. The success and efficiency for any set of supramolecular reactions can be judged by the frequency of occurrence of desired intermolecular interactions and connectivities in the resulting solid. The probability that a certain motif will appear in a crystalline lattice is, in many ways, a measure of the *yield of a supramolecular reaction*. Just as a covalent synthetic chemist searches for ways in which a specific reaction can be promoted or prevented, a supramolecular chemist tries to identify the experimental regime where a synthon prevails despite competition from other non-covalent forces.

### 2.5.3.6    Heteromeric Interactions are Better than Homomeric Interactions

A survey of hydrogen bonded co-crystals in the CSD [124] reveals that most of them have been prepared using strategies that utilize suitable combinations of chemical entities (or functional groups) located on different molecules, such that they would prefer to interact and bind heteromerically, Scheme 2.5.18, rather than with themselves [71, 75, 77, 78, 125].

The most widely used synthons for the directed assembly of binary co-crystals have contained a carboxylic acid in combination with a suitable N-containing heterocycle. For example, there are currently three co-crystals in the CSD with pyrazine [126], seven with phenazine[127], sixteen with 4,4'-bipyridine [128], one with pyrimidine [129], and nine co-crystals with either azapyrine, quinoline, phenanthroline [130] and a benzoic acid-based counterpart. In every case, the expected/intended carboxylic acid$\cdots$N(heterocycle), O–H$\cdots$N, hydrogen bond is present. For slightly more complex heterocycles (i.e. with added substituents capable of hydrogen bonding) the results are still very consistent; 11 of 12 carbox-

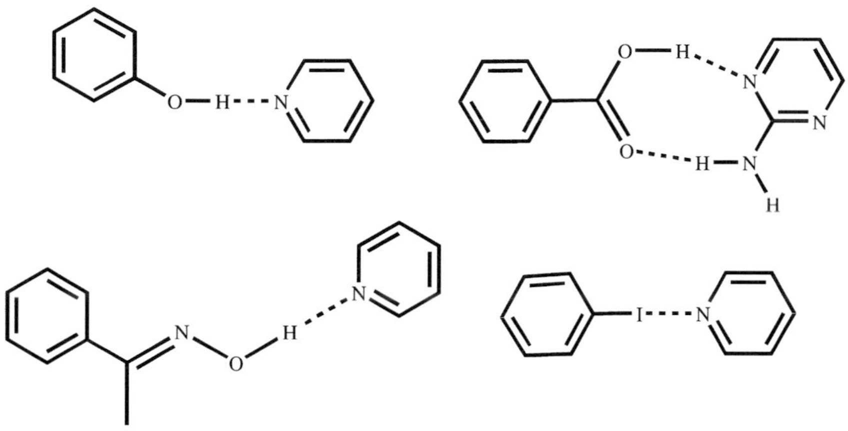

**Scheme 2.5.18** Four dimeric motifs constructed via heteromeric intermolecular interactions.

ylic acid: isonicotinamide co-crystals contain an acid· · ·pyridine interaction [131] – a good example of a high-yielding supramolecular reaction.

There are, in fact, very few occurrences of binary hydrogen-bond-based co-crystals that do not contain a primary intermolecular interaction between the two different molecules. Two such examples include 4 nitrobenzamide pyrazine carboxamide (1:1) [70] and 3,5-dintrobenzoic acid:4-(N,N-dimethylamino)benzoic acid (1:1) [132]. The latter is a rare example of a carboxylic acid· · ·carboxylic acid co-crystal that contains two homomeric dimers instead of one heteromeric dimer [133].

The overriding conclusion from the extensive data available in the CSD is clear. In order to convince two different discrete chemical species to coexist in a molecular co-crystal there needs to be some specific molecular-recognition based reason for their solid-state union [93]. Although individual structures that defy rationalization will appear from time to time, there is no doubt that the important 'big picture' reveals structural trends, patterns of behavior, and reproducible motifs that, when combined, can be developed into a library of high-yielding supramolecular reactions.

### 2.5.3.7 Do Polymorphic Compounds Make Good Co-crystallizing Agents [107]?

An effective co-crystallizing agent (CA) should engage in intermolecular interactions with the target molecule that are more favorable than any homomeric interactions that may exist [93]. The CA must also be able to tolerate a different molecule within the same crystalline lattice, which makes it reasonable to search for co-crystallizing agents amongst polymorphic compounds [134]. Polymorphism means that a compound is found in more than one crystalline manifestation, which [134] in turn indicates that such compounds display a degree of structural flexibility. In other words, the multidimensional potential-energy surface that describes the thermodynamics governing the molecular recognition processes is likely to contain many accessible energy minima.

The suggestion that polymorphic compounds are more likely to form co-crystals than compounds that never display polymorphism is difficult to quantify. However, from a practical perspective it would be extremely useful to have access to reliable guidelines for how a search for reliable co-crystallizing agents for a specific family of compounds can be very tightly focused. This challenge prompted an examination of four polymorphic compounds [107]: isonicotinamide **1**, 2-amino-5-nitropyrimidine [135] **2**, 4-chlorobenzamide [136] **3**, and maleic hydrazide [137, 138] **4**, and several co-crystals thereof. The reason for selecting **1–4** (apart from the fact that they are all polymorphic) was that they are all potentially capable of engaging in several well-defined and robust intermolecular hydrogen-bond interactions, Scheme 2.5.19.

Three of the four polymorphic compounds examined in this study [107] readily form co-crystals e.g. 2-hexeneoic acid:*iso*nicotinamide, 2-amino-5-nitropyrimidine:2-amino-3-nitropyridine, and 3-dimethylaminobenzoic acid:4-chlorobenzamide. The structural behavior of **1–3** supports the suggestion that polymorphs make good co-crystallizing agents (provided that solubility differences are not too large). However, no co-crystals with maleic hydrazide were obtained, despite the fact that it exists in three different polymorphs.

The failure to produce any co-crystals with **4** may cast doubts on the notion that polymorphic compounds make good co-crystallizing agents, unless there is a significant structural difference in the polymorphic behavior between **1–4**. A close examination of the known polymorphs of **1–4** using graph-set notation did, in fact, uncover possible reasons for the unwillingness of **4** to participate in molecular co-crystals. The two polymorphs of *iso*nicotinamide, **1 a** and **1 b**, display very different intermolecular connectivities. In **1 a** there is an amide···amide head-to-head interaction as well as the commonly occurring N–H···O link between adjacent dimers resulting in a $C(4)R_2^2(8)$ motif. However, the primary intermolecular forces in **1 b** are significantly different as there is a catemeric amide···amide hydrogen bond, $C(4)$, as well as an amide···pyridine interaction, $C(7)$. All three polymorphs

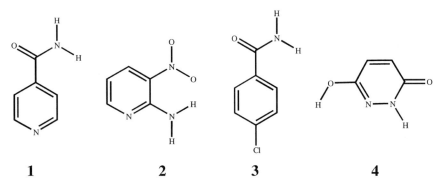

**1**     **2**     **3**     **4**

**Scheme 2.5.19** Molecules capable of forming a variety of hydrogen-bonded synthons; *iso*nicotinamide, **1**, 2-amino-3-nitro-pyridine, **2**, 4-chlorobenzamide, **3**, and maleic hydrazide, **4**.

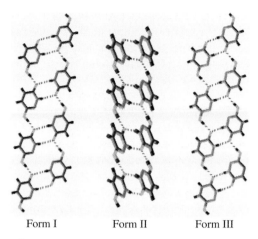

| Form I | Form II | Form III |

**Scheme 2.5.20** The primary hydrogen-bond motifs in the three known polymorphs of maleic hydrazide display the same connectivity.

of **2** have the expected intramolecular amine · · · nitro hydrogen bonds. **2 b** and **2 c** have identical intermolecular connectivities and the primary hydrogen-bond motifs can be with the same graph-set notion, $R_2^2(12)R_2^2(8)S(6)$. In **2 a**, however, there are infinite hydrogen-bonded ribbons, yielding the graph-set notation $R_2^2(10)\ R_2^2(8)S(6)$. In the structures of **3 a** and **3 c** the amide moieties form amide-amide catemers, $C(4)\ R_2^2(8)\ R_4^2(8)$. The difference between the two structures lies in the arrangement of aromatic rings with respect to one another. While the graph set notation for **3 b** is similar to **3 a** and **3 c**, the larger hydrogen-bonded ring creates a two-dimensional network, $C(4)\ R_2^2(8)\ R_6^4(16)$. In contrast, all three polymorphs of **4**, the one compound that would not readily produce co-crystals, contain identical hydrogen-bonded ladders: $R_4^2(14)R_2^2(8)$, Scheme 2.5.20.

The fact that the three polymorphs contain the same extended networks was noted in the original paper: *"Three forms of MH (maleic hydrazide) provide a unique example of polymorphic structures built of similar hydrogen-bonded aggregates. The same hydrogen-bonded supramolecular aggregates organize into different lattices of different symmetries...[138b]"*.

A detailed structural examination of **1–4** has enabled the identification of significant differences, in terms of molecular recognition patterns, between the different polymorphs that may provide practical guidelines for ranking candidates that are potentially suitable as co-crystallizing agents. An important consideration when attempting to prepare co-crystals is to choose a co-crystallizing agent that is already known to be polymorphic. However, structural flexibility alone is not always enough. It may be equally important to select molecules that can adopt alternative packing patterns as well as display *synthon flexibility* – an ability to participate in several different robust and well-defined intermolecular interactions that can satisfy the demands of multiple hydrogen-bond donors/acceptors on a variety of molecules.

### 2.5.3.8 Beyond Binary Co-crystals: The Need for Supramolecular Reagents

Through systematic structural studies of the effect of competing intermolecular interactions, it is becoming clear that under certain conditions it is possible to manipulate the way in which molecules recognize and bind by 'tuning' the strengths of site-specific complementary hydrogen-bond functionalities [139–141]. This knowledge about intermolecular forces is now also being forged into a practical tool for making ternary and higher-order co-crystals. Central to this approach is the availability of molecular building blocks that can provide supramolecular directionality, selectivity and reliability. Such building blocks, *supramolecular reagents*, contain two or more different binding sites, attachment points, that can be used to selectively attract, bind and organize two or more different molecules into a supermolecule with predictable connectivity and shape. The supramolecular reagent is the hub for the assembly process and will become part of the supermolecule – this heteromeric aggregate will also appear in the resulting crystalline solid enabling the synthesis of co-crystals of increasing complexity.

In order to establish how far this relatively simple modular supramolecular concept can be extended, we sought to design *ternary* supermolecules with predictable connectivity and stoichiometry [142], Scheme 2.5.21.

Since hydrogen bonds frequently form in a hierarchical fashion (best-donor to best-acceptor, second best-donor to second best-acceptor, etc.) [63, 64], the chances of producing a new binary co-crystal are greatly improved by positioning the best hydrogen-bond donor and the best hydrogen-bond acceptor on different molecular building blocks [70]. The assembly strategy was put into place in the systematic construction of ternary co-crystals using isonicotinamide as the supramolecular reagent [142]. As noted previously, this molecule readily recognizes and binds to carboxylic acids to form 1:1 binary co-crystals that contain a robust and reproducible heteromeric hydrogen-bonded motif, Fig. 2.5.1

The best donor (carboxylic acid) and the best acceptor (pyridine nitrogen) form an intermolecular O–H···N hydrogen bond, and the tetrameric supermolecule is completed through a self-complementary amide–amide interaction. Isonicotinamide is a good example of a supramolecular reagent suitable for constructing ternary co-crystals: it has two distinctly different, yet relatively strong, hydrogen-bond moieties, it shows good 'structural flexibility' since it is polymorphic, and it also displays 'synthon flexibility' as the two known polymorphs display different hydrogen-bond

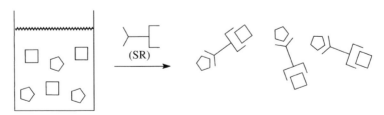

**Scheme 2.5.21** General description of a coupling reaction of two different molecules and a supramolecular reagent (SR) resulting in 1:1:1 ternary supermolecules.

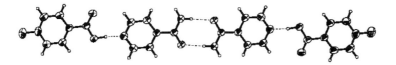

4-chlorobenzoic acid     Isonicotinamide     Isonicotinamide     4-chlorobenzoic acid

**Figure 2.5.1** Primary motif typical of the majority of
1:1 isonicotinamide:carboxylic acid co-crystals [93].

interactions. The primary hydrogen bond in this system is the acid–pyridine interaction and, since hydrogen bonds have large electrostatic components, the strength
of this interaction is governed by the acidity of the carboxylic acid donor [143].

The 'weaker' link in the tetrameric motif shown in Fig. 2.5.1 is the amide···
amide interaction. Since many reported structures contain heteromeric ami-
de···acid hydrogen bonds (in preference to the corresponding homomeric op-
tions) [144], the intention was to replace it with a more favorable heteromeric
acid···amide interaction, thus providing a method for bringing in the third
component in a specific manner. By offering two different carboxylic acids to
isonicotinamide, it was postulated that the stronger acid interact preferentially
with the best acceptor (the pyridine nitrogen atom) and the weaker acid with the
remaining amide moiety. This supramolecular design strategy was put to the test
by allowing equimolar amounts of a weaker acid, a stronger acid, and isonicotinamide to react in an aqueous solution.

Co-crystal **5**, 3-methylbenzoic acid:isonicotinamide:3,5-dinitrobenzoic acid
(1:1:1), contains the desired three-component supermolecule with the expected
connectivity. The stronger acid ($pK_a$=2.8) [145] interacts with the pyridine nitrogen
atom, and the weaker acid ($pK_a$=4.3) competes successfully for the amide moiety
and forms a heteromeric hydrogen-bonded motif, Fig. 2.5.2.

Co-crystal **6**, 4-($N,N$-dimethyl)aminobenzoic acid:isonicotinamide:3,5-dinitro-
benzoic acid (1:1:1), is also a ternary co-crystal with the intended three-component
supermolecule. The acid–pyridine nitrogen interaction persists, and the weaker
acid ($pK_a$=6.5) forms a heteromeric motif with the amide moiety, Fig. 2.5.3.

3,5-Dinitrobenzoic acid:isonicotinamide:4-hydroxy-3-methoxycinnamic acid **7**, is
also a 1:1:1 ternary co-crystal with the same primary supramolecular connectivity

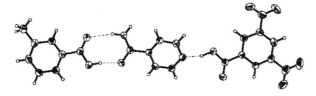

3-methylbenzoic acid     Isonicotinamide     3,5-dinitrobenzoic acid

**Figure 2.5.2** The ternary supermolecule in **5**.

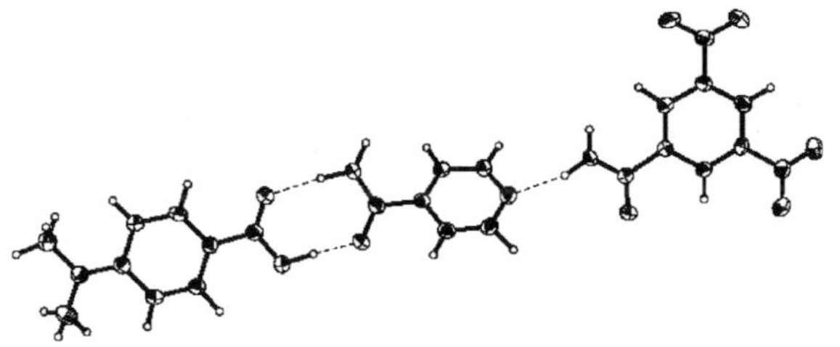

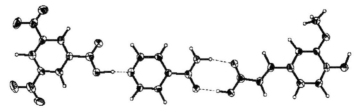

4-(*N,N*-dimethyl)aminobenzoic acid   Isonicotinamide   3,5-dinitrobenzoic acid

**Figure 2.5.3** The ternary supermolecule in **6**.

as in **5** and **6**, Fig. 2.5.4. The desired trimer persists even in the presence of the potentially disrupting OH-moiety on the weaker acid (p$K_a$=4.4).

All three structures, **5–7**, contain ternary supermolecules synthesized with the aid of a suitable supramolecular reagent that employs a hierarchy of hydrogen-bond interactions for the desired intermolecular organization and assembly. Although these forces are weaker than covalent interactions, it is clearly possible to assemble more than two different building blocks in a preconceived manner. To date about a dozen ternary co-crystals based upon the acid:isonicotinamide:acid combination have been reported; in each and every case the connectivity of the observed supermolecule is consistent with the underlying supramolecular synthetic strategy, and is readily rationalized through differences in p$K_a$ values – this reaction, with isonicotinamide as the supramolecular reagent proceeds with a very high supramolecular yield.

Despite its success, however, isonicotinamide is not sufficiently versatile to make it an ideal supramolecular reagent. First, it is capable of forming self-complementary amide···amide and amide···pyridine hydrogen bonds, which makes it inherently difficult to combine isonicotinamide with molecules that lack moieties that can compete successfully with the hydrogen-bonding capabilities of the amide functionality. Second, since the two binding sites on isonicotinamide are attached to the same backbone, it is impossible to tune the electronics of the two sites independently, which reduces the versatility. Consequently, there is a need for second-generation SRs that can be refined in such a way that they offer more

3,5-dinitrobenzoic acid      Isonicotinamide      4-hydroxy-3-methoxycinnamic acid

**Figure 2.5.4** The ternary supermolecule in **7**.

opportunities for modular supramolecular synthesis of ternary co-crystals through enhanced structural selectivity and specificity.

Against this background, a new family of SRs composed of asymmetric *bis*-heterocycles [146] was constructed, where two different binding sites (hydrogen-bond acceptor sites) are linked by a methylene bridge in order to provide increased solubility in a range of solvents. The two binding sites have significantly different basicities [147], which means that their ability to accept hydrogen bonds varies. They also lack strong hydrogen-bond donors and, consequently, homomeric inter-molecular interactions will be weak and less likely to prevent the desired hetero-meric interactions, Scheme 2.5.22.

Even though $pK_a/pK_b$ values do not provide direct measures of hydrogen-bond strength, hydrogen bond abilities and free energies of complexation have been correlated with $pK_a$ values and, within closely related classes of compounds, such comparisons frequently yield correct qualitative results [148, 149]. The basicity of each heterocycle can also be independently altered through suitable covalent substitution, which provides a practical handle for fine-tuning differences in intermolecular reactivity. Tunability is particularly important as it creates a versa-tile supramolecular reagent with the potential for a high degree of transferability. The ability of these SRs to form ternary supermolecules with predictable connec-tivity was put to the test by allowing each SR to react with pairs of different carboxylic acids in a 1:1:1 ratio [150]. The target in each case is a crystal structure containing 1:1:1 ternary supermolecule where the primary intermolecular inter-actions can be rationalized according to the best donor/best acceptor protocol.

The crystal structure of **8** contains two crystallographically unique ternary super-molecules with identical connectivity, Fig. 2.5.5.

The primary intermolecular interactions in this structure are the O–H$\cdots$N hydrogen bonds between the stronger acid (3,5-dinitrobenzoic acid) and the more basic nitrogen atom of the benzimidazol-1-yl ring, and the O–H$\cdots$N hydro-gen bonds from the weaker acid (4-nitrobenzoic acid) to the less basic nitrogen atom of the pyridyl moiety.

The crystal structure of **9** shows the presence of a ternary 1:1:1 supermolecule with the same connectivity as that found in **8**, Fig. 2.5.6.

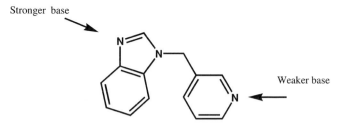

**Scheme 2.5.22** An asymmetric ditopic supramolecular reagent with two different binding sites (hydrogen-bond acceptors) that can be electronically modified independently through suitable covalent substituents.

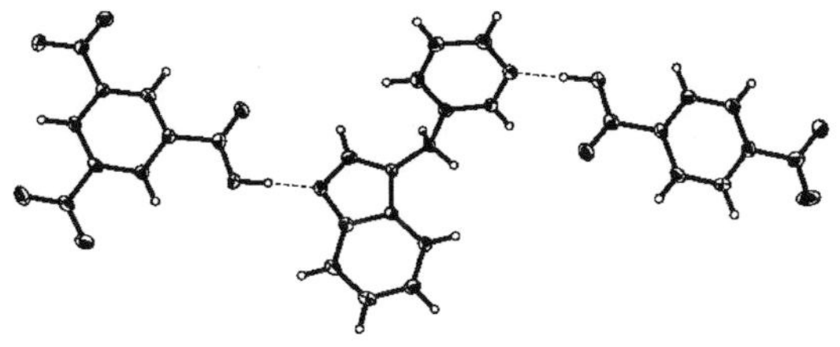

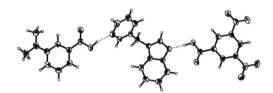

3,5-dinitrobenzoic acid    ditopic SR         4-nitrobenzoic acid

**Figure 2.5.5** One of the two ternary supermolecules in the crystal structure of **8**. The stronger acid binds to the stronger base (left-hand side), and the weaker acid binds to the second-best base (right-hand side).

The primary synthons comprise an O–H···N hydrogen bond between the stronger acid, 3,5-dinitrobenzoic acid, and the most basic heterocyclic moiety, and a second O–H···N interaction between the weaker acid, 3-*N,N*-dimethylaminobenzoic acid, and the less basic acceptor, the pyridyl moiety.

The crystal structure of **10** contains a 1:1:1 supermolecule with the desired connectivity, Fig. 2.5.7. The best acceptor, the benzimidazol-1-yl moiety, forms an O–H···N hydrogen bond with the best donor, the stronger acid. The second-best acceptor, the pyridyl moiety, binds to the weaker acid via an O–H···N hydrogen bond.

All three structures, **8–10**, contain ternary supermolecules constructed through the deliberate use of directional intermolecular synthetic operations. The supramolecular reagents each have two binding sites that differ primarily in their basicity, but neither site is otherwise biased or predisposed towards interacting preferentially with either of the two competing carboxylic acids. The differences in basicity are translated into supramolecular reactivity and selectivity that subsequently carry over into the solid state, which demonstrates that supramolecular assembly can be controlled by fine-tuning individual binding sites. This raises the possibility that a solution to the problem of making non-covalent one-pot synthesis "sequential" may be to devise modular assembly processes based upon a hierarchy of intermolecular interactions derived from molecular properties and structural trends.

3-(*N,N*-dimethyl)aminobenzoic acid   ditopic SR    3,5-dinitrobenzoic acid

**Figure 2.5.6** The ternary supermolecule in the crystal structure of **9**. The stronger acid binds to the stronger base (right-hand side), and the weaker acid binds to the weaker base (left-hand side).

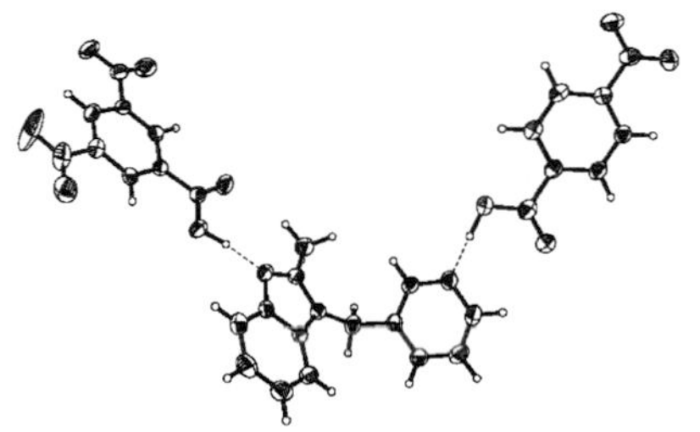

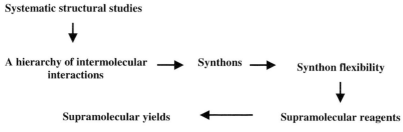

3,5-dinitrobenzoic acid     ditopic SR     4-nitrobenzoic acid

**Figure 2.5.7** The ternary supermolecule in the crystal structure of **10**. The stronger acid binds to the stronger base (left-hand side), and the weaker acid binds to the second-best base (right-hand side).

### 2.5.3.9  Codicil

Even though molecular recognition is typically associated with molecules in solution, such interactions are also responsible for organizing molecules in the solid state. Translating principles of molecular recognition to solid-state assembly of heteromeric molecular solids is of key importance to the development of versatile and reliable strategies for practical supramolecular synthesis. In this section we have tried to present some modular and transferable supramolecular design strategies based upon a hierarchy of intermolecular interactions. The starting point for these efforts is provided by a body of readily available structural information that subsequently allows the identification of suitable supramolecular reagents. The effectiveness of these compounds as active and reliable builders of heteromeric supramolecular aggregates can eventually be evaluated and quantified using the supramolecular yield, Scheme 2.5.23.

Inevitably, any supramolecular synthetic strategy will fail at some point, but through the power of covalent synthesis we have access to unlimited strategies for modulating the electronic and geometric details of each binding site on a supramolecular reagent such that a variety of chemical functionalities can be targeted for binding. By carefully examining the connections between molecular chemistry and supramolecular reactivity, it may be possible to build a library of SRs that, collectively, is capable of affecting the assembly of new supermolecules with a

**Systematic structural studies**

↓

**A hierarchy of intermolecular**          →          **Synthons**          →          **Synthon flexibility**
**interactions**

↓

**Supramolecular yields**          ←          **Supramolecular reagents**

**Scheme 2.5.23** Design, implementation, and evaluation of supramolecular synthesis of co-crystals.

high degree of specificity and reliability. If we can clear a path towards practical and transferable guidelines for versatile supramolecular synthesis through an improved awareness of the balance and competition between intermolecular interactions, then it is inevitable that much more complex supermolecules and higher-order co-crystals will be synthesized.

## References

1 J.-M. Lehn, *Angew. Chem. Int. Ed. Engl.* **1990**, *29*,1304; G. R. Desiraju, *Crystal Engineering: The Design of Organic Solids*, Elsevier, Amsterdam, 1989; C. V. K. Sharma, *J. Chem. Ed.* **2001**, *78*, 617; J. L. Atwood, J. E. D. Davies, D. D. MacNicol, F. Vögtle, K. S. Suslick (Eds.), *Comprehensive Supramolecular Chemistry*, Elsevier, Oxford, 1996; J. W. Steed, J. L. Atwood, *Supramolecular Chemistry: An Introduction*, J.Wiley & Sons, Chichester, 2000.

2 C. B. Aakeröy, *Acta Crystallogr., Sect. B* **1997**, *53*, 569.

3 G. R. Desiraju, *Nature* **2001**, *412*, 397; E. Coronado, J. R Galan-Mascaros, C. J. Gomez-Garcia, V. Laukhin, *Nature* **2000**, *408*, 447.

4 M. Muthuraman, R. Masse, J.-F. Nicoud, G. R. Desiraju, *Chem. Mater.* **2001**, *13*, 1473; O. König, H.-B. Burgi, T. Armbruster, J. Hulliger, T. Weber, *J. Am. Chem. Soc.* **1997**, *119*, 10632.

5 O. Kahn, *Acc. Chem. Res.* **2000**, *33*, 1; J. S. Miller, *Inorg. Chem.* **2000**, *39*, 4392.

6 Y. Aoyama, *Top. Curr. Chem.* **1998**, *198*, 131; D. M. Bassani, V. Darcos, S. Mahony, J.-P. Desvergne, *J. Am.Chem. Soc.* **2000**, *122*, 8795.

7 J.-M. Lehn, *Supramolecular Chemistry*, VCH, Weinheim, 1995.

8 E. J. Corey, *Chem. Soc. Rev.* **1988**, *17*, 111.

9 K. C. Nicolaou, E. J. Sorensen, *Classics in Total Synthesis*, Wiley-VCH, Weinheim, 1996.

10 B. Moulton, M. J. Zaworotko, *Chem. Rev.* **2001**, *101*, 1629.

11 G. R. Desiraju, *Acc. Chem. Res.* **2002**, *35*, 565.

12 K. T. Holman, A. M. Pivovar, J. A. Swift, M. D. Ward, *Acc. Chem. Res.* **2001**, *34*, 107; *Crystal engineering: From molecules and crystals to materials*, A NATO Advanced Study Institute and a Euroconference, D. Braga, F. Grepioni, A. G. Orpen (Eds.) Erice, Italy, 1999, *58*.

13 C. B. Aakeröy, A. M. Beatty, *Aus. J. Chem.* **2001**, *54*, 409.

14 M. W. Hosseini, *CrystEngComm.* **2004**, *6*, 318.

15 L. MacGillivray, *CrystEngComm.* **2004**, *6*, 77.

16 L. Brammer, *Chem. Soc. Rev.* **2004**, *33*, 476.

17 L. J. Prins, D. N. Reinhoudt, P. Timmerman, *Angew. Chem. Int. Ed.* **2001**, *40*, 2382; D. C. Sherrington, K. A. Taskinen, *Chem. Soc. Rev.* **2001**, *30*, 83; G. M. Whitesides, E. E. Simanek, J. P. Mathias, C. T. Seto, D. N. Chin, M. Mammen, D. M. Gordon, *Acc. Chem. Res.* **1995**, *28*, 37.

18 N. J. Turro, *Chem. Commun.* **2002**, 2279.

19 G. R. Desiraju, *Angew. Chem. Int. Ed. Engl.* **1995**, *34*, 2311.

20 A. Nangia, G. R. Desiraju, *Acta Crystallogr., Sect. A* **1998**, *54*, 934.

21 An extensive discussion on 'weaker' or unconventional hydrogen bonds is provided by T. Steiner, G. R. Desiraju, in *The Weak Hydrogen Bond in Structural Chemistry and Biology*, Oxford University Press, New York, 1999.

22 M. M. Conn, J. Rebek, Jr. *Chem. Rev.* **1997**, *97*, 1647; D. S. Lawrence, T. Jiang, M. Levett, *Chem. Rev.* **1995**, *95*, 2229; P. Stang, B. Olenyuk, *Acc. Chem. Res.* **1997**, *30*, 502; M. Fujita, *Chem. Soc. Rev.* **1998**, *27*, 417; B. Linton, A. D. Hamilton, *Chem. Rev.* **1997**, *97*, 1669.

23 C. T. Seto, G. M. Whitesides, *J. Am. Chem. Soc.* **1991**, *113*, 712; C. B. Aakeröy, K. R. Seddon, *Chem. Soc. Rev.* **1993**, *22*, 397; S. Subramanian, M. J. Zaworotko, *Coord. Chem. Rev.* **1994**, *137*, 357; J.-M.

Lehn, M. Mascal, A. DeCian, J. Fischer, *J. Chem. Soc., Perkin Trans. 2*, **1992**, 461.

**24** X. Zhao, Y.-L. Chang, F. W. Fowler, J. W. Lauher, *J. Am. Chem. Soc.* **1990**, *112*, 6627.

**25** M.W. Hosseini, *CrystEngComm.* **2004**, *6*, 318; L. R. MacGillivray, *CrystEngComm.* **2004**, *6*, 77; L. Brammer, *Chem. Soc. Rev.* **2004**, *33*, 476; D. L. Caulder, K. N. Raymond, *Acc. Chem. Res.* **1999**, *32*, 975; D. N. Reinhoudt, M. Crego-Calama, *Science* **2002**, *295*, 2403; J.-M. Lehn, *Science* **2002**, *295*, 2400.

**26** A. Marsh, M. Silvestri, J.-M. Lehn, *Chem. Commun.* **1996**, 1527.

**27** F. H. Beijer, H. Kooijman, A. L. Spek, R. P. Sijbesma, E. W. Mejer, *Angew. Chem. Int. Ed. Engl.* **1998**, *37*, 75.

**28** W. Jaunky, M. W. Hosseini, J. M. Planeix, A. De Cian, N. Kyritsakas, J. Fischer, *Chem. Commun.* **1999**, 2313.

**29** J. Toussaint, *Acta Crystallogr.* **1951**, *4*, 71; R. S. Miller, I. C. Paul, D. Y. Curtin, *J. Am. Chem. Soc.* **1974**, *96*, 6334; M. Colapietro, A. Domenicano, *Acta Crystallogr., Sect. B.* **1982**, *38*, 1953; H. Takazawa, S. Ohba, Y. Saito, *Acta Crystallogr., Sect. B.* **1989**, *45*, 432.

**30** K. Folting, W. M. Lipscomb, B. Jerslev, *Acta Crystallogr.* **1964**, *17*, 1263; E. Arte, J. P. Declercq, G. German, M. van Meerssche, *Bull. Soc. Chim. Belg.* **1980**, *89*, 155.

**31** Å. Kvick, S. S. Booles, *Acta Crystallogr., Sect. B.* **1972**, *28*, 3405.

**32** Å. Kvick, M. Backeus, *Acta Crystallogr., Sect. B.* **1974**, *30*, 474. A. Kvick, R. Thomas, T. F. Koetzle, *Acta Crystallogr., Sect. B.* **1976**, *32*, 224.

**33** B. R. Penfold, J. C. B White, *Acta. Crystallogr.* **1959**, *12*, 130.

**34** C. B. Aakeröy, J. Desper, E. Elisabeth, B. A. Helfrich, B. Levin, J. F. Urbina, *Z. Kristallogr.* **2005**, *220*, 325.

**35** M. Bailey, C. J Brown, *Acta. Crystallogr.* **1967**, *22*, 387.

**36** E. A. Burton, L. Brammer, F. C. Pigge, C. Aakeröy, D. Leinen, *New. J. Chem.* **2003**, *27*, 1084–1094.

**37** K. G. Jensen, *Acta. Chem. Scand.* **1970**, *24*, 3293.

**38** Y. Ducharme, J. D. Wuest, *J. Org. Chem.* **1988**, *53*, 5787.

**39** D. J. Duchamp, R. E. Marsh, *Acta. Crystallogr., Sect. B.* **1969**, *25*, 5.

**40** Y-H. Chang, M-A. West, F. W. Fowler, J. W. Lauher, *J. Am. Chem. Soc.* **1993**, *115*, 5991; S. Coe, J. J. Kane, T. L Nguyen, L. M. Toledo, E. Wininger, F. W. Fowler, J. W. Lauher, *J. Am. Chem. Soc.* **1997**, *119*, 86.

**41** T. L. Nguyen, A. Scott, B. Dinkelmeyer, F. W. Fowler, J. W. Lauher, *New J. Chem.* **1998**, *2*, 129.

**42** L. Leiserowitz, *Acta Crystallogr., Sect. B* **1976**, *32*, 775.

**43** O. Ermer, *J. Am. Chem. Soc.* **1998**, *110*, 3747.

**44** O. Saied, T. Maris, J. D. Wuest, *J. Am. Chem. Soc.* **2003**, *125*, 14956.

**45** X. Wang, M. Simard, J. D. Wuest, *J. Am. Chem. Soc.* **1994**, *116*, 12119; M. Simard, D. Su, J. D. Wuest, *J. Am. Chem. Soc.* **1991**, *113*, 4696.

**46** P. Metrangolo, H. Neukirch, T. Pilati, G. Resnati, *Acc. Chem. Res.* **2005**, *38*, 386; P. Metrangolo, G. Resnati, *Chem.–Eur. J.* **2001**, *12*, 2511.

**47** G. R. Desiraju, R. L. Harlow, *J. Am. Chem. Soc.* **1989**, *111*, 6757.

**48** A. D. Bond, J. Griffiths, J. M. Rawson, J. Hulliger, *Chem Commun.* **2001**, 2488.

**49** B. Ahrens, P. G. Jones, *Acta Crystallogr., Sect. C. Cryst. Struct. Commun.* **1999**, *55*, 1308.

**50** A. C. B Lucassen, M. Vartanian, G. Leitus, M. van der Boom, *Cryst. Growth Des.* **2005**, *5*, 1671.

**51** R. B. Walsh, C. W. Padgett, P. Metrangolo, G. Resnati, T. W. Hanks, W. T. Pennington, *Cryst. Growth Des.* **2001**, *1*, 165.

**52** A. Hinchliffe, R. W. Munn, R. G. Pritchard, C. J. Spicer, *J. Mol. Struct.* **1985**, *130*, 93.

**53** A. Maiga, Y. Haget, M. A. Cuevas-Diarte, *J. Appl. Crystallogr.* **1984**, *17*, 210.

**54** E. Estop, A. Alvarez-larena, A. Belaaraj, X. Solans, M. Labrador, *Acta Crystallogr., Sect. C*, **1997**, *53*, 1932.

**55** V. R. Thalladi, H-C. Weiss, D. Blaser, R. Boese, A. Nangia, G. R. Desiraju, *J. Am. Chem. Soc.* **1998**, *120*, 8702.

**56** J. D. Dunitz, *ChemBioChem.* **2004**, *5*, 614.

**57** M. Bremer, P. S. Gregory, P. V. R. Schleyer, *J. Org. Chem.* **1989**, *54*, 3796.

**58** V. R. Pedireddi, D. S. Reddy, B. S. Goud, D. C. Craig, A. D. Rae, G. R. Desiraju, *J. Chem. Soc. Perkin Trans. 2* **1994**, 2353.

**59** G. R. Desiraju, *Angew. Chem. Int. Ed. Engl.* **1995**, *34*, 2311.

**60** J. D. Dunitz, in *Perspectives in Supramolecular Chemistry: The Crystal as a Supramolecular Entity*, G. R. Desiraju (Ed.), Wiley, Amsterdam, 1995.

**61** M. C. Etter, D. A. Adsmond, *J. Chem. Soc., Chem. Commun.* **1990**, 589.

**62** M. C. Etter, P. W. Baures, *J. Am. Chem. Soc.***1988**, *110*, 639.

**63** M. C. Etter, *Acc. Chem. Res.* **1990**, *23*, 120.

**64** M. C. Etter, *J. Phys. Chem.* **1991**, *95*, 4601.

**65** G. R. Desiraju, *CrystEngComm*, **2003**, 5, 466; J. D. Dunitz, *CrystEngComm*, **2003**, 5, 506.

**66** This requirement may be deemed overly restrictive but it does offer an important distinction between solvates and co-crystals. However, in same cases, notably in the work by Boese and coworkers, it is clear that in their elegant work using low-temperature crystallizations they intentionally prepare co-crystals with a very clear and deliberate strategy (see e.g. M. T. Kirchner, R. Boese, A. Gehrke, D. Blaeser, *CrystEngComm*. **2004**, *6*, 360.)

**67** For information about classic charge-transfer compounds, see J. Rose, *Molecular Complexes*, Pergamon Press, Oxford, 1967.

**68** S. Shan, E. Batchelor, W. Jones, *Tetrahedron Lett.* **2002**, *43*, 8721; P. Vishweshwar, R. Thaimattam, M. Jaskolski, G. R. Desiraju, *Chem. Commun.* **2002**, 1830; P. Vishweshwar, A. Nangia, V. M. Lynch, *CrystEngComm.* **2003**, *5*; Ö. Almarsson, M. J. Zaworotko, *Chem. Commun.* **2004**, 1889.

**69** G. R. Desiraju, J. A. R. P. Sarma, *J. Chem. Soc., Chem. Commun.* **1983**, 45; F. Pan, W. S. Wong, V. Gramlich, C. Bosshard, P. Gunter, *J. Chem. Soc.,Chem. Commun.* **1996**, 2; V. R. Pedireddi, W. Jones, A. P. Chorlton, R. Docherty, *J. Chem. Soc., Chem. Commun.* **1996**, 997; S. H. Dale, M. R. J. Elsegood, M. Hemmings, A. L. Wilkinson, *CrystEngComm*, **2004**, *6*, 207; V. R. Pedireddi, J. PrakashaReddy, K. K. Arora, *Tetrahedron Lett.* **2003**, *44*, 4857.

**70** C. B. Aakeröy, J. Desper, B. A. Helfrich, *CrystEngComm* **2004**, *6*, 19.

**71** C. B. Aakeröy, A. M. Beatty, M. Nieuwenhuyzen, M. Zou, *Tetrahedron* **2000**, *56*, 6693.

**72** J.-M. Lehn, M. Mascal, A. DeCian, J. Fischer, *J. Chem. Soc., Chem. Commun.* **1990**, 479.

**73** R. D. Bailey Walsh, M. W. Bradner, S. Fleischman, L. A. Morales, B. Moulton, N. Rodriguez-Hornedo, M. J. Zaworotko, *Chem. Commun.* **2003**, 186.

**74** L. R. MacGillivray, J. L. Reid, J. A. Ripmeester, *J. Am. Chem. Soc.*, **2000**, *122*, 7817.

**75** J. A. Zerkowski, J. C. MacDonald, G. M. Whitesides, *Chem. Mater.* **1997**, *9*, 1933.

**76** J. J. Kane, R.-F. Liao, J. W. Lauher, F. W. Fowler, *J. Am. Chem. Soc.* **1995**, 117, 12003.

**77** P. Vishweshwar, A. Nangia, V. M. Lynch, *J. Org. Chem.* **2002**, *67*, 556.

**78** C. Huang, L. Leiserowitz, G. M. Schmidt, *J. Chem. Soc., Perkin Trans. 2* **1973**, 503.

**79** F. A. Allen, P. R. Raithby, G. P. Shields, R. Taylor, *J. Chem. Soc., Chem. Commun.* **1998**, 1043.

**80** O. Ermer, J. Neudörfl, *Helv. Chim. Acta.* **2001**, *84*, 1268; L. Leiserowitz, M. Tuval, *Acta Crystallogr., Sect. B* **1978**, *34*, 1230.

**81** M. C. Etter, G. M. Frankenbach, *Chem. Mater.* **1989**, *1*, 10.

**82** This observation was first made by J. Donohue *J. Phys. Chem.* **1952**, *56*, 502.

**83** P. Metrangolo, T. Pilati, G. Resnati, A. Stevenazzi, *Chem. Commun.* **2004**, 1492.

**84** A. De Santis, A. Forni, R. Liantonio, P. Metrangolo, T. Pilati, G. Resnati, *Chem. Eur. J.* **2003**, *9*, 3974.

**85** D. S. Reddy, D. C. Craig, G. R. Desiraju, *Chem. Commun.* **1994**, 1457.

**86** T. Dahl, O. Hassel, *Acta Chem. Scand.* **1970**, *24*, 377.

**87** J. A. R. P. Sarma, G. R. Desiraju, *J. Chem. Soc., Perkin Trans. 2* **1985**, 1905.

**88** T. Sugiyama, J. Meng, T. Matsuura, *J. Mol. Struct.* **2002**, *611*, 53.

**89** D. E. Lynch, G. Smith, D. Freney, K. A. Byriel, C. H. L. Kennard, *Aust. J. Chem.* **1994**, *47*, 1097.

**90** L. R. Nassimbeni, H. Su, E. Weber, K. Skobridis, *Cryst. Growth Des.* **2004**, *4*, 85.

**91** D. E. Lynch, T. Latif, G. Smith, K. A. Byriel, C. H. L. Kennard, S. Parsons, *Aust. J. Chem.* **1998**, *51*, 403.

**92** J. Zhu, J.-M. Zheng, *Chin. J. Struct. Chem.* **2004**, *23*, 417.

**93** C. B. Aakeröy, A. M. Beatty, B. A. Helfrich, *J. Am. Chem. Soc.* **2002**, *124*, 14423.

**94** C. B. Aakeröy, A. M. Beatty, M. Zou, *Cryst. Eng.* **1998**, *1*, 225.

**95** R. F. M. Lange, F. H. Beijer, R. P. Sijbesma, R. W.W. Hooft, H. Kooijman, A. L. Spek, J. Kroon, E. W. Meijer, *Angew. Chem., Int. Ed,* **1997**, *36*, 969.

**96** J. A. Zerkowski, C. T. Seto, G. M. Whitesides, *J. Am. Chem. Soc.* **1992**, *114*, 5473.

**97** M. Y. Antipin, A. I. Akhmedov, Y. T. Struchkov, E. I. Matrosov, M. I. Kabachnik, *Zh. Strukt. Khim.* **1983**, *24*, 86; D. E. Lynch, G. Smith, K. A. Byriel, C. H. L. Kennard, *Acta Crystallogr., Sect. C: Cryst. Struct. Commun.* **1993**, *49*, 718.

**98** M. C. Etter, T. W. Panunto, *J. Am. Chem. Soc.* **1988**, *110*, 5896.

**99** C. Glidewell, G. Ferguson, R. M. Gregson, A. J. Lough, *Acta Crystallogr., Sect. C: Cryst. Struct. Commun.* **1999**, *55*, 2133.

**100** A. Ranganathan, V. R. Pedireddi, C. N. R. Rao, *J. Am. Chem. Soc.* **1999**, *121*, 1752.

**101** A. De Santis, A. Forni, R. Liantonio, P. Metrangolo, T. Pilati, G. Resnati, *Chem.–Eur. J.* **2003**, *9*, 3974.

**102** R. B. Walsh, C. W. Padgett, P. Metrangolo, G. Resnati, T. W. Hanks, W. T. Pennington, *Cryst. Growth Des.* **2001**, *1*, 165.

**103** F. Garcia-Tellado, S. J. Geib, S. Goswami, A. D. Hamilton, *J. Am. Chem. Soc.*, **1991**, *113*, 9265.

**104** M. Tremayne, C. Glidewell, *Chem. Commun.* **2000**, 2425.

**105** O. Ermer, A. Eling, *J. Chem. Soc., Perkin Trans. 2* **1994**, 925.

**106** C. B. Aakeröy, J. Desper, B. Leonard, J. F. Urbina, *Cryst. Growth Des.* **2005**, *3*, 865.

**107** C. B. Aakeröy, A. M. Beatty, B. A. Helfrich, M. Nieuwenhuyzen, *Cryst. Growth Des.* **2003**, *3*, 159.

**108** T.-J. M. Luo, G. T. R Palmore, *Cryst. Growth Des.* **2002**, *2*, 337.

**109** V. R. Pedireddi, S. Chatterjee, A. Ranganathan, C. N. R. Rao, *J. Am. Chem. Soc.* **1997**, *119*, 10867.

**110** T. L. Nguyen, F. W. Fowler, J. W. Lauher, *J. Am. Chem. Soc.* **2001**, *123*, 11057.

**111** J. Xiao, M. Yang, J. W. Lauher, F. W. Fowler, *Angew. Chem., Int. Ed.* **2000**, *39*, 2132.

**112** F. H. Beijer, R. P. Sijbesma, J. A. J. M. Vekemans, E. W. Meijer, H. Kooijman, A. L. Spek, *J. Org. Chem.* **1996**, *61*, 6371.

**113** D. Q. Ma, P. Coppens, *Chem. Commun.* **2003**, 2290.

**114** P. Vishweshwar, A. Nangia, V. M. Lynch, *Cryst. Growth Des.* **2003**, *3*, 783.

**115** S. G. Fleischman, S. S. Kuduva, J. A. McMahon, B. Moulton, R. D. B. Walsh, N. Rodriguez-Hornedo, M. J. Zaworotko, *Cryst. Growth Des.* **2003**, *3*, 909.

**116** M. C. Etter, J. C. MacDonald, J. Bernstein, *Acta Crystallogr., Sect. B.* **1990**, *46*, 256.

**117** J. Bernstein, R. E. Davis, L. Shimoni, N.-L.Chang, *Angew. Chem., Int. Ed., Engl.,* **1995**, *34*, 1555.

**118** C. P Brock, J. D. Dunitz, *Acta Crystallogr., Sect. A* **1991**, *47*, 854.

**119** P. Kronenbusch, W. B. Gleson, D. Britton, *Cryst. Struct. Commun.* **1976**, *5*, 17.

**120** D. Britton, J. Konnert, S. Lam, *Cryst. Struct. Commun.* **1977**, *6*, 45.

**121** H. Quast, Y. Gorlach, E.-M. Peters, K. Peters, H. G. von Schnering, L. M. Jackman, G. Ibar, A. J. Freyer, *Chem. Ber.* **1986**, *119*, 1801.

**122** Charge-transfer complexes are not included in this discussion.

**123** Several co-crystals have been constructed from heterocycles like bipyridin or pyrazine and activated iodo-containing compounds; see refs. [51] and [102].

**124** F.A. Allen, *Acta Crystallogr., Sect. B* **2002**, *58*, 380.

**125** F. Pan, W. S. Wong, V. Gramlich, C. Bosshard, P. Gunter, *Chem. Commun.* **1996**, 2; J. A. Zerkowski, J. C. MacDonald, G. M. Whitesides, *Chem. Mater.* **1997**, *9*, 9; R.-F. Liao, J. W. Lauher, F. W. Fowler, *Tetrahedron*, **1996**, *52*, 3153; N. Shan, A. D. Bond, W. Jones, *Cryst. Eng.* **2002**, *5*, 9; V. R. Pedireddi, W. Jones, A. P. Chorlton, R. Docherty, *Chem. Commun.* **1996**, 997; P. Vishweshwar, A. Nangia, V. M. Lynch, *J. Org. Chem.* **2002**, *67*, 556.

**126** CSD codes: OCAYUM, OCASAT, and POFPAB.

**127** CSD codes: GURVOE, GURVUK, LUDFUL, UNECAR, ZUPKUQ, and ZUPLAX.

**128** CSD codes: BEQWAV, FIHYEA, FIJCIK, FIJCUW, HUPPOX, MUFNIK, NOPXIZ, OJENIA, PULWUO, RAPHAR, SUXVOW, UDUZIC, UDOZOI, UHELUO, XUNGIW, and XUNGOC.

**129** CSD code: POFPEF.

**130** CSD codes: CUFSET, LUSWEB, PANYIM, PANZEJ, PAPDIT, PAPFOB, UNEBOE, WUKREZ, and WUKROJ.

**131** CSD codes: AJAKEB, AJAKIF, ASAXOH, ASAXUN, BUDWEC, BUDZUV, BUFBIP, BUFQAU, LUNMEM, LUNMIC, LUNMOW, and VAKTOR.

**132** C. V. K. Sharma, K. Panneerselvam, T. Pilati, G. R. Desiraju, *J. Chem. Soc., Perkin Trans. 2* **1993**, 2209.

**133** V. R. Pedireddi, J. PrakashReddy, *Tetrahedron Lett.* **2002**, *43*, 4927.

**134** J. Bernstein, *Polymorphism in Molecular Crystals*, Oxford University Press, New York, 2002.

**135** C. B. Aakeröy, M. Nieuwenhuyzen, S. L. Price, *J. Am. Chem. Soc.* **1998**, 120, 8986.

**136** Forms I and III of 4-chlorobenzamide: T. Taniguch, K. Nakata, Y. Takaki, K. Sakurai, *Acta Crystallogr. Sect. B* **1978**, *34*, 2574; Form II: T. Hayashi, K. Nakata, Y. Takaki, K. Sakurai, *Bull. Chem. Soc. Jpn.* **1980**, *53*, 801.

**137** P. D. Cradwick, *Nature*, **1975**, 258, 774.

**138** A. Katrusiak, *Acta Crystallogr., Sect. C: Cryst. Struct. Commun.* **1993**, 49, 36; A. Katrusiak, *Acta Crystallogr., Sect. B: Struct. Sci.* **2001**, *57*, 697.

**139** C. B. Aakeröy, A. M. Beatty, D. S. Leinen, *CrystEngComm.* **2002**, *4*, 310; J. Valdés-Martínez, M. Del Rio-Ramirez, S. Hernández-Ortega, C. B. Aakeröy, B. A. Helfrich, *Cryst. Growth Des.* **2001**, *2*, 485; C. B. Aakeröy, A. M. Beatty, M. Tremayne, D. M. Rowe, C. C. Seaton, *Cryst. Growth Des.* **2001**, *2*, 377.

**140** C. B. Aakeröy, A. M. Beatty, D. S. Leinen, *Cryst. Growth Des.* **2000**, *1*, 47; C. B. Aakeröy, A. M. Beatty , M. Nieuwenhuyzen, M. Zou, *J. Mater. Chem.* **1998**, *8*, 1385.

**141** A. D. Burrows, *Struct. Bond.*, **2004**, *111*, 55.

**142** C. B. Aakeröy, A. M. Beatty, B. A. Helfrich, *Angew. Chem., Int. Ed.* **2001**, *40*, 3240.

**143** If the acid is too strong, the proton will be transferred to the nitrogen atom resulting in an ionic compound and not a co-crystal. Some examples of crystal engineering of *ionic* systems include e.g., J. A. Swift, A. M. Pivovar, A. M. Reynolds, M. D. Ward, *J. Am. Chem. Soc.*, **1998**, 120, 5887; F. Xue, T. C. W. Mak, *J. Phys. Org. Chem.* **2000**, *13*, 405; P. Cudic, J.-P. Vigneron, J.-M. Lehn, M. Cesario, T. Prangé, *Eur. J. Org. Chem.* **1999**, 2479; C. V. K. Sharma, A. Clearfield, *J. Am. Chem. Soc.* **2000**, *122*, 4394; S. Ferlay, R. Holakovsky, M. W. Hosseini, J.-M. Planeix, N. Kyritsakas, *Chem. Commun.***2003**, 1224; C. B. Aakeröy, P. B. Hitchcock, K. R. Seddon, *J. Chem. Soc., Chem. Commun.* **1992**, 553; C. B. Aakeröy, D. P. Hughes M. Nieuwenhuyzen, *J. Am. Chem. Soc.* **1996**, *118*, 10134; A. M. Beatty, K. E. Granger, A. E. Simpson, *Chem.–Eur. J.* **2002**, *8*, 3254; A. M. Beatty, C. L. Schneider, A. E. Simpson, J. L. Zaher, *CrystEngComm.* **2002**, *4*, 282; D. Braga, L. Maini, M. Polito, F. Grepioni, *Struct. Bond.* **2004**, *111*, 1.

**144** L. Leiserowitz, F. Nader, *Acta Crystallogr., Sect. B: Struct. Sci.* **1977**, *33*, 2719.

**145** All $pK_a/pK_b$ values from G. Kartum, W. Vogel, K. Andrussov, *Dissociation Constants of Organic Acids in Aqueous Solution*, Butterworth & Co., London, 1961.

**146** C. B. Aakeröy, J. Desper, J. F. Urbina, *Chem. Commun.* **2005**, 2820.

**147** An estimate for the differences in basicity between the pyridyl and the benzimidazol-1-yl moieties were obtained by calculating $pK_a$ values for the conjugated acids of the two hydrogen-bond acceptors in each SR. 3-(benzimidazol-1-yl)methylpyridine: $pK_a$ = 4.71 ± 0.10 and 5.72 ±0.30 for pyridyl and benzimidazol-1-yl moieties, respectively. 3-(2-methylbenzimidazol-1-yl)methylpyridine: $pK_a$ = 4.72 ±0.10 and 5.85 ± 0.18 for pyridyl and benzimidazol-1-yl moieties, respectively. Calculations were performed using Advanced Chemistry Development (ACD/Labs) Software Solaris V4.76 (© 1994–2005 ACD/Labs).

**148** M. H. Abraham, *Chem. Soc. Rev.* **1993**, *22*, 73.

**149** S. Shan, S. Loh, D. Herschlag, *Science*, **1996**, *272*, 97.

**150** 3,5-dinitrobenzoic acid ($pK_a$=2.8); 4-nitrobenzoic acid ($pK_a$= 3.44); 3-*N*,*N*-(dimethylamino)benzoic acid ($pK_a$= 4.30).

# 3
# Characterizations and Applications

## 3.1
### Diffraction Studies in Crystal Engineering
*Guillermo Mínguez Espallargas and Lee Brammer*

### 3.1.1
### Introduction

Crystal engineering concerns the design and synthesis, characterisation and applications of crystalline materials. The focus of this chapter is on diffraction methods, but it is valuable first to place this in a wider context of characterisation of crystalline materials.

Synthetic endeavors in molecular organic or inorganic chemistry generally result in discrete (molecular) species that can be examined in the first instance by spectroscopic methods readily applied to the solution phase, including NMR, IR and UV/visible spectroscopy and mass spectrometry. Crystal synthesis by contrast necessarily leads to crystalline products, i.e. solid phase periodic materials. Thus, characterisation methods suited to such materials are needed. Central among these techniques is diffraction, from which detailed, accurate and precise information about the composition and geometric characteristics of the material can often be obtained. However, any practitioner of crystal engineering should have at their disposal an array of physical techniques. Thus, whereas single crystal diffraction provides very accurate and precise structural information it is always vital to recognise that accurate determination of the structure of a single crystal is different from characterisation of the product as a whole. Powder diffraction studies permit assessment of phase homogeneity, i.e. assessment of whether the single crystal structure determined is the only product phase or one of many from a given crystal synthesis. Elemental analysis of the product (most commonly for carbon, hydrogen, and nitrogen content, but potentially for other elements) is also important in assessing overall product composition, but of course cannot distinguish between two crystalline phases of the same composition (i.e. polymorphs – see Chapter 3.3). Spectroscopic methods are still valuable tools for studying crystalline solids. Solid state NMR spectroscopy (Chapter 3.2) can be particularly valuable in understanding structural transformations in the solid state, and has an advantage over

*Making Crystals by Design*. Edited by Dario Braga and Fabrizia Grepioni
Copyright © 2007 WILEY-VCH Verlag GmbH & Co. KGaA, Weinheim
ISBN: 978-3-527-31506-2

diffraction methods of permitting such processes to be followed even when amorphous phases are produced and prohibit diffraction studies. However, NMR spectroscopy cannot provide accuracy and precision of internuclear distances comparable to diffraction methods in crystalline materials. IR spectroscopy is particularly useful for identifying the presence of non-covalent interactions involving particular functional groups. Carboxylic acids and amides are common functional groups used in designing hydrogen-bonded materials (Chapter 2.5) and carboxylates are widely used in the synthesis of metal-organic framework materials (Chapter 2.4). These groups provide an excellent spectroscopic handle for identification by IR methods. Thermal analysis of solids includes a range of techniques, foremost among them being differential scanning calorimetry (DSC) and thermal gravimetric analysis (TGA). DSC permits both the temperature and enthalpy of a structural transformation to be quantified. Such transformations include phase transitions between polymorphs and between solid and liquid phases or can involve loss of trapped guest molecules (e.g. in solvates). In the latter area TGA becomes important and more broadly, given the importance of hydrates in studies of pharmaceutical solids and the design and function of porous materials for gas storage, separations and other forms of molecular entrapment. TGA permits the change in mass of a solid to be measured accurately as a function of temperature and thus can be invaluable in assessing the ability of solids with entrapped molecules to retain the guests and quantifying the temperature at which the guest molecules are released. TGA is also very important in assessing the overall thermal stability of a solid and providing insight into its means of thermal decomposition, particularly if coupled with other means of analysis such as mass spectrometry or IR spectroscopy. Quantifying such properties is essential for developing applications of any new crystalline material.

Many other techniques may be used to assess properties pertinent to specific applications of designed crystalline materials, such as magnetic or conductivity measurements. Electron microscopy has been applied to studies of crystal growth and even to solid-state reactivity (Chapter 2.1) and theoretical calculations are vital to providing a better understanding of intermolecular interactions (Chapter 1.2), and understanding nucleation and crystal packing (Chapter 1.1). However, the linchpin technique in crystal engineering is most often diffraction, not only for the study of individual crystal structures but also for the collective information that can be derived via crystallographic databases on typical geometries.

### 3.1.2
### Scope

It is not within the scope of this chapter to provide tuition in crystallography. For this the reader is referred to many valuable texts [1] or to intensive crystallography schools such as those run by the British Crystallographic Association and the American Crystallographic Association [2]. The aim of the following sections is to provide a brief overview of what diffraction methods can offer in the context of crystal engineering. Diffraction of X-rays and neutrons by single crystals and polycrystalline powders will be covered. In the later subsections some illustrative examples will be provided of diffraction experiments that move beyond structure determination from single crystal diffraction. Thus, diffraction studies at different temperatures and pressures will be examined as will structure determination from polycrystalline powders. Charge density analysis will be outlined. The intent here is bring to the attention of the reader a wider array of means by which diffraction methods can provide valuable input into crystal engineering.

## 3.1.3
## Single Crystal X-ray Diffraction

Single crystal X-ray diffraction is the predominant means of crystal structure determination and is central to crystal engineering. Key to its importance is the relative ease of use and the accurate and typically unambiguous structural characterisation it provides.

### 3.1.3.1   Experimental Method

Single crystal X-ray diffraction can be carried out using a laboratory X-ray diffractometer comprising a sealed tube or rotating anode X-ray source, a goniometer to orient the crystal and a detector to measure diffraction intensities from which the crystal structure is to be determined. Such instruments are designed to operate at an X-ray wavelength characteristic of the anode chosen, e.g. Mo-K$\alpha$ (0.71073 Å) or Cu-K$\alpha$ (1.54056 Å). Alternatively, synchrotron X-ray sources provide a much higher flux (ca. $10^7$ times sealed tube flux and ca. $10^6$ times rotating anode flux) and brilliance of X-rays and the capability of tuning the wavelength over a wide range (e.g. 0.3–2.0 Å). The X-ray goniometer and detector are usually comparable to those used for laboratory diffractometers.

The most recent generation of CCD X-ray detectors enable laboratory structure determination on crystals of dimensions of 100 μm and in favorable cases smaller still. The use of synchrotron radiation may permit crystals of a few tens of microns in size to be studied. Typical intensity data collection time is usually of the order of a few hours (shorter for high flux synchrotron sources), but, with some compromise in precision, structure determination can be accomplished in some instances using data collection of less than an hour.

CCD and other area detectors such as image plates permit large numbers of intensities to be measured simultaneously, in contrast to their forerunner scintillation detectors for which intensities were measured sequentially. Area detectors therefore facilitate multiple measurements of each diffraction intensity (or symmetry-related intensity) within a relatively short time period (hours). Averaging over the crystal symmetry then yields a unique set of intensities of greater accuracy than where only individual measurements have been made. Such extensive data sets containing multiple measurements of each intensity also permit excellent estimation and empirical correction for systematic errors such as absorption, and instrumental/ experimental errors such as minor misalignments of the instrument or crystal.

Area detectors also offer the great advantage over point detectors that the unit cell dimensions and orientation of the crystal need not be determined prior to data collection, although such a step is an advantageous preliminary step in any diffraction experiment. The absence of this requirement allows data collection to be undertaken straightforwardly for samples that are twinned, i.e. contain domains or crystallites of substantially different orientations. Crystallographic software has advanced in parallel with the widespread use of area detectors over the past 10 years such that detecting and modeling twinned structures is now feasible. In many cases this allows structure solution and refinement and thus valuable structural information to be obtained on samples that would have proven intractable in most cases when using a point detector.

Single crystal diffractometers can be equipped with open flow cryostats capable of reaching temperatures as low as 90 K quite routinely, assuming a readily available supply either of liquid nitrogen or dry nitrogen gas. Commercially available helium-based cryostats also becoming more widely used allowing temperatures below 30 K to be reached and structure determination under application of modest pressure is a developing area of potential relevance to crystal engineering. The application to crystal engineering of these methods will be discussed in Sections 3.1.5 and 3.1.6.

### 3.1.3.2 What Information Can Be Obtained and How Reliable Is It?

Crystal structure determination requires the development of a suitable model for periodic electron density distribution in the crystal that is related by Fourier transform to the structure factors that can be derived from the experimentally measured diffraction intensities. The model, in practice, is atomistic and consists of coordinates and atom types for all symmetry independent atoms as well as parameters that are used to describe the (mean square amplitude of) displacement of the atoms about their mean positions, which arises due to molecular motions and small variations in the mean position across the collection of unit cells that comprise the crystal.

The quality of the structure determination depends of course upon data quality and also on the quality of the model used to fit the data. Measures of data quality and the quality of the structure refinement are numerous. A detailed and thorough means of assessment has been implemented in an automated form by the International Union of Crystallography (IUCr). The assessment is based upon the crystallographic information file (CIF), now the standard means to archive the results of a crystallographic experiment. Submission of a CIF via the IUCr web site [3] results in the generation of a detailed report that provides an excellent guide to researchers regarding any potential problems with the structure determination. This assessment has now been adopted as a standard guide to authors by all major publishers of scientific journals in which crystal structures are reported.

A complete structure determination of course reveals the chemical composition of the crystal studied, but most importantly it provides accurate geometric information. This includes bond lengths, bond angles, torsion angles, interplanar distances and the geometries of intermolecular interactions and an estimate of the uncertainty (precision) of these values. The precision of intramolecular distances is usually between 0.001 and 0.01 Å, while for bond angles and torsion angles uncertainties are often 0.01–0.1°. However, it should be recognised that geometries of chemically identical bond distances or angles can vary by much larger amounts between compounds or between crystal structures. Martín and Orpen have examined this question in some detail [4].

In the context of crystal engineering, rather than the use of crystallography for molecular structure determination, single crystal diffraction is often vital for establishing chemical composition. It is then intermolecular interactions, supramolecular synthons, network topology or perhaps pore or cavity dimensions rather than intramolecular geometries that are the most important results derived form a crystal structure determination. However, the latter should of course always be checked as part of the overall assessment of quality of the structure determination..

### 3.1.3.3 **Limitations**

Although providing highly accurate and precise structural information, the limitations of single crystal X-ray diffraction should also be recognised. A most important limitation in the context of crystal design and synthesis bears repetition: a single crystal structure determination is just that. In other words it represents a structure determination only of the single crystal studied, not of the crystalline product of the synthesis. Analysis of the entire product, preferably by powder diffraction (Section 3.1.7) is required to establish if this structure is representative of the product as a whole.

Since X-ray diffraction involves scattering of X-rays by the electron distribution of the crystal, atoms of greater electron density (atomic number) contribute substantially more to the diffraction intensities than do lighter atoms. In the extreme case the contribution of hydrogen atoms is small, which consequently makes the location by X-ray diffraction difficult or imprecise. Location of hydrogen atoms is also systematically inaccurate and leads to apparently shortened covalent bonds and therefore overestimation of hydrogen bond lengths $[(D)H \cdots A]$ and other intermolecular contacts involving hydrogen atoms. Since hydrogen atoms have only one electron, which is inevitably involved in formation of a covalent bond (C–H, N–H, O–H, etc) their electron density is centered not about the hydrogen atom nucleus but at a point displaced towards the covalently bonded atom (e.g. C, N, O, etc). In a conventional crystallographic model used in structure determination each atom is represented by the electron density of a gas-phase spherical atom convoluted with the probability density function associated with the displacement parameter model. Hence, derived atomic positions are associated with the sites of maximum electron density. Thus hydrogen atom positions are determined systematically displaced by ca. 0.1–0.15 Å from the nuclear position [5]. Hydrogen atom positions are also routinely assigned using a standard geometric calculation based upon the hybridisation of the covalently attached atom. This is a reliable means of assigning hydrogen atom positions except in cases of torsional uncertainty, such as for methyl groups or hydroxyl groups.

Accurate location of hydrogen atom positions requires the use of neutron diffraction (Section 3.1.4). When X-ray diffraction is used, therefore, it should be recognised that for calculation of intermolecular geometries involving hydrogen atoms, most prominently for hydrogen bonds, the experimentally determined (or assigned) hydrogen atom position should be adjusted to the anticipated nuclear position by displacement along the X–H vector to a suitable internuclear distance (viz. C–H 1.083, N–H 1.01, O–H 0.96 Å [5]).

Another consequence of the relationship between X-ray diffraction and the electron density is that X-ray diffraction can lead to difficulties in distinguishing atoms of similar atomic number assuming that their coordination geometry does not provide a clear distinction. This is not uncommon for transition metal compounds and when diffraction data quality is poor difficulty in distinguishing carbon from nitrogen atoms in organic compounds or ligands can arise. Although X-rays are sensitive to electron density they are highly insensitive to electron spin and thus largely uninformative on magnetic structure. There are also practical considerations when undertaking X-ray diffraction experiments at very low temperature or at elevated pressures wherein closed sample vessels, typically made of metals for strength, are used. The high absorption and large background scattering of X-rays by these sample containers make such studies more difficult.

Fortunately a number of the limitations highlighted here can be addressed by turning to single crystal neutron diffraction.

### 3.1.4
### Single crystal neutron diffraction

Single crystal neutron diffraction is in many ways a complementary technique to X-ray diffraction. In neutron diffraction scattering by the atomic nuclei rather than the electron density gives rise to diffraction. However, neutrons have a spin and polarisation of the neutron beam can be used to undertake diffraction experiments to map the distribution of *unpaired* electrons (the spin density) in a crystal.

#### 3.1.4.1    Experimental Method
Unlike X-ray diffraction, all neutron diffraction experiments are conducted at major national or regional facilities since a nuclear reactor or an accelerator facility, known as a spallation neutron source, is required to generate a suitable neutron beam. Nuclear reactor sources, such as at the Institut Laue Langevin in Grenoble, France, have traditionally used a monochromator crystal to select a single wavelength from the broad spectrum of neutrons. However, the recent development of image plate area detectors for neutrons has led to the introduction of Laue or quasi-Laue diffraction experiments in which a range of wavelengths are used simultaneously. At spallation neutron sources Laue diffraction is also used, but in time-of-

flight mode (TOF), whereby neutrons of different energies (wavelengths) arrive at the sample crystal at different times following a pulse of neutrons departing from a heavy atom target that is bombarded with a beam from a proton synchrotron. The TOF method allows good temporal resolution of diffraction events resulting from neutrons of different wavelengths. Laue methods add complexity to the measurement of diffraction intensities and to the application of corrections for absorption and extinction, which are wavelength dependent. However, neutron diffraction is highly flux limited compared to X-ray diffraction and Laue methods are important in increasing the effective flux and thus decreasing the time required for the data collection or permitting the use of smaller crystals. Most materials have a low absorption cross-section for neutrons compared to X-rays. Thus, construction of robust sample containers that can be evacuated, for very low temperature studies, or subjected to high pressures is straightforward.

Despite advances in Laue diffraction the much lower flux of neutron sources even compared to the X-ray flux of a sealed tube source necessitates the use of large crystals, typically with linear dimensions in excess of 1 mm. However, the latest generation of neutron sources and instruments currently being designed and constructed in the USA, Europe and Asia are anticipated to facilitate single crystal diffraction studies on crystals of dimensions of ca. 0.5 mm. The limited availability of instruments for single crystal neutron diffraction and the need for at least somewhat larger crystals dictates that X-ray diffraction is far more widely used and indeed neutron diffraction studies are usually conducted on compounds for which the structure has been previously determined by X-ray diffraction. Indeed, crystal structures are rarely solved directly from neutron diffraction data since the common presence of both atoms with positive scattering lengths and negative scattering lengths (phase shifted by 180°) makes structure solution by direct methods difficult to implement.

### 3.1.4.2 What Information Can Be Obtained and How Reliable Is It?

A key difference between X-ray and neutron diffraction is that for the latter there is no systematic increase in atomic scattering power with atomic number (Fig. 3.1.1). Furthermore, isotopes of the same element have different scattering powers. Thus, not only neighboring elements, but isotopes (significantly H and D) can readily be distinguished. In addition to this, hydrogen (and deuterium) have scattering powers comparable to most other elements and thus contribute significantly to all diffraction intensities. Consequently hydrogen atom positions can be located more precisely than by X-ray diffraction. Hydrogen atom location is also more accurate since nuclear positions rather than centroids of electron density constitute the structural model. Thus, a common application of neutron diffraction is obtaining accurate metrical information on hydrogen bonds [6].

A more specialist application of neutron diffraction involves the use of polarised neutrons, which can interact with the unpaired electron spins (magnetic scattering) as well as the atomic nuclei (nuclear scattering). The difference between the total neutron scattering and the nuclear scattering can be used to establish the

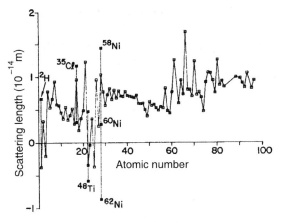

**Figure 3.1.1** Neutron scattering lengths.

magnetic scattering and from the magnetic scattering factors the spin density distribution can be modeled [7]. This is particularly important for magnetic materials, the design of which represents a growing interface within crystal engineering [8].

### 3.1.5
### Single Crystal Diffraction Studies at Low Temperatures

In the previous sections single crystal diffraction has been described and an overview of the typical information one may wish to obtain from such an experiment has been presented. This includes chemical composition, atomic positions and molecular geometries, atomic displacements, and, in the context of crystal engineering, the identification of supramolecular synthons, network topologies, cavity sizes, etc. The study of non-covalent interactions is one of the main research

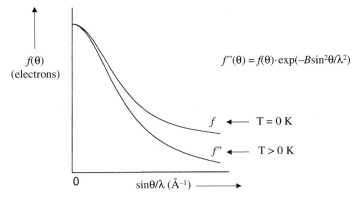

**Figure 3.1.2** Dependence of atomic scattering factor, $f$, upon scattering angle $\theta$ and temperature.

areas in crystal engineering, and in order to identify weak non-covalent interactions it is of great importance to obtain accurate positions of the atoms. Use of low temperature data collection is often the simplest way of enhancing the accuracy (and precision) of this information. A valuable exposition on the advantages and applications of low temperature X-ray diffraction has recently been published [9].

Reduction of the temperature of the crystal decreases the thermal motion of the atoms, which can be expressed in terms of the temperature factor (*B*). A reduction in *B* enhances the atomic scattering factors (Fig. 3.1.2) and thus the corresponding structure factors and measured intensities (Table 3.1.1), particularly at higher Bragg angles.

**Table 3.1.1** Expected increase in observed Bragg intensities based on reduction of overall *B* values from 4.0 Å$^2$ (300 K) to 1.5 Å$^2$ (100 K) to 0.75 Å$^2$ (20 K). (Adapted from Ref. [10].)

| $\sin\theta/\lambda (\text{Å}^{-1})$ | $I_{hkl}$ (100 K) / $I_{hkl}$ (300 K) | $I_{hkl}$ (20 K) / $I_{hkl}$ (100 K) |
|---|---|---|
| 0.7 | 11.6 | 2.1 |
| 1.0 | 149.6 | 4.5 |
| 1.3 | 4675 | 12.6 |

An illustration of the influence of temperature on diffraction intensities is provided in Fig. 3.1.3, which shows two frames collected with a CCD area detector using a 0.3° scan over a 10 s time interval. Each has been collected for the compound

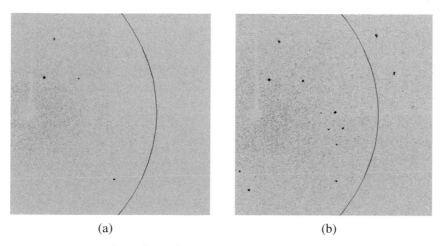

(a)                  (b)

**Figure 3.1.3** Intensity data collected for a crystal of (4-bromo-pyridinium)$_2$[CoCl$_4$] in the same orientation but at two different temperatures, (a) 240 K, and (b) 30 K, using Mo-K$_\alpha$ radiation [11b]. The solid line corresponds to a resolution of *d* = 1 Å.

(4-bromopyridinium)$_2$[CoCl$_4$] [11] using the same crystal and orientation. It can be clearly seen that the number of reflections, especially at high Bragg angles, has considerably increased by reducing the temperature from 240 K to 30 K.

Low temperatures are normally achieved using open flow cryostats based on nitrogen gas, which permit the studies to be undertaken with good temperature control ($\pm$ 0.1 K) at temperatures as low as 80 K for the best cryostats [9]. Open flow cryostats based upon other gases have also become more widely commercially available to the crystallographic community in the past few years. The lowest achievable temperature is usually in the range 10–30 K depending upon the design of the cryostat [9, 12]. Lower temperatures and far lower helium consumption can be achieved using closed-cycle cryostats, which are at present uncommon on X-ray diffractometers [13]. The two-stage models that are typically in use are capable of delivering sample temperatures of around 10 K (or slightly lower), whereas three-stage models and other related closed-cycle models in use at a number of neutron diffraction installations can achieve temperatures of ca. 2 K, albeit with some compromise in re-orientation of the sample.

Temperature not only has an effect on the quality of the data, but also affects the lattice parameters of the crystal and the intermolecular geometries present in a structure. It is normally expected that unit cell lengths decrease with temperature, with the smallest variations occurring in the directions where the stronger inter-actions are found. However, this is not always the case and for some compounds one or more of the unit cell axes expands with reduction of the temperature [14]. The temperature dependence of the distance between a pair of atoms is expected to be inversely related to the strength of the interaction, since the potential well is deeper for a strong interaction, meaning that a small variation in the distance produces a big increase in the energy. Conversely, for a weak interaction, which has a shallow potential well, variation of the distance does not cost much in energy, permitting greater deformation with change in temperature. Thus valuable infor-mation on the strength of non-covalent interactions can be obtained by under-taking crystal structure determinations over a range of temperatures.

The remainder of this section includes some illustrative examples of the use of low temperature crystallography of relevance to crystal engineering.

### 3.1.5.1 Hydrogen Bonds

Hydrogen bonds are a particularly important class of non-covalent interactions in crystal engineering and have been the subject of widespread study across many areas of chemistry. Most early studies on the effect of temperature on hydrogen bonds were made using neutron diffraction since the positions (and displacement parameters) of the hydrogen atoms cannot be accurately obtained by X-ray diffrac-tion [6]. This type of experiment has been very important in the study of the nature of both strong and weak hydrogen bonds, allowing the estimation of bond energies and force constants, and also permitting a better understanding of the role of hydrogen bonds in determining the structures and properties of the hydrogen-bonded systems.

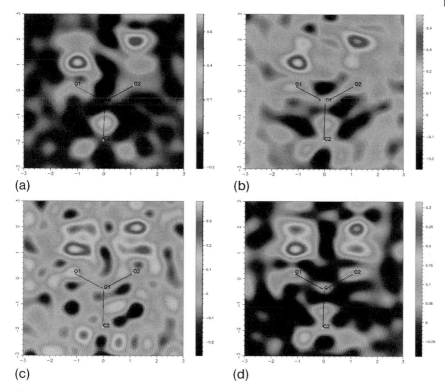

**Figure 3.1.4** Proton migration upon increasing temperature in the O–H···O hydrogen bonds between carboxyl groups in 2,4,6-trimethylbenzoic acid. X-ray difference Fourier maps showing the electron density associated with the carboxyl protons at 100 (a), 170 (b), 240 (c) and 290 K (d). Figure courtesy of Prof. Chick C. Wilson and Dr. Andrew Parkin, University of Glasgow.

Many hydrogen bonded systems exhibit temperature-dependent properties, such as proton migration or hydrogen atom disorder [15], and variable temperature neutron diffraction is the diffraction technique of choice to address these issues [16]. Nevertheless, some recent experiments using variable temperature X-ray diffraction have successfully enabled distinction between static and dynamic disorder of the H-bond proton (Fig. 3.1.4) [17]. Very recently a combination of both neutron and X-ray diffraction with variable temperature has been applied in the study of temperature-dependence of the hydrogen bond present in potassium hydrogen phthalate, and shows no evidence of disorder or proton migration dependence with temperature (Fig. 3.1.5) [18].

### 3.1.5.2 Other Non-covalent Interactions

Recently the change in the length of halogen bonds, O···I–C (in bpNO·F$_4$dIb) and N···X–C (X = I in bpe·F$_4$dIb and X = Br in bpe·F$_4$dBrb; Scheme 3.1.1) has been

O4
O2
C7
O1
30K
100K
300K

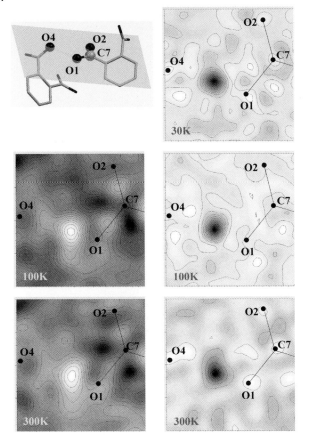

**Figure 3.1.5** Difference Fourier maps are shown for the plane defined by C7, O1 and O4 (as labelled) from neutron diffraction data (left) and X-ray diffraction data (right) showing the temperature-independence of the proton position in the O–H···O⁻ hydrogen bond in the crystal structure of potassium hydrogen phthalate. Note that hydrogen atom peaks are negative in neutron diffraction. Reproduced from Ref. [18] with permission of Elsevier.

studied as a function of temperature using X-ray diffraction (Fig. 3.1.6) [19]. Refinement of the crystal structures using data collected at 90, 145, 200 and

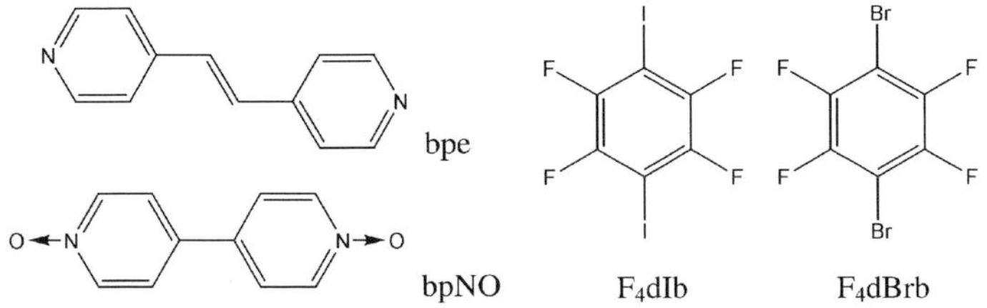

bpe
bpNO
$F_4dIb$
$F_4dBrb$

**Scheme 3.1.1** Molecular building blocks used in the study of the temperature dependence of halogen bonds.

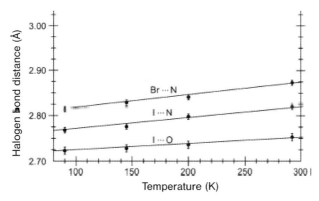

**Figure 3.1.6** Variation of X···O/N halogen bond distances (Å) with temperature (K) (error bars ±3σ). Reproduced from Ref. [19] with permission of the American Chemical Society.

292 K showed only small differences in the rate of change of the X–C interactions [X = I 2.8(3) ×10$^{-4}$ Å K$^{-1}$; X = Br 3.0(2) ×10$^{-4}$ Å K$^{-1}$] but significantly less variation associated with the stronger O···I–C halogen bond [1.5(2) ×10$^{-4}$ Å K$^{-1}$]. Thus, there is a qualitative agreement between the less easily deformed interactions and the accepted order of interaction strengths O···I–C > N···I–C and N···I–C > N···Br–C [20].

### 3.1.5.3   Other Applications
Being able to conduct diffraction studies at different temperatures is important when correlations between structural changes and other temperature-dependent physical properties are to be studied. Examples include the study of structures that undergo phase transitions. These may be structural phase transitions or other types of transitions such as magnetic order/disorder or onset of superconductivity transitions.

Recent studies have shown how low-temperature X-ray and neutron diffraction studies can be used to identify the location and population of gas molecules in metal–organic framework materials, providing insight into an important area of application for materials designed by a crystal engineering approach [21].

### 3.1.6
## Single Crystal Diffraction Studies at Increased Pressures

### 3.1.6.1   Experimental Method and Applications
High-pressure crystallography has long been of interest in mineralogy, but has only recently begun to be taken up by those whose interests are in molecular crystals. An important analogy can be drawn between increasing pressure and reducing temperature as discussed in the previous section. Both typically lead to decreases in volume of the crystal (assuming isobaric conditions for the temper-

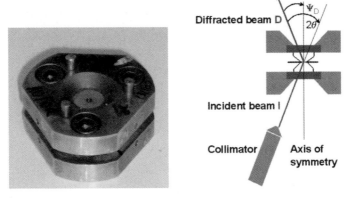

**Figure 3.1.7** Photograph (left) and schematic drawing (right) of a diamond-anvil cell (see Ref. [25]). Diagram courtesy of Dr. David R. Allan, University of Edinburgh.

ature reduction and isothermal conditions for the increase in pressure). Thus, both methods provide a valuable probe of contractions of the non-covalent interactions, especially the weaker ones and, by inference, at least qualitative information on interaction strengths. However, the obvious practical and physical limitations on the lowering of temperature are less apparent for high-pressure studies. Hence, greater volume reductions can be achieved upon application of pressure to the sample than upon temperature reduction. Elevated pressure can also be valuable in a crystal engineering context permitting reduction of the cavities within crystal structures, and allowing studies of phase transitions including polymorphic transitions. Crystal growth under pressure has also recently been applied in the quest to identify new polymorphs of pharmaceutical compounds and materials that are liquids at ambient pressure and temperature [22].

A common experimental approach to high-pressure studies of molecular crystals involves the use of a diamond-anvil cell (Fig. 3.1.7) [23], which consists of two diamond faces positioned either side of a hole drilled in a tungsten gasket. The sample, placed between the diamond faces, can be either a single crystal surrounded by a hydrostatic fluid (e.g. 1:1 pentane:isopentane mixture) which ensures isotropic application of the pressure or can be loaded in the hole as a liquid or in solution from which the growth of a single crystal is induced by application of pressure. This type of cell enables samples to be studied at pressures of up to 15 GPa. The pressure is applied by simply tightening a set of screws on the cell. The pressure within the cell is calibrated by measuring the fluorescence of a very small chip of ruby included inside the cell with the sample [23]. A problem associated with the use of diamond-anvil cells is the limitation on data quality and quantity. The accessible fraction of reciprocal space is limited due to shading by the steel body of the pressure cell and the data quality is compromised by the presence of diffraction from the diamonds and the beryllium windows. The hydrostatic fluid also leads to high background counts. An alternative approach using a thick-walled quartz capillary permits ease of viewing the crystal optically and removes angular restrictions placed upon the diffracted beams by the (diamond-anvil) cell. However, such capillaries are limited to pressures up to $10^5$ kPa [24].

Since the pressures applied to the crystals are, to a first approximation, unlikely to affect significantly the intramolecular geometry of the molecules, in the crystallographic refinements conducted in high pressure studies the intramolecular geometry is often constrained to that obtained from a suitable low temperature study. Limits on completeness of the data set also often lead to isotropic rather than anisotropic models for atomic displacements being necessary. In this context, the most useful information that can be obtained is on systematic trends in the behavior of different intermolecular interactions.

### 3.1.6.2  Intermolecular Interactions

The variation of hydrogen bonds with pressure has been recently reviewed by Boldyreva [26]. In this subsection some illustrative examples of studies of the change in metrics of intermolecular interactions upon increasing pressure are provided.

The structure of cyclobutanol, $C_4H_7OH$, at ambient pressure comprises pseudo-threefold hydrogen-bonded catemers which lie parallel to the crystallographic $a$-axis (Fig. 3.1.8(a)), forming a chain with a graph-set $C_2^2(4)$. Upon application of a pressure of 1.3 GPa, a change in space group from $Aba2$ to $Pna2_1$ occurs in tandem with an 11.2 % compression of the molecular volume relative to the 220 K structure at ambient pressure. The hydrogen-bonded catemers also change, adopting pseudo-twofold $C(2)$ chains at high pressure [22c].

The hexagonal polymorph of L-cystine contains hydrogen-bonded layers which consist of $R_4^4(16)$ hydrogen-bonded ring motifs, connected on one side by the disulfide bridges within the cystine molecules, and on the other by $N–H\cdots O$

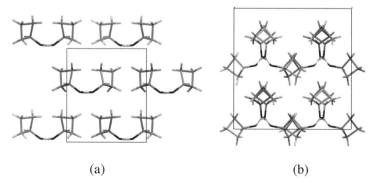

(a)                                              (b)

**Figure 3.1.8**  (a) View looking down the pseudo-threefold hydrogen-bonded molecular chains in the low-temperature ($Aba2$) crystal structure of cyclobutanol. (b) The wave-like pseudo-twofold hydrogen-bonded chains of the high-pressure ($Pna2_1$) phase of cyclobutanol [22 c].

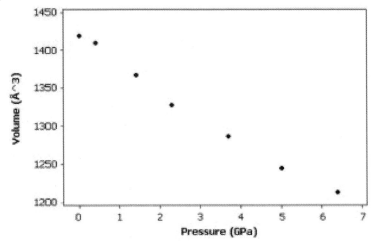

**Figure 3.1.9.** Variation of the volume of hexagonal L-cystine as a function of pressure. Reproduced from Ref. [27] with permission of the International Union of Crystallography.

hydrogen bonds to other layers. Application of pressure (studied up to 3.7 GPa [27]) pushes the layers closer together, compressing approximately equally both the regions of the interlayer hydrogen bonds and the disulfide bridges, which vary the torsion angles allowing the cystine molecules to behave like springs. The increase in pressure leads to the closing-up of voids in the structure, a decrease in the N–C–C–O and C–S–S–C torsional angles, shortening of the N–H···O hydrogen bonds by 0.10–0.60 Å and compression of the interlayer S···S contact from 3.444(4) Å at ambient pressure to 3.264(4) Å at 3.7 GPa. The variation in unit cell volume with pressure up to 6.4 GPa is shown in Fig. 3.1.9.

Other recent studies of intermolecular interactions under pressure include the study of Cl···Cl contacts in $CH_2Cl_2$ [28] and S···S contacts in $CS_2$ [29] by Katrusiak.

### 3.1.6.3    Polymorphism

The importance of polymorphism within crystal engineering is substantial and in the area of pharmaceutical crystals has proven to be of great importance financially (Chapter 3.3). Studies using pressure as a variable have been applied recently to the studies of pharmaceutical or related compounds to explore more widely potential polymorphism in such compounds. The examples of glycine and paracetamol are discussed below.

Glycine has three known polymorphs under ambient conditions, namely α, β and γ. The α-polymorph is stable with respect to the application of pressures up to 23 GPa [30], whereas both the β- and the γ-polymorphs undergo phase transitions leading to two new forms. β-glycine transforms into δ-glycine at a pressure of about 0.76 GPa [31], while a powder of the γ-polymorph (single crystals do not survive the application of pressure), which is the most stable of the polymorphs of glycine at ambient conditions, suffers an irreversible phase transition at about 3 GPa into what has been called ε-glycine. Two alternative structures have been proposed for this phase [31, 32]. In all the different forms, the application of pressure produces similar effects, which are the reduction of the voids in the structure as well as an increase in the number and shortening of the C–H···O contacts. However, the different responses of these polymorphs with pressure have been argued to be due to the orientational relationship between the molecules before and after the phase transitions.

The analgesic and antipyretic drug paracetamol exists in two principal polymorphic forms, monoclinic form I, which requires additives for tablet manufacture and the orthorhombic form II, which can be formed directly into tablets. The existence of a third form (form III) has also been the subject of a number of investigations. A recent study of the application of pressure to both forms I and II revealed no polymorphic transformations, but only a reversible anisotropic structural distortion was observed for the single crystals of the monoclinic form I ($P2_1/n$) at pressures up to 5 GPa [33]. However, for powder samples of form I, an irreversible partial transformation into the orthorhombic form II (*Pbca*) was observed at pressures below 2 GPa. Forms I and II showed similar bulk compressibilities (Fig. 3.1.10) but different anisotropic behavior arising from differences in intermolecular interactions. Maximum compression is observed in the direction normal to the hydrogen-bonded layers in both polymorphs. Transitions between the polymorphs induced by pressure were found to be poorly reproducible and depended strongly on the sample and on the procedure of increasing/decreasing pressure. Polymorph II transformed partly into the polymorph I during grinding.

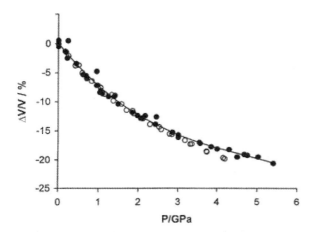

**Figure 3.1.10** Relative volume changes with pressure change for form I (open circles) and form II (shaded circles) of paracetamol. Reproduced from Ref. [33] with permission of Kluwer Academic Publishers.

### 3.1.7
### Powder Diffraction

Powder diffraction is an important tool in crystal engineering and can be applied in a number of different ways to yield valuable structural information for polycrystalline materials. Recent reviews provide a useful entry to the literature, giving an overview of the technique and, in particular, its use in structure determination [34] and describing a wider range of applications including *in situ* studies, following reactions in solids including obtaining kinetic data, and investigating metastable structures (polymorphs) [35]. X-ray powder diffraction is taking on a growing importance in crystal structure determination due to the requirements of pharmaceutical and other industries (e.g. pigments for paints) to establish structures of commercially important products and explore their potential polymorphs and solvates. It is also expected to grow in importance for structure determination of organometallic catalysts, which are often available only as small amounts of crystalline powder.

In its simplest but most widely used application, powder diffraction is used as a means for establishing phase homogeneity of crystalline products. This is an important part of any crystal engineering study where the aim is to design and synthesise a new crystalline material. Assuming that a crystal structure of the product can be obtained by single crystal diffraction, it is vital to establish whether or not this structure is representative of the entire product or whether more than one polymorph or solvate has formed, or indeed whether more than one product of different composition is present. Comparison of the powder pattern calculated from the single crystal data with that of the measured powder diffraction pattern is used to make this determination. However, detection of additional phases that constitute only a few percent of the entire sample can be difficult with a conventional laboratory instrument, due to inadequate signal-to-noise or too great a peak width. Thus, additional effort may be required in designing the experiment or alternatively synchrotron radiation may be needed.

#### 3.1.7.1 Crystal Structure Determination

Structure solution from X-ray powder diffraction began as a means of characterising inorganic products of solid state synthesis [36] but has been more widely used in recent years for the study of organic compounds [34b, 37], particularly of pharmaceutical importance. There have been some notable recent successes in the area of organometallic compounds [38], but structure determination of coordination framework materials is very much in its infancy [39].

Early approaches to structure solution involved extraction of intensities from the powder pattern and solution (using the very limited data set obtained) by Patterson or direct methods. More common at present are direct space global optimization approaches, which include grid search, simulated annealing, Monte Carlo and genetic algorithm methods. Once the lattice parameters and space group have been established, this approach is used to generate possible crystal structures whose powder patterns are calculated and compared against the experimental pattern. Accurate peak positions from a well-calibrated diffractometer are of course essential for accurate lattice parameter determination. The limited number of extracted Bragg intensities can, however, lead to ambiguities in space group determination. Rietveld refinement [40], in which the solved structure is refined using the whole powder pattern, gives the best overall structure refinement. In contrast to single crystal structure refinement, typically a structure model with a number of constraints or restraints is needed. The level of success in structure determination depends on the quality of the powder data, which is associated with the nature of the sample and the source used for data collection. Synchrotron X-ray sources provide significant advantages due to their high intensity radiation and the narrow peak width obtained. However, the small number of facilities worldwide ensures that it is more practical to use a conventional laboratory X-ray source in many cases. Indeed the number of crystal structures solved and refined successfully using laboratory X-ray data has been rising in recent years. Some illustrative examples of structure determination follow.

### 3.1.7.2 Crystal Structures of Organic Compounds

Florence et al. have conducted an extensive study using simulated annealing methods for the structure determination of 35 organic compounds in which they have progressively introduced more degrees of freedom, thus making the systems more complicated each time [37]. They found that at the lower end of the complexity scale the structures were solved with excellent reproducibility and high accuracy, while the more complex cases were a significant challenge to global optimization procedures. The inclusion of constraints derived from the CSD permitted the correct solution by simplifying the problem. The accuracy of the results was measured by comparison with single crystal structures (Scheme 3.1.2).

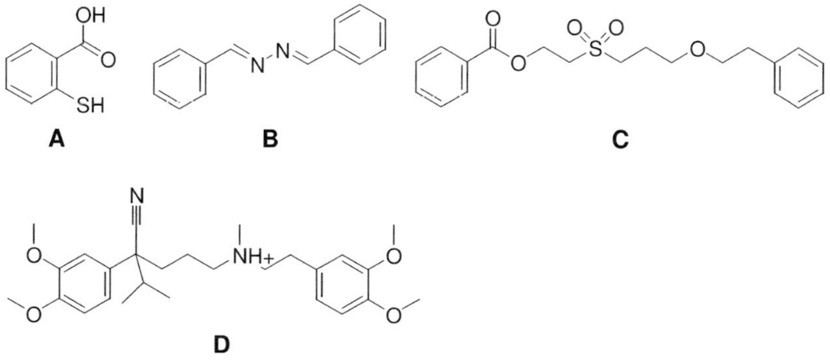

**Scheme 3.1.2** Some structures determined by laboratory X-ray powder diffraction. RMS distance between atomic coordinates for structures solved by powder diffraction and single crystal diffraction were **A** 0.026 **B** 0.026, **C** 0.165, **D** 0.204 Å.

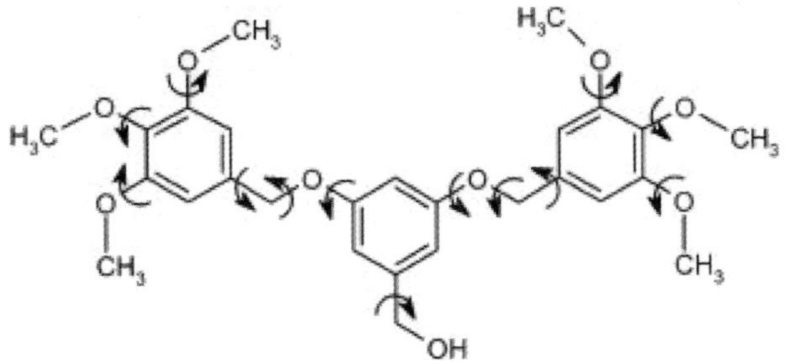

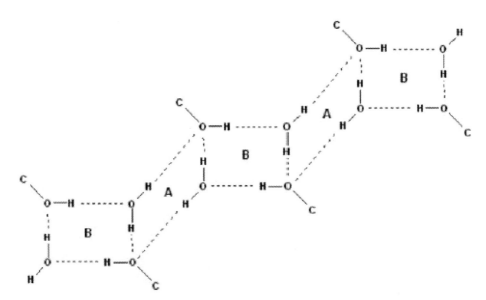

**Scheme 3.1.3** Molecular structure of BTBA, 3,5-bis((3,4,5-trime-thoxybenzyl)oxy)benzyl alcohol depicting the 13 variable torsion angles used in the structure solution calculation. Reproduced from Ref. [41] with permission of the American Chemical Society.

Harris and colleagues have demonstrated the power of a genetic algorithm approach in the solution of the crystal structure of the highly flexible structure of BTBA, 3,5-bis((3,4,5-trimethoxybenzyl)oxy)benzyl alcohol, as its monohydrate (Scheme 3.1.3) [41]. Pertinent to crystal engineering is the extended hydrogen-bonding network that encompasses the alcohol and water moieties. Several different permutations could be proposed for the arrangement of hydrogen bonds within the network, such that each O⋅⋅⋅O edge is occupied by an O–H⋅⋅⋅O hydrogen bond. However, consideration of the geometry of the oxygen framework and the assignment of O–H⋅⋅⋅O hydrogen bonds that are as close to linear as possible

**Scheme 3.1.4** Proposed hydrogen-bonding scheme for BTBA showing two independent circuits, A and B. Reproduced from Ref. [41] with permission of the American Chemical Society.

within the network of fixed oxygen atom positions, has permitted the hydrogen-bonding scheme shown in Scheme 3.1.4 to be proposed as the most probable.

### 3.1.7.3  Crystal Structures and Reactions of Metal–Organic Compounds

Masciocchi and colleagues have reported the crystal structure of the coordination polymer bis(pyrazylato)copper(II) hydrate, $Cu(pz)_2 \cdot H_2O$ (Fig. 3.1.11(a)) in which the water molecules are very weakly associated with the copper centres in axial sites relative to the square planar coordination of the pyrazylate ligands, $\beta$-$Cu(pz)_2$ [39c]. Upon gentle heating, water is lost to give an anhydrous coordination polymer in which the square planar coordination geometry is maintained (Fig. 3.1.11(b)). This contrasts with a previously characterised polymorph in which the anhydrous form has tetrahedral coordination geometry at the copper sites, $\alpha$-$Cu(pz)_2$ (Fig. 3.1.11(c)) [42]. The compound also rehydrates upon exposure to water vapor or indeed absorbs other small molecules that can occupy the same coordination sites as water, e.g. $NH_3$, $CH_3CN$.

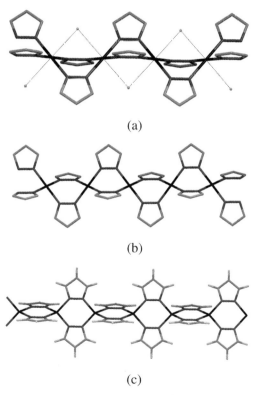

(a)

(b)

(c)

**Figure 3.1.11**  (a) Crystal structure of bis(pyrazylato)copper(II) hydrate, $Cu(pz)_2 \cdot H_2O$. (b) Crystal structure following removal of water upon heating, $\beta$-$Cu(pz)_2$. (c) Crystal structure of known polymorph of anhydrous coordination polymer, $\alpha$-$Cu(pz)_2$.

In our own work we have studied the loss of HCl by the salts (3-XpyH)$_2$[CuCl$_4$] (3-XpyH = 3-halopyridinium) in which anions and cations are linked via N–H$\cdots$Cl–Cu hydrogen bonds and Cu–Cl$\cdots$X–C halogen bonds [43]. Polycrystalline powders of these yellow salts undergo slow loss of HCl under ambient conditions (more rapidly under heating) to yield blue coordination compounds [CuCl$_2$(3-Xpy)$_2$] that retain the halogen bonds. The process involves a change in the ligands coordinated at the Cu centre and a change in coordination geometry from distorted tetrahedral to square planar. Remarkably, the reaction is fully reversible upon exposure of the coordination compounds to vapour from a concentrated HCl solution. Crystal structures of the products of both the forward and reverse reaction have been determined by powder diffraction using simulated annealing methods.

## 3.1.8
### Charge Density Studies

Charge density analysis is an advanced method of X-ray diffraction in which a more sophisticated model for the electron density than that adopted for crystal structure determination is used [44]. Instead of using spherical gas-phase free-atom electron densities an atom-centered model that accounts for the asphericity of the electron distribution about each atom is adopted. The electron density of each atom is represented as a combination of a spherical core density and an aspherical valence electron distribution, which arises due to the formation of chemical bonds (including intermolecular interactions). The valence shell electron density is modeled using parameters that comprise a sum of spherical harmonic functions, the contribution of each being refined as part of the crystallographic least-squares model. The radial description of the valence charge density is also determined. In order for charge density models to be refined successfully, X-ray data measured to much higher resolution than for a structure determination are required (typically $d_{min} < 0.5$ Å) and high redundancy of data is required for very accurate and precise determination of X-ray intensities. Data are collected at low temperature to reduce atomic motions and improve accuracy and deconvolution of the charge density from these motions. Most commonly temperatures of 100–150 K have been used but increasingly much lower temperatures are being applied due to the greater availability of helium cryostats. Careful correction for systematic errors such as absorption and extinction is also crucial. Where hydrogen atoms are involved it is helpful to be able to determine accurately the hydrogen atom positions and to describe the atomic displacements. This can be accomplished by a parallel neutron diffraction experiment.

Charge density analyses can provide experimental information on the concentration of electron density around atoms and in intra- and intermolecular bonds, including the location of lone pairs. Transition metal d-orbital populations can be estimated from the asphericity of the charge distribution around such metal centers. A number of physical properties that depend upon the electron density distribution can also be calculated. These include atomic charges, dipole and higher moments, electric field gradients, electrostatic potentials and interaction energies. Although charge density studies have not made a major impact in the field of crystal engineering to date, as the method becomes more widely accessible, with the continued development of rapid accurate data collection and more user-friendly software, there are a number of areas of potential impact. In particular, the qualitative and quantitative evaluation of intermolecular interactions and the study of electronic/magnetic properties of designed molecular materials can benefit from charge density studies.

Using the approach pioneered by Bader for the analysis of (initially theoretically determined) charge densities [45], quantitative correlations between geometric measures of hydrogen bonds and the topology of their electron density have been established. These geometric and topological measures include the positions of saddle points in electron density (known as critical points) between the hydrogen atom and acceptor group (A) of a D–H$\cdots$A hydrogen bond (D = C, N, O; A = O), the curvature of the electron density along the H$\cdots$A bond path, and even the energy density at the hydrogen bond critical point, from which estimates of the hydrogen bond energy can be made [46]. Charge density studies of zeolites and related porous materials [47], including determination of electrostatic potentials within pores and of framework–guest interaction energies, provide a good indication that porous metal–organic frameworks, a vibrant field within crystal engineering, could valuably be studied by this approach. Indeed, recently the first charge density study of such a framework material has been reported in which investigation of metal–metal bonding allowed inferences regarding the magnetic super-exchange pathway to be made [48].

## 3.1.9
## Conclusions

Diffraction is the key experimental technique in crystal engineering. It provides the means of accurately characterising the product of a crystal synthesis endeavor. However, the applications of diffraction are much wider than simply structure determination. An overview of the contribution of crystallography to many areas of chemistry is provided in a recent issue of the journal *Chemical Society Reviews* dedicated to crystallography [49]. The purpose of this chapter has been to take a broader view of diffraction studies and their present and future potential to play an important role in the continued development of all aspects of crystal engineering.

## References

1 For example, see W. Clegg, A. J. Blake, R. O. Gould, P. Main, *Crystal Structure Analysis: Principles and Practice*, Oxford University Press, Oxford, **2002**.
2 For ACA crystallography school(s) see http://www.hwi.buffalo.edu/aca/. For BCA crystallography school see http://www.crystallography.org.uk/
3 Presently at http://journals.iucr.org/ services/cif/checking/checkfull.html
4 A. Martín, A. G. Orpen, *J. Am. Chem. Soc.* **1996**, *118*, 1464.

**5** F. H. Allen, *Acta Crystallogr.,Sect.B* **1986**, *42*, 515.

**6** (a) F. Takusagawa, T. F. Koetzle, *Acta Crystallogr.,Sect. B* **1978**, *34*, 1149.

**7** B. Gillon, C. Mathoniere, E. Ruiz, S. Alvarez, A. Cousson, T. M. Rajendiran, O. Kahn, *J. Am. Chem. Soc.* **2002**, *124*, 14433.

**8** (a) S. R. Batten, K. S. Murray, *Coord. Chem. Rev.* **2003**, *246*, 103; (b) J. S. Miller, J. L. Manson, *Acc. Chem. Res.* **2001**, *34*, 563.

**9** A. E. Goeta, J. A. K. Howard, *Chem. Soc. Rev.* **2004**, *33*, 490.

**10** F. K. Larsen in *The Application of Charge Density Research to Chemistry and Drug Design*, G. A. Jeffrey, J. F. Piniella (Eds.), *NATO ASI:Physics*, Vol. 250, p.187, Plenum, New York 1991.

**11** (a) G. Mínguez Espallargas, L. Brammer, P. Sherwood, *Angew. Chem. Int. Ed.* **2006**, *45*, 435; (b) G. Mínguez Espallargas, L. Brammer, D. R. Allan, C. R. Pulham, N. Robertson, **2006**, manuscript in preparation.

**12** M. J. Hardie, K. Kirschbaum, A. Martin, A. A. Pinkerton, *J. Appl. Crystallogr.* **1998**, *31*, 815.

**13** (a) K. Hendriksen, F. K. Larsen, S. E. Rasmussen, *J. Appl. Crystallogr.* **1986**, *19*, 390; (b) R. C. B. Copley, A. E. Goeta, C. W. Lehmann, J. C. Cole, D. S. Yufit, J. A. K. Howard, J. M. Archer, *J. Appl. Crystallogr.* **1997**, *30*, 413; (c) M. Messerschmidt, M. Meyer, P. Luger, *J. Appl. Crystallogr.* **2003**, *36*, 1452.

**14** T. A. Mary, J. S. O. Evans, T. Vogt, A. W. Sleight, *Science* **1996**, *272*, 90.

**15** G. A. Jeffrey, *An Introduction to Hydrogen Bonding*, Oxford University Press, New York 1997, Ch. 2.

**16** (a) C. C. Wilson, N. Shankland, A. J. Florence, *Chem. Phys. Lett.* **1996**, *253*, 103; (b) C. C. Wilson, N. Shankland, A. J. Florence, *J. Chem. Soc., Faraday Trans.* **1996**, *92*, 5051; (c) T. Steiner, I. Majerz, C. C. Wilson, *Angew. Chem. Int. Ed.* **2001**, *40*, 2651.

**17** (a) P. Gilli, V. Bertolasi, L. Pretto, A. Lycka, G. Gilli, *J. Am. Chem. Soc.* **2002**, *124*, 13554; (b) C. C. Wilson, A. E. Goeta, *Angew. Chem. Int. Ed.* **2004**, *43*, 2095; (c) P. Gilli, V. Bertolasi, L. Pretto, V. Ferretti, G. Gilli, *J. Am. Chem. Soc.* **2004**, *126*, 3845; (d) A. Parkin, S. M. Harte, A. E. Goeta, C. C. Wilson, *New J. Chem.* **2004**, *28*, 718; (e) P. Gilli, V. Bertolasi, L. Pretto, L. Antonov, G. Gilli, *J. Am. Chem. Soc.* **2005**, *127*, 4943.

**18** S. M. Harte, A. Parkin, A. E. Goeta, C. C. Wilson, *J. Mol. Struct.* **2005**, *741*, 93.

**19** A. Forni, P. Metrangolo, T. Pilati, G. Resnati, *Cryst. Growth Des.* **2004**, *4*, 291.

**20** P. Metrangolo, H. Neukirch, T. Pilati, G. Resnati, *Acc. Chem. Res.* **2005**, *38*, 386.

**21** (a) J. L. C. Rowsell, E. C. Spencer, J. Eckert, J. A. K. Howard, O. M. Yaghi, *Science* **2005**, *309*, 1350; (b) E. C. Spencer, J. A. K. Howard, G. J. MacIntyre, J. L. C. Rowsell, O. M. Yaghi, *Chem. Commun.* **2006**, 278.

**22** (a) I. D. H. Oswald, D. R. Allan, G. D. Day, W. D. S. Motherwell, S. Parsons, *Acta Crystallogr., Sect. B* **2005**, *61*, 69; (b) I. D. H. Oswald, D. R. Allan, G. D. Day, W. D. S. Motherwell, S. Parsons, *Cryst. Growth. Des.* **2005**, *5*, 1055; (c) P. A. McGregor, D. R. Allan, S. Parsons, C. R. Pulham, *Acta Crystallogr., Sect. B* **2005**, *61*, 449.

**23** L. Merrill, W. A. Bassett, *Rev. Sci. Instrum.* **1974**, *45*, 290.

**24** D. S. Yufit, J. A. K. Howard, *J. Appl. Crystallogr.* **2005**, *38*, 583.

**25** University of Edinburgh Chemical Crystallography Group: http://www.crystal.chem.ed.ac.uk/research/hpcg.php?page=3

**26** E. V. Boldyreva, *J. Mol. Struct.* **2004**, *700*, 151

**27** S. A. Moggach, D. R. Allan, S. Parsons, L. Sawyer, J. E. Warren, *J. Synchrotron Rad.* **2005**, *12*, 598.

**28** M. Podsiadło, K. Dziubek, A. Katrusiak, *Acta Crystallogr., Sect. B* **2005**, *61*, 595.

**29** K. F. Dziubek, A. Katrusiak, *J. Phys. Chem. B* **2004**, *108*, 19089.

**30** C. Murli, S. M. Sharma, S. Karmakar, S. K. Sikka, *Physica B* **2003**, *339*, 23.

**31** A. Dawson, D. R. Allan, S. A. Belmonte, S. J. Clark, W. I. F. David, P. A. McGregor, S. Parsons, C. R. Pulham L. Sawyer, *Cryst. Growth Des.* **2005**, *5*, 1415.

**32** E. V. Boldyreva, H. Ahsbahs, H.-P.Weber, *Z. Kristallogr.* **2003**, *218*, 231.

**33** E. V. Boldyreva, T. P. Shakhtshneider, H. Ahsbahs, H. Sowa, H. Uchtmann, *J. Therm. Anal. Calorim.* **2002**, *68*, 437.

**34** (a) K. D. M. Harris, E. Y. Cheung, *Chem. Soc. Rev.* **2004**, *33*, 526; (b) K. D. M. Harris, M. Tremayne, B. M. Kariuki, *Angew. Chem. Int. Ed.* **2001**, *40*, 1626; (c) W. I. F. David, K. Shankland, L. B. McCusker, C. Baerlocher, *Structure Determination from Powder Diffraction Data*, IUCr Monographs in Crystallography, no. 13, Oxford University Press, Oxford, **2002**.

**35** J. S. O. Evans, I. R. Evans, *Chem. Soc. Rev.* **2004**, *33*, 539.

**36** (a) A. Clearfield, L. B. McCusker, P. R. Rudolf, *Inorg. Chem.* **1984**, *23*, 4679; (b) P. R. Rudolf, A. Clearfield, *Inorg. Chem.* **1985**, *24*, 3714. (c) P. Rudolf, A. Clearfield, *Acta Crystallogr., Sect. B* **1985**, *41*, 418.

**37** A. J. Florence, N. Shankland, K. Shankland, W. I. F. David, E. Pidcock, X. Xu, A. Johnston, A. R. Kennedy, P. J. Cox, J. S. O. Evans, G. Steele, S. D. Cosgrove, C. S. Frampton, *J. Appl. Crystallogr.* **2005**, *38*, 249.

**38** G. Mínguez Espallargas, L. Brammer, Structure and Bonding in Organometallic Compounds: Diffraction Methods, in *Comprehensive Organometallic Chemistry 3*, Vol. 1, G. Parkin (Ed.), Elsevier, 2006, in press.

**39** (a) N. Masciocchi, A. Sironi, *J. Chem. Soc., Dalton Trans.* **1997**, 4643; (b) E. Barea, J. A. R. Navarro, J. M. Salas, N. Masciocchi, S. Galli, A. Sironi, *Inorg. Chem.* **2004**, *43*, 473; (c) A. Cingolani, S.Galli, N. Masciocchi, L. Pandolfo, C. Pettinari, A. Sironi, *J. Am. Chem. Soc.* **2005**, *127*, 6144.

**40** H. M. Rietveld, *J. Appl. Crystallogr.* **1969**, *2*, 65.

**41** Z. Pan, E. Y. Cheung, K. D. M. Harris, E. Constable, C. E. Housecroft, *Cryst. Growth Des.* **2005**, *5*, 2084.

**42** (a) M. K. Ehlert, S. J. Rettig, A. Storr, R. C. Thompson, J. Trotter, *Can. J. Chem.* **1989**, *67*, 1970; (b) Structure redetermined in M. K. Ehlert, A. Storr, R. C. Thompson, F. W. B. Einstein, R. J. Batchelor, *Can. J. Chem.* **1993**, *71*, 331.

**43** G. Mínguez Espallargas, L. Brammer, J. van de Streek, K. Shankland, A. J. Florence, H. Adams, *J. Am. Chem. Soc.* **2006**, *128*, 9584.

**44** (a) P. Coppens, *X-ray Charge Densities and Chemical Bonding*, Oxford Science Publishing, 1997; (b) T. S. Koritsanszky, P. Coppens, *Chem. Rev.* **2001**, *101*, 1583; (c) P. Macchi, A. Sironi, *Coord. Chem. Rev.* **2003**, *238–239*, 383; (d) P. Coppens, B. Iversen, F. K. Larsen, *Coord. Chem. Rev.* **2005**, *249*, 179; (e) C. Lecomte, E. Aubert, V. Legrand, F. Porcher, S. Pillet, B. Guillot, C. Jelsch, *Z. Kristallogr.* **2005**, *220*, 373.

**45** R. F. W. Bader, *Atoms in Molecules – A Quantum Theory*, International Series of Monographs in Chemistry 22, Oxford University Press, Oxford, **1990**.

**46** (a) E. Espinosa, M. Souhassou, H. Lachekar, C. Lecomte, *Acta Crystallogr., Sect. B* **1999**, *55*, 563 ; (b) E. Espinosa, C. Lecomte, E. Molins, *Chem. Phys. Lett.* **1999**, *300*, 745 ; (c) E. Espinosa, E. Molins, C. Lecomte, *Chem. Phys. Lett.* **1998**, *285*, 170.

**47** E. Aubert, F. Porcher, M. Souhassou, C. Lecomte, *J. Phys. Chem. Solids* **2004**, *65*, 1943.

**48** (a) R. D. Poulsen, A. Bentien, M. Chevalier, B. B. Iversen, *J. Am. Chem. Soc.* **2005**, *127*, 9156. (b) See also S. Pillet, M. Souhassou, C. Mathoniere, C. Lecomte, *J. Am. Chem. Soc.* **2004**, *126*, 1219.

**49** *Whole issue devoted to crystallography: Chem. Soc. Rev.*, **2004**, *33*(8), 463–565.

## 3.2
## Solid State NMR
*Roberto Gobetto*

### 3.2.1
### Introduction

Solid state NMR (SSNMR) is becoming a well established tool for molecular structure determination in a wide variety of systems. Advances in commercial solid state hardware including high power amplifiers, high field superconducting magnets, stable probes allowing fast spinning speeds of the samples and the development of new pulse sequences have been of paramount importance for overcoming the earlier resolution problems and yield spectra of linewidths comparable to those of liquids.

Solid state NMR appears to be particularly attractive in solid systems lacking order or homogeneity for crystallographic examination. SSNMR spectroscopy has therefore emerged as a powerful technique for structural studies of a great number of technologically important materials, which are either insoluble or difficult to crystallize. Unlike X-ray diffraction, SSNMR is able to study largely disordered molecular systems such as, for example, non-crystalline proteins [1, 2] or polymers [3]. NMR is a localized technique having the capacity to investigate the microscopic environment at each individual site in an amorphous material.

Also, where X-Ray data are available, the technique acts as a bridge between crystallographic information and solution NMR spectra [4–6]. In addition to the high sensitivity to the local anisotropic parameters (quadrupolar, chemical shift anisotropy and dipolar interactions), which directly report on the geometric and electronic structure of the material, NMR experiments give new insights into the dynamic behavior of functional groups or single atoms in the solid state. A wealth of information available from the spectra makes SSNMR an attractive and powerful method for the complete characterization of a solid material. In the recent years SSNMR has been successfully applied to organic and inorganic complexes, polymers, resins, zeolites and aluminosilicates, mesoporous and microporous solids, minerals, biological molecules, glasses, ceramics, cements, food products, wood, bones, archaelogical specimens, semiconductors, surfaces and catalytic studies, etc. [4–6].

This chapter gives a short introduction to solid state NMR and then presents some recent results where the development of new pulse sequences have brought some insights into various aspects of the characterization of solid state materials. Finally attention will be focused on the information that the technique can provide in the field of crystal engineering, in the detection of polymorphism and in the dynamic processes occurring in the solid state.

3.2.2
**The Fundamentals**

In a general case the Hamiltonian governing the energy levels involved in the NMR experiment [5] is:

$$H = H_Z + H_Q + H_D + H_{CS} + H_J$$

where $H_Z$ is the Zeeman Interaction, $H_Q$ is the quadrupolar interaction, $H_D$ is the nuclear dipolar interaction, $H_{CS}$ is the chemical shielding interaction and $H_J$ is the scalar interaction. All these Hamiltonians must be considered as tensor interactions, depending on their relative orientation with respect to the magnetic field direction, and therefore are responsible for the broadening of the spectral resonances. In solution, where all these terms are reduced to their isotropic averages by rapid molecular tumbling, the observed signals are much sharper than the analogous resonances in the solid state. Due to the wealth of information to be gained by using solid state NMR, several methods have been developed to reduce these interactions and thereby improve the quality of the observed signals.

### 3.2.2.1 Quadrupole Interaction

Nuclei with spin number $I > 1/2$ possess an electric quadrupolar moment that represents a measure of their difference from spherical symmetry. The quadrupolar effect on the NMR spectrum arises from the interaction of the nuclear charge with this non-spherically symmetric electric field gradient. The asymmetric charge distribution in the nucleus is described by the nuclear electric quadrupole moment, $eQ$, which is an intrinsic property of the nucleus, and is the same regardless of the environment. Such interaction can be of the same order of magnitude as the Zeeman interaction and usually, when a quadrupolar nucleus is present, dipolar interactions, anisotropic chemical shifts and scalar spin–spin coupling interactions can be neglected to a first approximation. In the case of $^2H$, where $I=1$, the energy level diagram is given in Fig. 3.2.1.

The levels $m=1$ and $m=-1$ are affected in the same direction by the quadrupolar interaction whereas the $m=0$ level is affected in the opposite direction. If the electric

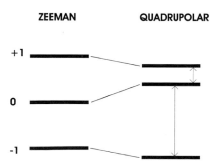

**ZEEMAN**        **QUADRUPOLAR**

+1

0

-1

**Figure 3.2.1** Energy levels for a nucleus having $I = 1$.

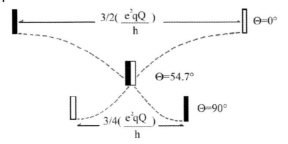

**Figure 3.2.2** Peak separation as a function of the θ angle.

field gradient is symmetric about the principal axis (i.e. deuterium), the energy shift due to the quadrupolar interaction is

$$\Delta E_Q = (3/8)(e^2qQ/\hbar)(3\cos^2\theta - 1)$$

For a single deuteron in a crystalline material the spectrum consists of a doublet with peak separation equal to:

$$\Delta v = (3/4)(e^2qQ/\hbar)(3\cos^2\theta - 1)$$

The peak separation is then dependent on the orientation, expressed by the θ angle, of the electric field gradient with respect to the magnetic field (Fig. 3.2.2).

In a powder spectrum, where all the orientations are present, the superimposition of two symmetric powder distributions reversed in sign forms the well known Pake doublet (Fig. 3.2.3) [6].

When molecules undergo reorientation on a timescale of the order of the quadrupole interaction or faster (rapid exchange limit, correlation time of the motion < $10^{-7}$ s), the observed frequencies will be averaged and the resulting lineshape can be analysed to determine the type and the rate of the motion [7].

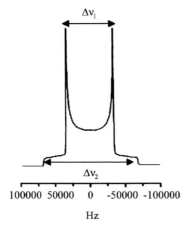

**Figure 3.2.3** Pake doublet for a powder spectrum of a single deuterium site.

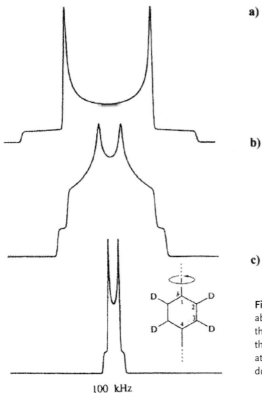

a)

b)

c)

100 kHz

**Figure 3.2.4** The effect of motion about the molecular symmetry axis, either by two-fold or n-fold ($n>3$) flips, on the deuterium spectrum of a deuterated para-substituted benzene. Reproduced with permission from Ref. [10].

Temperature dependent lineshape variation of the Pake doublet represents then an experimental evidence of a motion occurring in the solid state. $^2$H NMR investigations on selectively deuterated compounds, therefore, are important tools for the recognition of molecular motions in solids. This method has been applied to the characterization of molecular motions in molecular solids [8] and inclusion compounds [9]. As an example in Fig. 3.2.4 is reported the effect of the motion about the molecular symmetry axis, either by two-fold or n-fold ($n>3$) flips, on the deuterium spectrum of a deuterated para-substituted benzene [10].

Since molecular motions also affect the ordering of molecules, dynamic NMR methods can also be used for the determination of the ordering features in solids (e.g. orientational order or disorder).

### 3.2.2.2 Dipolar Interaction

The most important solid state interaction is the direct dipolar broadening of the observed spin by its neighboring spins. The dipolar interaction is independent of the magnetic field strength, $B_o$, and has a short range order which is proportional to $r^{-3}$ [5].

The magnetic field experienced at a specified nucleus can be calculated from:

$$B_{loc} \propto \pm \mu_s \, (3\cos^2\theta - 1) \, / \, r^3$$

where $\mu_s$ is the magnetic moment of the neighboring nuclei, $r$ is the internuclear distance between the nuclei and $\theta$ is the angle formed between the internuclear vector and the applied magnetic field. When rare spins are considered, this broadening arises from the nearby abundant spins, frequently protons.

Relatively few NMR studies have been reported for single crystals, mainly because of the need for a large size crystal to overcome the sensitivity problems. In the standard experiment an oriented single crystal is placed in the spectrometer and the spectra are measured as a function of the orientation of the crystal in the magnetic field. Pake reported [11] a single crystal NMR study of gypsum, $CuSO_4 \cdot 2\,H_2O$, in which the proton pairs of the water molecules are relatively isolated. The orientation of the two non-equivalent water molecules in the unit cell was determined by fitting the orientation dependence of the spectral components to the dipolar angular dependence (Fig. 3.2.5). Pake was able to determine a proton separation in the water molecules of 1.58 Å with an accuracy of 2%.

For an isolated pair of nuclei in a polycrystalline system the band shape shows a Pake doublet. When the number of interacting nuclei becomes larger and a polycrystalline material is analyzed, the resulting spectrum is much more complex and much less informative. Nevertheless an expression for the dipolar second moment (or mean square width) of the NMR spectrum in a polycrystalline sample has been suggested by Van Vleck [12]:

$$M^2(\text{Homonuclear}) = \left(\frac{3}{4}\gamma^2\hbar\frac{\mu_0}{4\pi}\right)^2 \sum_k \frac{\left(1 - 3\cos^2\theta_{jk}\right)^2}{r_{jk}}$$

The second moment of the proton spectra is related to the sum of $(1 - 3\cos2\theta jk)^{2r}{}_{jk}{}^{-6}$ over all nuclear pairs $j,k$ in the crystal and the $\theta_{jk}$ is the angle between the field direction and the internuclear vector. The dipolar interaction that dominates the

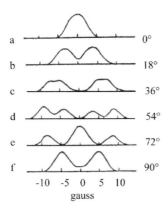

Figure 3.2.5 Proton single crystal NMR spectra measured for gypsum, $CuSO_4 \cdot 2\,H_2O$, as a function of the crystal orientation in the magnetic field. Reproduced with permission from Ref. [11].

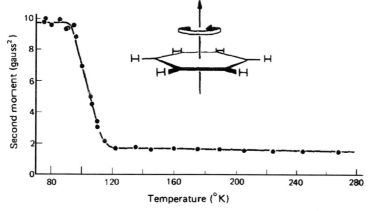

**Figure 3.2.6** Temperature dependence of the proton second moment for benzene. Reproduced with permission from Ref. [14].

linewidth of the proton spectrum is then related to the crystal structure and can afford structural information concerning molecules, groups and ions [13]. Any form of molecular motion which significantly changes the nuclear dipolar inter- actions will cause spectral narrowing. The presence of molecular motion in the solids can be shown by a significant reduction of the linewidth of the NMR spectra since there is a partial or total average of the $(1 - 3\cos2\theta jk)$ term. The study of the dependence of the second moment as a function of temperature allows to obtain information on the molecular motion (Fig. 3.2.6) [14].

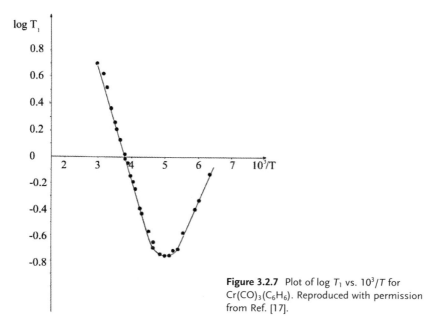

**Figure 3.2.7** Plot of log $T_1$ vs. $10^3/T$ for $Cr(CO)_3(C_6H_6)$. Reproduced with permission from Ref. [17].

Actually the measurements of the temperature dependence of spin–lattice and spin–spin relaxation times is preferred since it allows the exploitation of a wider range of frequencies for the reorientational motions occurring in the solid state. The quantitative comprehension of the effect of molecular motion on relaxation times $T_1$ and $T_2$ is described by the Bloenbergen, Purcell and Pound (BPP) theory [15]. For example if one considers a specific motion that contributes to proton relaxation, the Kubo and Tomita equation [16]

$$1/T_1 = C[(\tau_c/(1+\omega^2\tau_c^2)+4\tau_c/(1+4\tau_c^2w^2)]$$

can be applied, where $\omega$ is the Larmor frequency, $\tau_c$ is the correlation time following a simple Arrhenius-type activation law $\tau_c=\tau_0 \exp (E_a/RT)$. By plotting the log $T_1$ values versus $10^3/T(K)$ the slope of the curve affords the activation energy $(E_a)$ associated with the dynamic process.

An example of this method can be seen in Fig. 3.2.7 for $Cr(CO)_3(C_6H_6)$ [17]. The curve of the proton relaxation time versus $1000/T$ shows a minimum at around 190 K and allows the calculation of an activation energy of 17.6 kJ mol$^{-1}$ for the free rotation of the benzene ring around the principal molecular axis.

A particularly interesting example of the use of wideline spectra is represented by the cis- and trans- isomers of $Fe_2Cp_2(CO)_4$, (Fig. 3.2.8), obtainable separately in the solid state by careful crystallization with appropriate temperature and solvents [18].

In the cis-isomer the two Cp rings are seen to be dynamically and crystallographically independent, whereas the trans-isomer shows equivalence not only of the two Cp rings but also of the two bridging CO groups and the two terminal CO groups. The relaxation times of the two isomers were investigated [18] over the temperature range from –128 °C to 25 °C (Fig. 3.2.9). The trans-isomer shows $T_1$ values decreasing with increasing temperature, with a calculated activation energy of 10.48 kJ mol$^{-1}$. The cis-isomer shows a much more complex curve with a minimum at –63.3 °C and a second minimum that cannot be reached. The calculated activation energy values for the two types of Cp rings (7.22 and 15.76 kJ mol$^{-1}$, respectively) are in agreement with the observation of different anisotropic displacement parameters ("thermal ellipsoids"), and this indicates a variation in the two Cp ring dynamics.

The potential energy barriers associated with reorientational processes of rigid molecular fragments can also be evaluated by means of the pairwise atom–atom potential energy method [19], by using the information in the crystal structure as determined by diffraction techniques. The combined use of spectroscopic and crystallographic methods to investigate dynamic processes provides possible models and a better understanding of the intimate nature of the motion under investigation [20].

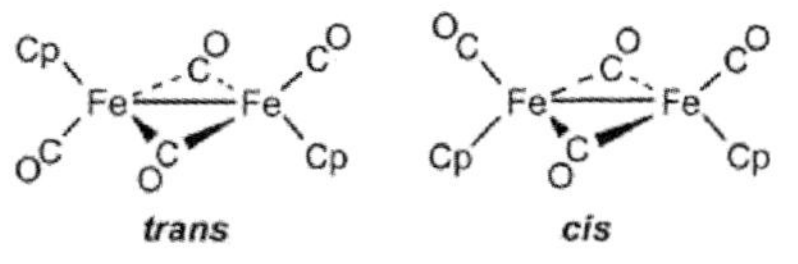

**Figure 3.2.8** Cis- and trans-isomers of $Fe_2Cp_2(CO)_4$.

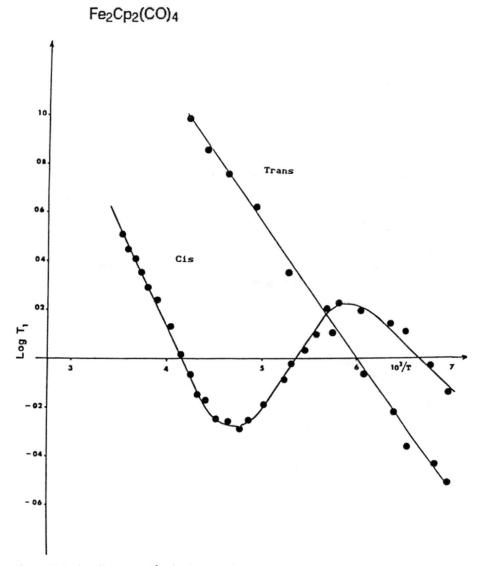

Fe$_2$Cp$_2$(CO)$_4$

**Figure 3.2.9** Plot of log $T_1$ vs. $10^3/T$ for the cis- and trans-isomers of Fe$_2$Cp$_2$(CO)$_4$. Reproduced with permission from Ref. [18].

The frequencies of motion to which NMR is sensitive in the solid state have been substantially extended to slow motion, by measuring the rotating-frame relaxation time $T_{1\varrho}$ [21] and the dipolar relaxation time $T_{1D}$ [22]. This allows NMR studies to cover a range of molecular motions from $10^{-1}$ to $10^{11}$ Hz.

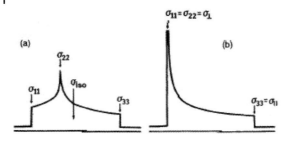

**Figure 3.2.10** Schematic representation of theoretical powder line shapes for the chemical-shift tensor, (a) – asymmetric shift anisotropy, (b) axially symmetric shift anisotropy.

### 3.2.2.3 The Shielding Term ($H_S$) or Chemical Shift Anisotropy (CSA)

The shielding of the magnetic field felt at the nucleus is altered by the presence of electrons. If the electrons have directional character the chemical shift is also directional. Subsequently chemical shift values will be dependent on the orientation of the molecule with respect to the applied field and therefore be regarded as anisotropic. The chemical shift was found to be dependent on $(3\cos^2\theta - 1)/2$, where $\theta$ is the angle between the magnetic field and the molecule-fixed direction.

As a consequence, in a sample where there are many differently oriented molecules there will be a range of chemical shift values giving a very wide signal (Fig. 3.2.10).

The overlapping CSA patterns often hide the isotropic information and then the main NMR parameter cannot be extracted from solid state spectra.

### 3.2.2.4 Line Narrowing in the Solid State

The dipolar, chemical shift anisotropy and quadrupolar (at least at the first order level) interactions have one angular dependence in common, which can be summarised by the expression $3\cos^2\theta - 1$, where $\theta$ is the angle between the interaction vector and the static magnetic field. In order to remove the linebroadening effects it is then necessary to orient the molecules along a particular angle able to make this expression equal to zero. In a powder sample, where the molecules are randomly oriented it is possible to fulfil the need by rotating the sample at a value of $\theta = 54.7°$, commonly known as the magic angle (MA). In order to remove all the effects, the sample should have a spinning speed larger than the magnitude of the different interactions. The speed limit (determined by the rotor diameter) cannot exceed a few tens of KHz, therefore the magic angle spinning is often insufficient to completely remove interactions that can be of the order of hundreds of KHz or even MHz. The remaining part of such interactions (dipolar, CSA or quadrupolar) is seen in the spectra as a distortion of the signal or by the appearance of spinning sidebands. In the case of quadrupolar interaction the signals appear severely distorted and often the centre of gravity is changed. When the CSA is relatively large, therefore greater than the spinning frequency, the phenomenon of spinning

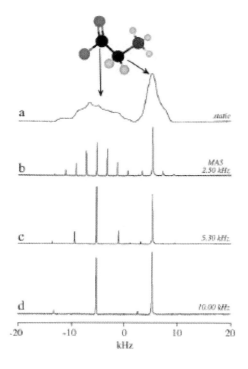

a       *static*

b       *MAS 2.50 kHz*

c       *5.30 kHz*

d       *10.00 kHz*

-20    -10    0    10    20

kHz

**Figure 3.2.11** The experimental $^{13}C$ spectra of a powder of 99%-$^{13}C_2$-labeled glycine, acquired at a magnetic field of 9.4 T. High-power proton decoupling was employed during the acquisition. (a) The spectrum of a static sample. (b–d) Spectra for a sample rotating at the magic angle and recorded at a spinning frequency equal to (b) 2.50, (c) 5.30, and (d) 10.00 kHz. Reproduced with permission from Ref. [23].

side bands (SSB) is observed. The spinning side band manifold appears (Fig. 3.2.11) as a series of sharp lines spaced at the rotational frequency and centred on the isotropic (averaged) chemical shift [23].

### 3.2.2.5 Dipolar Decoupling

The strong broadening due to the dipolar interaction can be easily removed by heteronuclear decoupling. The method consists in applying a continuous pulse at the proton resonance frequency in a direction perpendicular to the applied field $B_o$, while detecting the heteronuclear nucleus (e.g. $^{13}C$). Not only the heteronuclear direct dipolar coupling is averaged to zero by high-power decoupling but also the heteronuclear indirect coupling (since $J \ll D$).

In the cases of strong homonuclear dipolar interactions, normally MAS spinning rates are too slow to provide efficient decoupling: in this case multipulse techniques can be used, which consist of a cycle of radio frequency (rf) pulses repetitively applied to the spin system and able to selectively averaging different interactions. For example, a special pulse sequence known as CRAMPS (combined rotation and multipulse sequence) [24, 25] may be constructed in such a way that at certain points the effect of the dipolar interaction on the nuclear magnetization is zero. If the magnetization is detected only at these points, the effects of dipolar coupling are removed from the spectrum. CRAMPS has been successfully applied to $^1H$ and $^{19}F$ studies [26]. Using very high sample spinning frequencies (rotation speed

≥ 30 KHz ) proton line narrowing can sometimes be sufficient to reach the necessary resolution for detecting separate signals and measuring proton–proton distances. A variable field and spinning speed study up to 25 T and 40 kHz shows that the homogeneous line broadening is inversely proportional to the product of magnetic field strength and spinning speed [27]. However, such spin rates can cause substantially higher increases in sample temperature, unless controlled.

### 3.2.3
### CPMAS

In the solid state the restricted molecular motions may cause relaxation times to be relatively long. This means that the experiment cannot be repeated as fast as in solution and a smaller number of accumulations is possible in a given time. The signal can be enhanced by a double resonance technique that is called *cross polarization* (CP) that solves the problem of slow signal accumulation caused by the long longitudinal relaxation times $T1$ of heteronuclei in the solid state. The polarization required for the experiment comes from the protons. Thermal equilibrium polarization of the protons is restored with the longitudinal relaxation time of the protons, which is much faster than that of heteronuclei. Most commonly, the CP technique is combined with MAS and is denoted CP/MAS [28, 29]. Today this is the predominant method for $^{13}$C solid state NMR spectroscopy, but is not restricted only to this isotope.

In the CP experiment the magnetization of the abundant nucleus $I$ ($^{1}$H) is rotated at 90° to the y-axis and then "spin-locked" along the y-axis by an on-resonance "spin-locking" pulse applied for a time $t$ (Fig. 3.2.12). During this period a strong pulse is applied to the low abundant spin $S$ (for example $^{13}$C) in order to orient these spins along the same y-direction. When both spins are locked along the y-direction and if the spin locking fields fulfil the Hartmann–Hahn condition:

$$\gamma_I B_{1I} = \gamma_s B_{1s}$$

where $\gamma$ is the gyromagnetic ratio and $B_1$ the applied field, the two spin species, each in its own rotating frame, have the same energy splittings. Now "flip-flop" transitions, at a constant total energy, tend to equalize the spin temperatures of

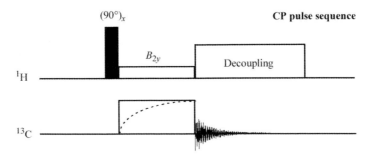

**Figure 3.2.12** Cross polarization pulse sequence.

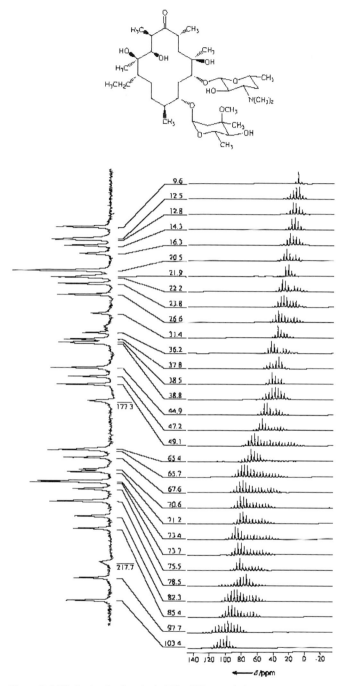

**Figure 3.2.13** Isotropic chemical shift ±CSA correlation spectrum of erythromycin A measured with the FIREMAT technique. Reproduced with permission from Ref. [30].

both the abundant $I$ ($^1$H) and rare $S$ ($^{13}$C) spin systems. After this period, defined as the contact time, the carbon field is switched off and the carbon FID recorded. The $^1$H field is kept on for heteronuclear decoupling. Finally there is a delay time of about 3 times the longitudinal relaxation time ($T_1$) of the $^1$H spins (significantly shorter than the longitudinal relaxation time of the $^{13}$C spins).

The development of new 1D and 2D pulse sequences enables the spectroscopist to obtain structure and dynamic information about systems that were previously very hard to study. As an example is reported in Fig. 3.2.13 the 2D spectrum of erythromycin A measured with the FIREMAT (FIve p Replicated Magic Angle Turning ) technique [30]. The slow spinning speed of 390 Hz produces a spinning sideband pattern for each peak in one dimension, whereas a multipulse sequence in combination with a special processing method produces isotropic lines in the second dimension.

### 3.2.4
### Advantages and Disadvantages of Solid State NMR Spectroscopy

With respect to other analytical techniques, solid state NMR spectroscopy offers several advantages. First, it is a bulk method, it is non-invasive and non-destructive and it does not usually require any special sample preparation. It allows one to focus on a single nucleus (i.e., $^{31}$P, $^{13}$C $^{15}$N, etc) and the isotopic labeling can increase the sensitivity on the specific site of interest. SSNMR spectroscopy can analyze crystalline and amorphous substances and it can, therefore, be applied to the active substance as well as to the formulations at each production step to monitor changes in the physical form. Formulations can be analyzed directly without the need to extract the single components. Since the technique is highly sensitive to changes in the local environment, it is possible to identify polymorphs and even amorphous material can be analyzed. Furthermore, if the technique is performed properly, the intensity of the peaks is directly proportional to the number of magnetically equivalent nuclei, and quantitation of the different components of a mixture can be achieved.

The main disadvantages of SSNMR are related to the relatively high cost of the spectrometers and to the fact that the technique is nonroutine and requires a highly trained operator. Furthermore, because of the low sensitivity of the most interesting nuclei such as $^{13}$C or $^{15}$N, a reasonable signal-to-noise ratio can only be achieved with a long acquisition time. Selective isotopic enrichment of certain sites in the molecule shortens the acquisition time by increasing the sensitivity; however, the expense of the analysis increases dramatically.

### 3.2.5
### Polymorphism

Conformational and packing effects and polymorphism in molecular crystals are all of great interest to chemists attempting to produce 'designed' solids with specific properties. Polymorphism is defined as the ability of a substance to exist

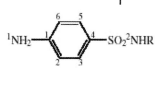

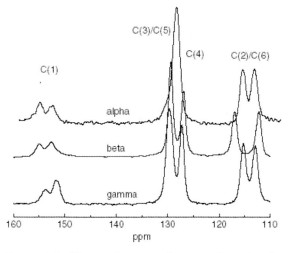

**Figure 3.2.14** $^{13}$C CPMAS spectra of various polymorphs of sulfanilamide. Reproduced with permission from Ref. [33].

in two or more crystalline forms that differ in the arrangement and/or conformation of the molecules in the crystal lattice. Many organic solids exhibit conformational polymorphism, i.e. they possess the same chemical structure, but differ in molecular conformation. Solid state NMR can be a complementary technique with respect to the well known diffractometric approach for the investigation of polymorphism, since the technique is often able to identify the number of crystallographically inequivalent sites in the unit cell and to understand the molecular structure. In many cases the presence of polymorphs is detected by their different chemical shift in one or more chemical sites. However the small chemical shift differences of polymorphs are not always observable due to the larger linewidth in the solid state compared to that in the solution state.

Conformational differences can result in variations in local electron density and NMR can be an ideal probe for this type of behavior, via the chemical shielding tensor [31, 32]. For example detailed inter- and intramolecular conformations can be described at high resolution. Proton sites and hydrogen bonds, which are usually averaged in solution state NMR and not easily assessed in crystallographic studies, can be directly determined by solid state NMR. In Fig. 3.2.14 is presented a typical example in which the different polymorphs of sulfanilamide can be distinguished by their $^{13}$C CPMAS spectra [33].

It is worth noting that solid state NMR is one of the preferred techniques for the detection of changes in the polymorphs with time or temperature. Also, the occurrence of internal motions or the presence of phase transitions can be easily assessed. In several cases (e.g. sulfatiazole [34]) SSNMR was able to resolve different polymorphs that were erroneously described by other techniques, probably because of compound stability and changing hydrated form. For all these reasons solid state NMR has been widely employed for the characterization of pharmaceutical solids where the preparation and detection of polymorphs is essential in quality control. Polymorphic forms of steroids (testosterone, prednisolone, cortisone and 4-azasteroids (finasteride)) have been detected [35–38].

The proportion of the different polymorphs can be quantified if one takes into consideration the relaxation behavior of each polymorph. This property is essential, for example, in all cases where drug-storage requirements play a central role.

Unfortunately, even in moderately complex organic solids, overlapping spectral features make chemical shift analysis of 1D MAS spectra nearly impossible. In this case several approaches have been suggested for separating the isotropic and anisotropic parts of the spectra.

2D solid-state NMR techniques, such as the 2D-TOSS pulse sequence [39], can allow the separation of isotropic and anisotropic chemical shift information over two dimensions, making it possible to distinguish individual carbon atoms under moderate magic angle spinning (MAS) conditions.

Smith and others reported [40] the study of conformational polymorphism of 5-methyl-2-[2-nitrophenyl)amino]-3-thiophenecarbonitrile by using a 2D-TOSS experiment for separation of isotropic and anisotropic part of the spectra with heavy overlapped systems (Fig. 3.2.15).

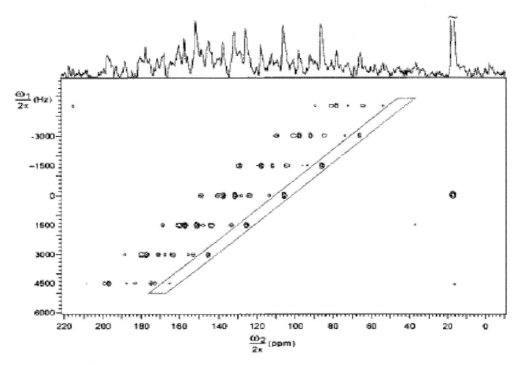

**Figure 3.2.15** Separation by 2D-TOSS experiment of the spinning sideband pattern of 5-methyl-2-[2-nitrophenyl)amino]-3-thiophenecarbonitrile. Reproduced with permission from Ref. [40].

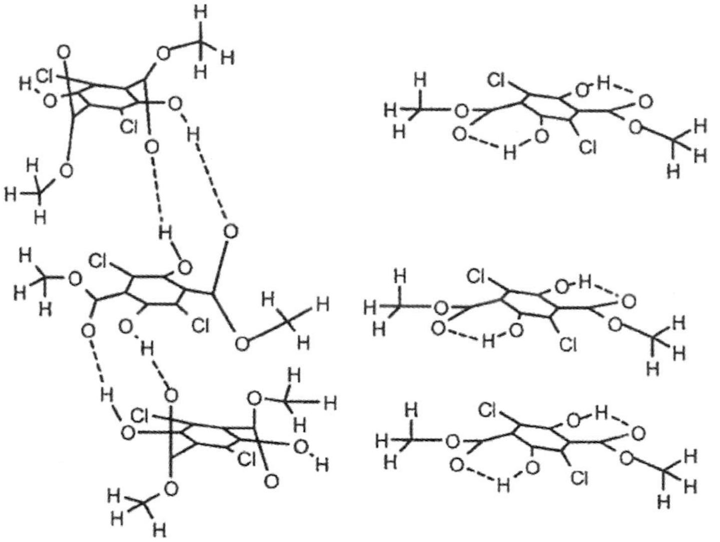

**Figure 3.2.16** Scheme of the hydrogen bonding pattern in dimethyl 3,6-dichloro-2,5-dihydroxylterephthalate.

In the last few years several examples have been published showing that new methodologies, mainly based on two-dimensional (2D) techniques, can afford new insights in the solid state analysis of polymorphs. More sophisticated pulse sequences have been used by Grant and coworkers [41]: the FIREMAT pulse sequence was employed in the investigation of the polymorphism in dimethyl 3,6-dichloro-2,5-dihydroxyl-terephthalate (Fig. 3.2.16).

The main difference between the two polymorphs is the hydrogen bonding pattern. The isotropic chemical shifts of the ester carbons differ by only 1 ppm, whereas the $\delta_{11}$ and $\delta_{22}$ components of the chemical shift tensor show a significant change of 10 ppm in opposite directions, allowing a clear distinction of the polymorphs.

It seems obvious that polymorphs with different lattice arrangements can experience different dynamic parameters. For example in the case of glycine, the carboxyl group in the $\alpha$ and $\gamma$ polymorphs shows a significant difference between the $^{13}C$ $T_1$ relaxation times (11.9 s and 61.0 s respectively) [42].

Due to efficient spin diffusion, homogeneous crystalline solids present a unique relaxation time, regardless of chemical differences. Then different polymorphs can be differentiated by using pulse sequences able to discriminate substances on the basis of their different relaxation properties. In particular cases it has been possible to extract the individual subspectra from a mixture of polymorphs, by modifying the inversion recovery pulse sequence to decompose the spectra [43].

An interesting case of polymorphism has been reported by Harbison and coworkers for the complex of pentachlorophenol with 4-methylpyridine (4MPPCP) [44]. The substitution of hydrogen with deuterium at the hydrogen-bonded position leads to an entirely different crystal polymorph (Fig. 3.2.17). In particular the

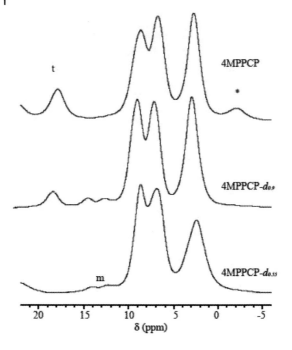

t

4MPPCP

*

4MPPCP-$d_{0.9}$

m

4MPPCP-$d_{0.55}$

20    15    10    5    0    -5

δ (ppm)

**Figure 3.2.17** Effect of deuteration on the solid state proton spectrum of pentachlorophenol with 4-methylpyridine. Reproduced with permission from Ref. [44].

protonated crystals correspond to a triclinic unit cell, whereas crystals of the 90% deuterated form gives a monoclinic cell.

### 3.2.6
### Resolution of Enantiomers by Solid State NMR

In principle, solid state NMR is able to distinguish enantiomers and racemates since their crystals often belong to different point groups and then a small, observable difference in their isotopic chemical shifts may be observed also in powdered samples. The property is of particular interest since in the pharmaceutical industry it is well known that enantiomeric and racemic drugs often show different physiological properties [45]. Better results in the enantiomer discrimination can be obtained by using nuclei with a larger chemical shift range like $^{31}$P. By using this technique Jakobsen and coworkers [46] were able to determine the enantiomeric purity of a series of organophosphorus compounds. Solid state NMR has also been used in many cases for extracting information about enantiomeric excess.

An interesting method for distinguishing enantiomers and racemates has been proposed by Tekely and coworkers [47] for the determination of enantiomers in P-chiral oxazaphosphorine derivatives (Fig. 3.2.18).

The pulse sequence used by these authors is ODESSA (one-dimensional exchange spectroscopy by sideband alternation) [48]. The experiment is based on the fact that intermolecular distances between chemically equivalent nuclei differ significantly in racemates and enantiomers according to slight but substantial

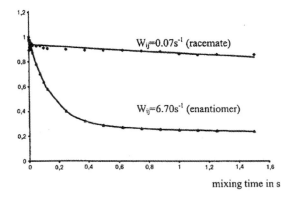

**Figure 3.2.18** Structure of cyclophosphamide.

differences in molecular symmetry. ODESSA allows one to distinguish the enan-tiomeric excess on the base of the intensity changes of the $^{31}$P spinning sideband pattern as a function of mixing time (Fig. 3.2.19).

## 3.2.7
### Distances Determined by SSNMR

The strength of the dipolar coupling between two nuclei is proportional to the inverse cube of the internuclear distance according to the following equation:

$$D_{IS} = (\mu_0/4\,\pi)\ \gamma_I \gamma_S \hbar / r^3{}_{IS}$$

where $D_{IS}$ is the dipolar coupling between nuclear spins $I$ and $S$, $\mu_0$ is the permeability of free space, $\gamma_I$ and $\gamma_S$ are the gyromagnetic ratio of spins $I$ and $S$, respectively, $\hbar$ is the Planck's constant divided by $2\,\pi$ and $r_{IS}$ is the internuclear distance. Therefore, if one can measure the magnitude of the dipolar coupling between two nuclei, their internuclear distance can be determined.

Several methods have been proposed for determining the distances by solid state NMR, since high resolution can be reached, typically ±0.05 nm and often better. Unfortunately sample spinning is often able to average the dipolar coupling interaction, since under MAS, the value of the heteronuclear dipolar coupling, changes both intensity and sign. Over a single rotor period, the integrated value of the heteronuclear dipolar coupling is zero, so the dipolar coupling is effectively removed under MAS. New pulse sequences have been developed for reintroducing and determining $D_{IS}$, and consequently $r_{IS}$. By applying pulses at the proper

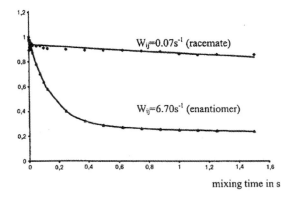

**Figure 3.2.19** Intensity changes of the $^{31}$P spinning sideband pattern as a function of mixing time in the racemic and enantiomeric forms of cyclophosphamide. Reproduced with permission from Ref. [47].

intervals, it is possible to counteract the effect of MAS on the heteronuclear dipolar coupling. Indeed the dipolar interaction is suitably recoupled by the application of rotor synchronized pulses on the dipolar coupled $I$ spins, applied twice per rotor period, with a single $S$ spin pulse applied halfway during the sequence. This two π-pulse technique, known as Rotational Echo DOuble Resonance (REDOR) [49, 50], is a powerful tool for determining dipolar couplings between two heteronuclei, like $^{13}C$ and $^{15}N$.

The scheme of the REDOR pulse sequence is shown in Fig. 3.2.20. After cross polarization from protons, the magnetization on the $^{13}C$ nuclei is allowed to evolve for a period $\tau_m$ before detection. During this time two $^{15}N$ π-pulses are applied every rotor period, in order to reintroduce $^{13}C$–$^{15}N$ heteronuclear dipolar coupling. The same experiment is repeated without the $^{15}N$ pulses and the spectra are compared. The decay of the $^{13}C$ magnetization when $^{13}C$–$^{15}N$ dipolar coupling is reintroduced, is dependent on the dephasing effect of the heteronuclear dipolar coupling. The normalized decay of the magnetization, obtained as a function of $\tau_m$, allows one to obtain a REDOR decay curve (Fig. 3.2.20). The technique has been used in many different cases mainly with isotopic labelled compounds. In the case of a sample of 10% 1–$^{13}C$,$^{15}N$ glycine, the fitted line from the experimental dephasing curve allows one to determine a $^{13}C$–$^{15}N$ dipolar coupling constant of 195 Hz [51]. The estimated value for the $^{15}N$–$^{13}C$ internuclear distance is then 2.47 Å.

The dipolar coupling allows one to obtain distance information that can be used to confirm signal assignment or to elucidate phase separation. Solid state NMR is then a powerful method for investigating molecular conformation and geometry in large molecular systems. The technique is able to provide information about protein structure of membrane, biomaterials and polymers that are not accessible with X-Ray or liquid state NMR spectroscopy.

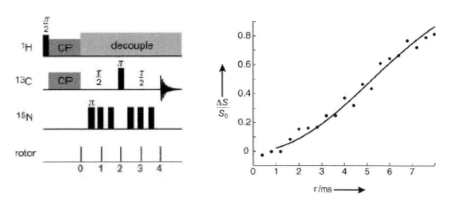

**Figure 3.2.20** Scheme of REDOR pulse sequence. Dephasing curve: $^{13}C$ magnetization in $^{13}C$- and $^{15}N$-labeled glycine as a function of the length of the dephasing time τ. $\Delta S/S_0$ (the REDOR difference) is the normalized difference between the $^{13}C$ signal intensities with and without the $^{15}N$ pulses. Reproduced with permission from Ref. [51].

Homonuclear dipolar recoupling can be accomplished by DRAMA (Dipolar Recovery At the Magic Angle) [52] or by RR (rotational resonance) [53] pulse sequence. During the DRAMA pulse sequence, two $\pi/2$ pulses per rotor period move the magnetization from the $z$-axis to the $y$-axis and back again. This two pulse sequence interrupts the averaging of the chemical shift anisotropy under MAS; to remove the CSA a $\pi$-pulse is introduced. Distance determination via DRAMA is accomplished in the same way as REDOR. During the mixing period, $\tau_m$, the magnetization of the recoupled nuclei will decay due to the dephasing effects of the homonuclear dipolar coupling. By measuring the decay of the magnetization with and without the DRAMA sequence, it is possible to determine the dipolar couplings and the internuclear distances.

In RR the recoupling can be achieved by setting the spinning rate $\omega_r$ to multiples of the frequency difference between the NMR resonances.

Although the origin of this recoupling is complex, it can generally be understood by considering the rotor to be an additional source of energy for the system to bridge the energy gap between the nuclear transitions. Magnetization transfer occurs during the mixing time, reducing the spectral intensity. The study of the decay curve as a function of the mixing time also allows in this case the quantification of the dipolar interaction.

## 3.2.8
## Hydrogen Bond

Strong, selective, and directional hydrogen bonding has been noted as a most powerful organizing force in molecular assembly. Due to its importance not only in crystal engineering but in many others fields and due to the difficult in detecting and characterising it, several techniques have been used for the detection and characterization of the hydrogen bond.

The various techniques have some intrinsic limitations and cannot be universally applied, for example in the case of a polycrystalline powder obtained from solvent-free reactions such as those occurring in the solid state between molecules, or between two solids, or between a solid and a gas. For these kinds of reaction, which are important from both the environmental and topochemical viewpoints, the alternative investigation techniques are mainly IR and NMR spectroscopy.

Since in the solid state the signals are not averaged by solvent effects or by rapid exchange processes often present in solution, the solid state NMR approach allows a more accurate study.

There are different parameters that can provide insights into the hydrogen bond interaction in the solid state. Solid state $^{13}$C-NMR studies of protonated and deprotonated carboxylates in amino acids have shown that the values of the principal elements of the nuclear shielding tensor change significantly with the protonation state of the carboxylic groups [54–59]. The orientation of the three CSA tensors for the COOH and COO$^-$ groups in a peptide is depicted in Fig. 3.2.21.

$\delta_{11}$ lies in the plane of symmetry of the carboxylic group and it is directed along the C–C axis in the deprotonated form and perpendicularly to the C=O group in the

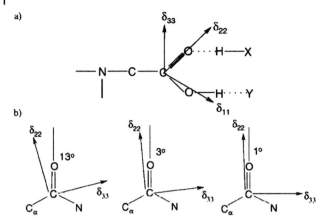

a)

b)

**Figure 3.2.21** The orientation of the $^{13}$C chemical shift tensors with respect to the molecular frame of the carboxylic group of a peptide.

protonated case, $\delta_{22}$ lies perpendicular to the plane of symmetry of the C=O and is the most diagnostic parameter, reflecting the strength of the hydrogen bond. $\delta_{33}$, the most shielded tensor, is perpendicular to the plane of symmetry.

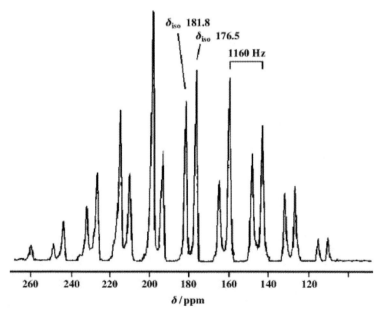

**Figure 3.2.22** $^{13}$C CPMAS NMR spectrum (carboxylic region) recorded at 67.94 MHz at a spinning speed of 1160 Hz. Reproduced with permission from Ref. [60].

The carbon chemical shift tensors of the COOH group obtained from the sideband intensity of low-speed spinning NMR spectra, provide a reliable criterion for assigning the protonation state of compounds. A typical example has been reported in the case of the hydrogen bonded supramolecular adducts between the diamine 1,4-diazabicyclo–[2.2.2]octane (DABCO) and dicarboxylic acids of variable chain length [60]. The CPMAS $^{13}$C spectrum of the [N(CH2CH$_2$)$_3$N]-H-[OOC(CH$_2$)$_5$COOH] adduct obtained at low spinning speed (Fig. 3.2.22) is indicative of the difference in the sideband pattern between the COOH ($\delta_{iso}$=176.5) and the COO$^-$ groups ($\delta_{iso}$=181.8 ppm).

In the [N(CH$_2$CH$_2$)$_3$N]-H-[OOC(CH$_2$)$_n$COOH] (n=1–7) adducts the chemical shift tensors, obtained by computer simulation of the spectrum measured at low spinning speed with the Herzfeld–Berger method [61], demonstrate that $\delta_{11}$ values change from 242±2 ppm (carboxylate form) to 257±4 ppm (carboxylic form), $\delta_{22}$ values change from 177±10 ppm for the COO$^-$ group to 155±20 ppm for the COOH moiety, whereas $\delta_{33}$ is usually not very sensitive to the protonation state of the carboxylic group. The isotropic value $\delta_{iso}$ increases in shielding upon protonation, but unfortunately this information is limited by the fact that $\delta_{11}$ and $\delta_{22}$ change their values in the opposite direction (Fig. 3.2.23).

The $^{15}$N chemical shift is also a useful parameter for the location of the hydrogen in hydrogen bond systems involving nitrogen and oxygen atoms. It is expected that the $^{15}$N chemical shift will be more sensitive to the presence of the hydrogen bond than that of $^{13}$C, due to the wider chemical shift range of the former. Intermolecular hydrogen bonds produce high or low frequency shifts in the $^{15}$N values, according to the type of nitrogen atom and to the type of interaction.

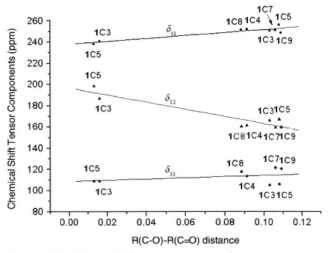

**Figure 3.2.23** Chemical shift tensor components for [N(CH$_2$CH$_2$)$_3$N]–H–[OOC(CH$_2$)$_n$COOH] (n=1, 1C3; n=2, 1C4; n=3, 1C5; n=5, 1C7; n=6, 1C8, n=7, 1C9) plotted as a function of the difference between C–O and C=O bond lengths obtained from the crystallographic data. Reproduced with permission from Ref. [60].

For the $^{15}N$ isotropic chemical shift the protonation-induced shifts are of the order of 100 ppm towards lower frequencies for aromatic amines, and about 25 ppm towards higher frequencies for aliphatic amines [62]. For example, in the case of glycine residues in solid oligopeptides, a relationship between the N–O distance and the $^{15}N$ chemical shift tensor in $C=O\cdots H–N$ hydrogen bonds has been proposed [63].

The possibility of spinning the sample at a speed higher than 20 kHz allows the direct measurement of chemical shifts of protons in HB, leading to an increasing number of $^1H$ studies [64–66].

Frey and Sternberg [67, 68] have observed direct relationships between proton chemical shift ($\delta_{1H}$) and hydrogen-bond strength, and between $\delta_{1H}$ and X–H distance, respectively, for different classes of hydrogen-bonded compounds. Correlation between the $\delta_{1H}$ and the heteroatom separation supports this interpretation [69, 70].

The proton transfer reaction along a hydrogen bond between aliphatic dicarboxylic acids and DABCO, investigated by $^1H$ solid state NMR (Fig. 3.2.24), has shown that intramolecular $O–H\cdots O$ and intermolecular $N\cdots H–O$ hydrogen bonds are strong interactions, with proton chemical shifts of around 16±1.5 ppm, and N–O and O–O bond lengths of around 2.55–2.60 Å, while intermolecular $N^+–H\cdots O^-$ interactions are weaker and are characterised by a $\delta_{1H}$ of about 12.3 ppm and by an N–O bond length of about 2.7 Å [71, 72].

It has been shown [73–75] that the plot of the chemical shifts vs. bond lengths is useful for obtaining the positions of the hydrogen atoms. Due to the intrinsic difficult of the X-ray technique in detecting the position of the hydrogen atoms, an

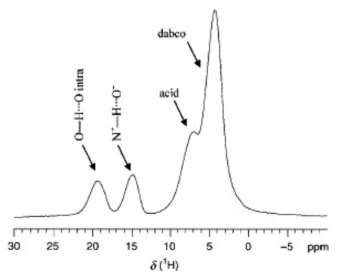

**Figure 3.2.24** $^1H$ MAS NMR spectrum of dabco–maleate 1:2 at 298 K obtained at 499.7MHz. Reproduced with permission from Ref. [71].

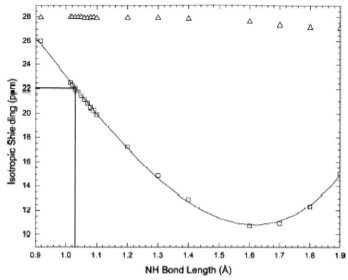

**Figure 3.2.25** DFT computed isotropic shielding constant (on an absolute scale) of the hydrogen-bonded proton in the white form of methylnitroacetanilide, plotted as a function of the N–H distance for the dimer. The curve represents an empirical fit to a cubic equation. The horizontal and vertical lines show the derivation of the N–H distance from the observed shielding. Reproduced with permission from Ref. [76].

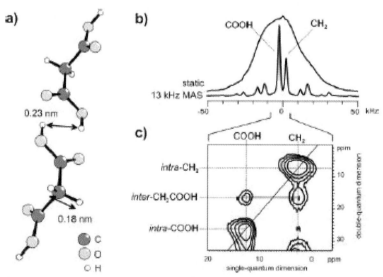

**Figure 3.2.26** (a) Dimeric arrangement of the carboxylic acid groups in the crystal structure of malonic acid. The interproton distances obtained from $^1$H SQ and DQ MAS spectra are also shown. (b) $^1$H one-pulse spectra of malonic acid under static conditions and under MAS at 13 kHz. (c) $^1$H DQ MAS spectrum of malonic acid under MAS at 13 kHz at a $^1$H Larmor frequency of 500 MHz. The subspectrum at the first-order MAS sideband in the DQ dimension is displayed. Reproduced with permission from Ref. [77].

effective method that combines $^1$H fast MAS NMR and DFT calculations for the determination of the hydrogen bond O–H distance has been proposed (Fig. 3.2.25) [71, 76].

In the last few years the combination of high spinning speed with 2D multiple quantum spectroscopy has opened up new possibilities in the area of hydrogen bond investigation by $^1$H solid state NMR. The field has been exhaustively reviewed by Schnell and Spiess [77]. In Fig. 3.2.26 is reported, for example, the information available by 1D and 2D double quantum proton spectra in the case of malonic acid. Each signal can be unequivocally assigned to the respective proton, and the chemical shift of the carboxylic protons can be correlated to the O–O distance. Furthermore the intermolecular correlation between $CH_2$ and COOH protons can be used as a constraint for establishing the conformation and array structure in the solid state.

## References

1 Y. Tomita, E. J. O'Connor, A. E. McDermott, *J. Am. Chem. Soc.* 1994, *116*, 8768.

2 X. Feng, P. J. E. Verdegem, Y. K. Lee, D. Sandstro'm, M. Ede'n, P. Bovee-Guerts, W. J. De Grip, J. Lungtenborg, H. J. M. de Groot, M. H. Levitt , *J. Am. Chem. Soc.* 1997, *119*, 6853.

3 K. Schmidt-Rohr, *Macromolecules* 1996, *29*, 3975.

4 C. A.Fyfe, *Solid State NMR for Chemists*, CFC, Guelph, Canada 1984.

5 M. Mehring, *Principles of High Resolution NMR in Solids*, Springer Verlag, Berlin 1983.

6 M. Duer *An Introduction to Solid-State NMR*, Blackwell Science Ltd, Oxford 2004.

7 H. W. Spiess, *Colloid Polym. Sci.* 1983, *261*, 193.

8 A. G. Stepanov, T. O. Shegai, M. V.Luzgin, N. Essayem, H.Jobic, *J. Phys. Chem. B.* 2003, *107*, 12438.

9 A. G. Stepanov, M. M Alkaev, A. A Shubin, M. V. Luzgin, T. O. Shegai, H. Jobic, *J. Phys. Chem. B* 2002, *106*, 10114.

10 J. A. Ripmeester, C. I. Ratcliffe, *Compr. Supramol. Chem.* 1996, *8*, 323.

11 G. E. Pake, *J. Chem. Phys.* 1948, *16*, 127.

12 J. H. Van Vleck, *Phys. Rev.* 1948, *74*, 1168.

13 H. S. Gutowsky, G. B. Kistiakoowsky, G. E. Pake, E. M. Purcell, *J.Chem. Phys.* 1949, *17*, 72.

14 E. R. Andrew, R. G. Eades, *Proc. R. Soc. London, Ser. A* 1953, *218*, 537.

15 N. Bloenbergen, E. M. Purcell, R. V. Pound, *Phys. Rev.* 1948, *73*, 679.

16 R. Kubo, K. Tomita, *J. Phys. Soc.* 1954, *9*, 888.

17 P. Delise, G. Allegra, E. R. Mognaschi, A. Chierico, *J. Chem. Soc., Faraday Trans. 2,* 1975, *71*, 207.

18 S. Aime, M. Botta, R. Gobetto, A. Orlandi, *Magn. Res. Chem.* 1990, *28*, 552.

19 A. J. Pertsin, A. I. Kitaigorodsky, *The Atom-Atom Potential Method*, Springer-Verlag, Berlin 1987.

20 S. Aime, D. Braga, L. Cordero, R. Gobetto, F. Grepioni, S. Righi, S. Sostero, *Inorg. Chem.* 1992, *31*, 3054

21 C. P. Slichter, D. Ilion, *Phys. Rev. A* 1964, *135*, 1099.

22 J. Jeener, P. Broekaert, *Phys. Rev.* 1967, *157*, 232.

23 M. Ede'n, *Concepts Magn. Reson., Part A,* 2003, *17*, 117.

24 B. C. Gerstein, C. Clor, R. G. Pembleton, R. C. Wilson, *J. Phys. Chem.* 1977, *81*, 565.

25 C. E. Broniman, B. L. Hawkings, M. Zhang, G. E. Maciel, *Anal. Chem.* 1988, *60*, 1743.

26 R. K. Harris, P. Jackson, *Chem Rev.* 1991, *91*, 1427.

27 A. Somoson, T. Tuherm, Z. Gan, *Solid State Nucl. Magn. Reson.* 2001, *20*, 130.

28 A. Pines, M. G. Gibby, J. S. Waugh, *J. Phys. Chem.* **1972**, *56*, 1776.

29 J. Schaefer, E. O. Stejskal, *J. Am. Chem. Soc.* **1976**, *98*, 1030.

30 D. W. Alderman, G. McGeorge, J. Z. Hu, R. J. Pugmire, D. M. Grant, *Mol. Phys.* **1998**, *95*, 1113.

31 E. A. Christopher, R. K. Harris, R. A. Fletton, *Solid State NMR*, **2002**, *1*, 93.

32 E. D. Smith, R. B.Hammond, M. J. Jones, K. J. Roberts, J. B. O. Mitchell, S. L. Price, R. K. Harris, D. C. Apperley, J. C. Cherryman, R. Docherty, *J. Phys. Chem. B* **2001**, *105*, 5818.

33 A. Portieri, R. K. Harris, R. A. Fletton, R. W. Lancaster, T. L. Threlfall, *Magn. Reson. Chem.* **2004**, *42*, 313.

34 D. C. Apperley, R. A. Fletton, R. K.Harris, R. W. Lancaster, S. Tavener, T. L. Threlfall, *J. Pharm. Sci.* **1999**, *88*, 1275.

35 A. Szyczewski, K. Holderna-Natkaniec, I. Natkaniec, *J. Mol. Struct.* **2004**, *E 698*, 41.

36 P. J. Saidon, N. S. Cauchon, P. A. Sutton, *Pharm. Res.* **1993**, *10*, 197.

37 R. K. Harris, *Solid State Sci.* **2004**, *6*, 1025.

38 J. W. Morzychi, I. Wawer, A. Gryszkiewicz, J. Maj, L. Siergiejczyk, A. Zaworska, *Steroids* **2002**, *67*, 621.

39 S. F. De Lacroix, J. J. Titman, A. Hagemeyer, H. W. Spiess, *J. Mag. Reson.* **1992**, *97*, 435.

40 J. Smith, E. MacNamara, D. Raftery, T. Borchards, S. J. Byrn, *J. Am. Chem. Soc.* **1998**, *120*, 11710.

41 M. Strohmeier, A. M. Orendt, D. W. Alderman, D. M.Grant, *J. Am. Chem. Soc.* **2001**, *123*, 1713.

42 M. J. Potrzebowski, P. Tekely, Y. Dusausoy, *Solid State Nucl. Magn. Reson.* **1998**, *11*, 253.

43 N. Zumbulyadis, B. Antalek, W. Windig, R. P. Scaringe, A. M. Lanzafame, T. Blanton, M. Helber, *J. Am. Chem. Soc.* **1999**, *121*, 11544.

44 J. Zhou, Y.-S. Kye, G. S. Harbison, *J. Am. Chem. Soc.* **2004**, *126*, 8392.

45 A. J. Hutt, S. C. Tan, *Drugs* **1996**, *52*, 1.

46 K.V. Andersen, H. Bildsoe, H. J. Jakobsen, *Magn. Reson. Chem.* **1990**, *28*, S47.

47 M. J. Potrzebowski, E. Tadeusiak, S. Misiura, W. Ciesielski, P. Tekely, *Chem. Eur. J.* **2002**, *8*, 5007.

48 V. Gerardy-Mountouillout, C. Malveau, P. Tekely, Z. Olender, Z. Luz, *J. Magn. Res. A* **1996**, *123*, 7.

49 T. Gullion, J.Schaefer, *J. Magn. Reson.* **1989**, *81*, 196.

50 T. Gullion, J. Schaefer, *Adv. Magn. Reson.* **1989**, *13*, 57.

51 D. D. Laws, H.-M. L. Bitter, A. Jerschow *Angew. Chem. Int. Ed.* **2002**, *41*, 3096.

52 R. Tycko, G. Dabbagh, *Chem. Phys. Lett.* **1990**, *173*, 461.

53 D. P. Raleigh, M. H. Levitt, R. G. Griffin, *Chem. Phys. Lett.* **1988**, *146*, 71.

54 Z.Gu, A. McDermott, *J. Am. Chem. Soc.* **1995**, *115*, 4262.

55 A. Naito, S. Ganapathy, K. Aakasaka, C. J. McDowell, *J. Chem. Phys.* **1981**, *74*, 3198.

56 R. Haberkom, R. Stark, H. van Willigen, R. Griffin, *J. Am. Chem. Soc.* **1981**, *103*, 2534.

57 N. James, S. Ganapathy, E. Oldfield, *J. Magn. Res.* **1983**, *54*, 111.

58 R. Griffin, A. Pines, S. Pausak, J. Waugh, *J. Phys. Chem.* **1975**, *65*, 1267.

59 W. Veeman, *Prog. NMR Spectrosc.* **1984**, *16*, 193.

60 D. Braga, L. Maini, G. de Sanctis, K. Rubini, F. Grepioni, M. R. Chierotti, R. Gobetto, *Chem. Eur. J.* **2003**, *1*, 5538.

61 J. Herzfeld, A. E. Berger, *Chem. Phys.* **1980**, *73*, 6021.

62 G. C. Levy, R. L. Lichter, *Nitrogen-15 Nuclear Magnetic Resonance Spectroscopy*, Wiley New York, 1979.

63 A. Fukutani, A. Naito, S. Tuzi, H. Saito, *J. Mol. Struct.* **2002**, *602*, 491.

64 J. Brus, J. Dybal, *Macromolecules* **2002**, *35*, 10038.

65 J. Brus, J. Dybal, P. Sysel, R. Hobzova, *Macromolecules* **2002**, *35*, 1253.

66 P. Lorente, I. G. Shenderovich, N. S. Golubev, G. S. Denisov, G. Buntkowsky, H. H. Limbach, *Magn. Reson. Chem.* **2001**, *39*, S18.

67 P. A. Frey, *Magn. Reson. Chem.* **2001**, *39*, S190.

68 U. Sternberg, E. Brumer, *J. Magn. Reson. Ser. A* **1994**, *108*, 142.

**69** A. McDermott, C. F. Ridenour, *Encyclopedia of NMR*, D. M. Grant, R. K. Harris (Eds.), Wiley, Chichester, 1996, p. 3820.

**70** A. S. Mildvan, T. K. Harris, C. Abeygunawardana, *Methods Enzymol.* **1999**, 219.

**71** R. Gobetto, C. Nervi, E. Valfrè, M. R. Chierotti, D. Braga, L. Maini, F. Grepioni, R. K. Harris, P. Y. Ghi, *Chem. Mater.* **2005**, *17*, 1457.

**72** R. Gobetto, C. Nervi, M. R. Chierotti, D. Braga, L. Maini, F. Grepioni, R. K. Harris, P. Hodgkinson, *Chem. Eur. J.* **2005**, *11*, 7461.

**73** B. Berglund, R. W. Vaughan, *J. Chem. Phys.* **1980**, *73*, 2037.

**74** K.Yamauchi, S. Kuroki, I. Ando, *J. Mol. Struct.* **2002**, *602*, 9.

**75** R. K. Harris, P. Jackson, L. H. Merwin, B. J. Say, G. Hagele, *J. Chem. Soc., Faraday Trans.* **1988**, *84*, 3649.

**76** R. K. Harris, P. Y. Ghi, R. B. Hammond, C. Y. Ma, K. J. Roberts, *J. Chem. Soc., Chem. Commun.* **2003**, 2834.

**77** J. Schnell, H. W. Spiess, *J. Magn. Reson.* **2001**, *151*, 153.

**3.3**

**Crystal Polymorphism: Challenges at the Crossroads of Science and Technology**

*Dario Braga and Joel Bernstein*

3.3.1

**Introduction**

The quest for, and identification and characterization of different crystal forms (polymorphs and solvates) of the same molecule or of aggregates of the same molecule with other molecules (co-crystals) is one of the most active and challenging research areas of modern solid state chemistry. The effort is by no means theoretical or academic. In the pharmaceutical field in particular, the existence of multiple crystal forms is relevant for the choice of the solid dosage form of an active pharmaceutical ingredient (API) most suitable for drug development and marketing, and has important implications both in terms of the drug ultimate efficacy and in terms of the protection of the intellectual property rights associated with the final pharmaceutical product.

3.3.1.1    **What Are Polymorphism and the Multiplicity of Crystal Forms?**

Materials are traditionally classified in three states of matter: gases, liquids and solids distinguished by their properties. However, the solid phase of a material can exhibit different structures, which in turn can show different properties. In addition to different crystal structures, called *polymorphs*, which are characterized by long-range order, a material may appear as an amorphous solid, characterized by the lack of long-range order. The *polymorphism* of calcium carbonate (calcite, vaterite and aragonite) was identified more than 200 years ago by Klaproth in 1788 [1], but formal recognition of the phenomenon is generally credited to Mitscherlich [2]. Diamond, graphite, fullerenes and nanotubes are polymorphic forms (denoted as allotropes for elements) of carbon all exhibiting very different properties. Cocoa butter can crystallize in at least five different ways, the various crystal structures affecting the perception of the epicurean quality of the prepared chocolate, although all forms are chemically identical.

In this chapter we discuss the different crystal forms of molecular crystals, in which a particular molecule crystallizes in different ways (polymorphs) and/or with solvent molecules (solvates). The general classification of these crystal forms is summarized in Fig. 3.3.1. Among practitioners in the field the discussion of both polymorphism and solvates is often encompassed under the umbrella of polymorphism [3]. As we will attempt to show, although the subject has been widely investigated, mainly in the field of solid state organic chemistry, the polymorphism of molecular crystals is still a fascinating phenomenon, and still represents a substantial scientific challenge to the very idea of rational design and construction of crystalline solids with predefined architectures and physical properties starting from the choice of the molecular components, which is the paradigm of molecular crystal engineering [4].

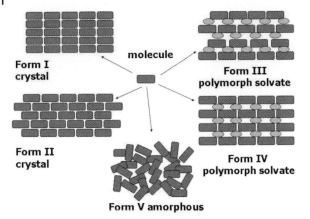

**Figure 3.3.1** Schematic representation of the structural relationship between "true" polymorphs, solvates, polymorphs of solvates and the amorphous phase.

From thermodynamic principles, under specified conditions only one polymorph is the stable form (except at a transition point) [5]. In practice, however, due to kinetic considerations, metastable forms can exist or coexist in the presence of more stable forms. Such is the case for diamond, which is metastable with regard to graphite, the thermodynamically stable form of carbon under ambient conditions. In practice, the relative stability of the various crystal forms and the possibility of interconversion between crystal forms, between crystals with a different degree of solvation and between an amorphous phase and a crystalline phase, can have very serious consequences on the life and effectiveness of a polymorphic product and the persistence over time of the desired properties (therapeutic effectiveness in the case of a drug, chromatic properties in the case of pigment, etc).

Conversions between different crystal forms are possible and often take place. Among the many possibilities for conversion, depending on variables such as temperature, pressure, relative humidity, etc. specific examples are:

1. a metastable form can convert to a thermodynamically more stable crystal form with very slow kinetics;
2. an unsolvated crystal form can form solvates and co-crystals with other "innocent" molecules which will nonetheless alter significantly the physical properties with respect to the "homomolecular" crystals;
3. an anhydrous crystalline form can be transformed into a crystal hydrate via vapor uptake from the atmosphere;
4. a solvate can, in turn, be transformed into another crystal form with a different degree of solvation up to the anhydrous crystal via stepwise solvent/water loss;
5. a (metastable) amorphous phase may transform into a stable crystalline phase over time.

The variety of phenomena related to polymorphism (hydration, solvation, amorphicity and interconversions) demonstrates the importance of acquiring a thorough mapping of the "crystal space" of a substance that is ultimately intended for some specific application.

### 3.3.1.2 Why Are Polymorphism and Multiple Crystal Forms Important?

The example of the polymorphs (allotropes) of carbon illustrate the key messages of this chapter: different crystal forms of a substance can possess very different properties and behave as different materials. This concept has important implications in all fields of chemistry associated with the production and commercialization of molecules in the form of crystalline materials (drugs, pigments, agrochemicals and food additives, explosives, etc). The producer, in fact, needs to know not only the exact nature of the material in the production and marketing process, but also its stability with time, the variability of its chemical and physical properties as a function of the crystal form, etc. In some areas, e.g. the pharmaceutical industry, the search for and characterization of crystal forms of the API has become a crucial step for the choice of the best form for formulation, production, stability and for intellectual property protection.

Table 3.3.1 summarizes some major possible differences in chemical and physical properties between crystal forms and solvates of the same substance. Different crystal forms are often recognized by differences in the color and shape of crystals. A striking example of these two properties is provided by the differences in color and form of the crystal forms of ROY (ROY = red, orange, yellow polymorphs of 5-methyl-2-[(2-nitrophenyl)amino]-3-thiop hene carbonitrile) [6] shown in Fig. 3.3.2.

**Table 3.3.1** Examples of chemical and physical properties that can differ among crystal forms and solvates of the same substance.

| | |
|---|---|
| Physical and thermodynamic properties | density and refractive index, thermal and electrical conductivity, hygroscopicity, melting points, free energy and chemical potential, heat capacity, vapor pressure, solubility, thermal stability |
| Spectroscopic properties | electronic, vibrational and rotational properties, nuclear magnetic resonance spectral features |
| Kinetic properties | rate of dissolution, kinetics of solid state reactions, stability |
| Surface properties | surface free energy, crystal habit, surface area, particle size distribution |
| Mechanical properties | hardness, compression, thermal expansion |
| Chemical properties | chemical and photochemical reactivity |

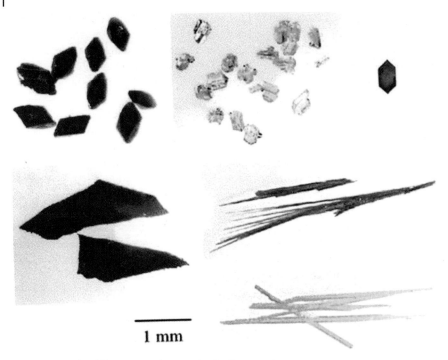

**1 mm**

**Figure 3.3.2** The differences in shape and color between the polymorphs of ROY (ROY = red, orange, yellow polymorphs of 5-methyl-2-[(2-nitrophenyl)amino]-3-thiophene carbonitrile) (Ref. [6], reprinted with permission).

## 3.2.2
### How Do We Detect and Characterize Multiple Crystal Forms?

In spite of the efforts of a great number of research groups worldwide, and of a familiarity with the experimental factors that can lead to multiple crystal forms, our ability to predict or control the occurrence of polymorphism is still embryonic. In many cases the crystallization of a new crystal form or of an amorphous phase of a given substance turns out to be the result of *serendipity* [7] rather than a process under complete human control.

The exploration of the "crystal form space" (polymorph screening) of a substance is the search of the polymorphs and solvates with a twofold purpose: (i) identification of the relative thermodynamic stability of the various forms including the existence of enantiotropic crystalline forms (that interconvert as a function of the temperature) or of monotropic forms (that do not interconvert) and of amorphous and solvate forms and (ii) physical characterization of the crystal forms with as many analytical techniques as possible. The relationships between the various phases and commonly used industrial and research laboratory processes are illustrated schematically in Fig. 3.3.3.

Polymorph assessment, on the other hand, is part of the system of quality control. It is necessary to make sure that the scale-up from laboratory preparation to industrial production does not introduce variations in crystal form. Polymorph assessment also guarantees that the product conforms to the guidelines of the appropriate regulatory agencies and does not infringe the intellectual property protection that may cover other crystal forms.

The polymorph pre-screening, screening and assessment are best achieved by the combined use of several solid-state techniques, among them (not exclusively or in any preferential order): microscopy and hot stage microscopy (HSM), differential scanning calorimetry (DSC), thermogravimetric analysis (TGA), infrared and Raman spectroscopy (IR and Raman), single crystal/powder X-ray diffraction (SCXRD, XRPD), solid state nuclear magnetic resonance spectroscopy (SSNMR) [See Chapter 3.2].

It is also important to mention in the context of this discussion the advantages offered by the possibility of determining the molecular and crystal structure of a crystal form by means of single crystal X-ray diffraction. This technique, although much more demanding than powder diffraction in terms of experiment duration and data processing, has the great advantage of providing detailed structural information on the molecular geometry, but more important for this discussion, it provides information on the packing of the molecules in the crystal and the nature and structural role of solvent molecules. Moreover, the knowledge of the single-crystal structure allows one to calculate the X-ray powder diffraction pattern that can be compared with that measured on the polycrystalline sample, as demonstrated in Fig. 3.3.4. Importantly, the calculated diffraction pattern is not affected by the typical sources of errors or the experimental powder diffraction (preferential orientation, mixtures, presence of amorphous) that often complicate

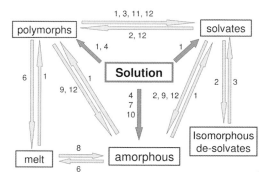

**Figure 3.3.3**  Some general relationships between polymorphs, solvates and amorphous phases and the type of research laboratory or industrial process for preparation and interconversion: 1, Crystallization; 2, Desolvation; 3, Exposure to solvent/vapor uptake, 4, Freeze drying; 5, Heating; 6, Melting; 7, Precipitation; 8, Quench cooling; 9, Milling; 10, Spray drying; 11, Kneading; 12, Wet granulation. Analogous relationships apply to polymorphic modifications of solvate forms. Note that the figure represents general trends rather than every possible transformation; the presence or absence of an arrow or number does not represent the exclusive existence or absence of a transformation.

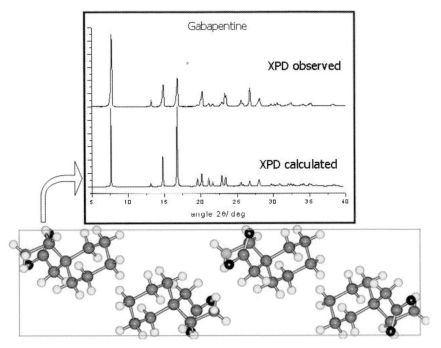

**Figure 3.3.4** Comparison between measured and calculated X-ray diffraction patterns for form II of gabapentine (single crystal data taken from Ref. [8]).

or render uncertain the interpretation of the measured powder diffractograms; hence the calculated powder pattern is often referred to as the "gold standard" pattern for a crystal form.

The search for various crystal forms requires that the behavior of a solid phase is investigated as a function of the variables that can influence or determine the outcome of the crystallization process, e.g. temperature, choice of solvents, crystallization conditions, rate of precipitation, interconversion between solid forms (from solvate to unsolvate and vice versa), pressure and mechanical treatment, absorption and release of vapor, etc. The most effective way to search for crystal forms is to evaluate the effect of the changes of one variable at a time. There has been a recent burst of activity in developing crystallization techniques and variables for obtaining new crystal forms [9], some of which are described in the next section.

The efficiency of screening protocols, whether *high throughput* automatic methods or manual procedures, can be considerably increased by carrying out preliminary HSM, DSC and variable temperature XRPD investigations for initial detection of multiple phases and the temperature ranges of their existence, as well as transformations among them. These observations can then be summarized with a semiempirical energy–temperature diagram [10] that can be helpful in designing protocols for screening for crystal forms. In particular, it is possible to determine if various phases are related enantiotropically (reversibly) or monotropically (non-reversibly). Once the thermodynamic screening of the crystalline product has been completed, the quest for new forms can extend to the investigation of the effect of changing the solvent or the mixture of solvents and/or to the temperature gradient, the presence of templates or seeds. Examples of the utilization of variable temperature diffraction methods (VT-XRPD) to investigate phase transitions between enantiotropic systems (in the case of anthranilic acid [11]) and desolvation proces-

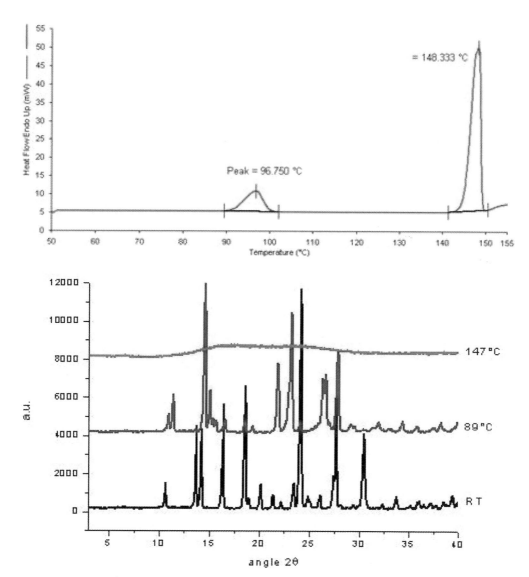

**Figure 3.3.5** DSC and VT-XRPD measurements applied to the investigation of phase transitions between Form I and Form II of anthranilic acid [11].

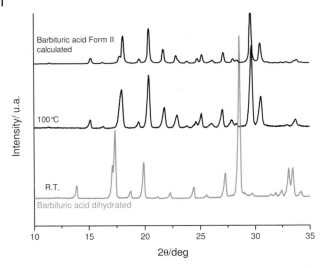

**Figure 3.3.6** Variable temperature XRPD measurements applied to the investigation of the dehydration of barbituric acid dihydrate with formation of Form I of barbituric acid [12].

ses (water removal from barbituric acid dihydrate [12]) are shown in Fig. 3.3.5 and Fig. 3.3.6, respectively.

### 3.3.3
### New Developments in Detecting and Characterizing Multiple Crystal Forms

The last decade has witnessed many developments in the generation and detection of new crystal forms. These have resulted from the increased awareness of the possibility of multiple crystal forms of a substance, the utility that may be derived by preparing a crystal form with enhanced properties and the potential intellectual property implications of new crystal forms. These factors, combined with the development of new technology [13], the attempts to design and control crystal structure [14], combined with some spectacular encounters with new (and undesired) crystal forms [15] (see next section) and some high profile pharmaceutical patent litigations [3b], have led to many new techniques for exploring the crystal form space of any particular substance. Some of these depend simply on an awareness of the older literature [16], the application of crystal engineering principles, based on hydrogen-bonding patterns, to the preparation of new multicomponent solids [17], the induction of crystal forms by incorporating a variety of functional groups onto a polymer backbone [18], the development of high throughput crystallization technology [19], the utilization of solid–solid and solid–gas reactions [20], solvent-free synthesis [21], the desolvation of solvated crystals [22] and crystallization from a supercritical solvent [23].

3.3.4
## Examples of Crystal Form Identification and Characterization

The familiarity with, and understanding of, a system exhibiting multiple crystal forms are best achieved by studying the system with a wide variety of analytical tools. In a manner similar to becoming acquainted with another human being, it is then possible to recognize particular forms and to prepare and utilize the desired ones, while avoiding those that are less desired. Of course, every system is different, and must be treated differently depending on its own idiosyncrasies and the resources available. Examples from recently studied systems [24, 25] are given here.

1,3-bis(*m*-nitrophenyl)urea, MNUP, was originally studied in 1899 [26] and summarized by Groth in 1906 [27] as exhibiting three concomitant forms: α, yellow prisms; β, white needles; and γ, yellow tablets as shown in Fig. 3.3.7. The system

**Figure 3.3.7** Photographs of crystals of 1,3-bis(*m*-nitrophenyl)urea, MNPU. From top to bottom: yellow prisms (α form), white needles (β/δ form) and yellow plates (γ form). Since it is difficult to visually distinguish between the β and δ forms, we will refer to them as the β/δ form.

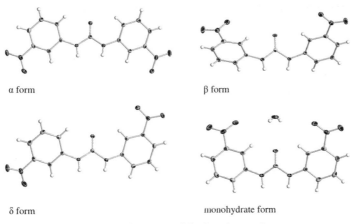

α form  β form

δ form  monohydrate form

**Figure 3.3.8** Molecular conformation of the four forms of MNPU. For ease of comparison of molecular conformation, all molecules are plotted on the place of the N–C(O)–N with the C=O bond vertical.

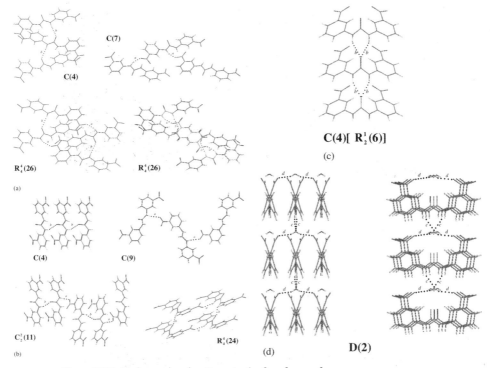

**Figure 3.3.9** Hydrogen bond patterns in the four forms of MNPU, showing four types of hydrogen bonds (labeled with lower case bold italic letters) and their graph set designations. (a) α form; (b) δ form; (c) β form; (d) monohydrate form (for clarity, the hydrogen bonds are shown from two different orientations).

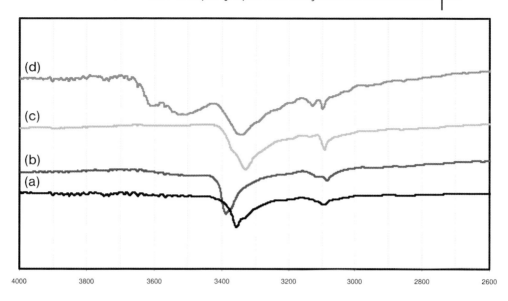

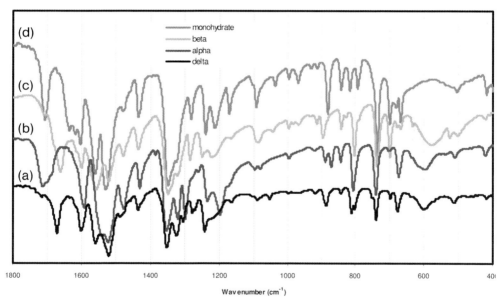

**Figure 3.3.10** Solid state FT-IR spectra (KBr disk) of δ(a) α (b), β (c) and monohydrate (d) forms. The intensities of the spectra were normalized to facilitate comparison.

was investigated by Etter in the early 1990 s [28, 29], with the methods of characterization including structure determination of the α and β forms along with solid state FT-IR, XRPD, thermal analysis, and determination of powder second-harmonic generation efficiencies.

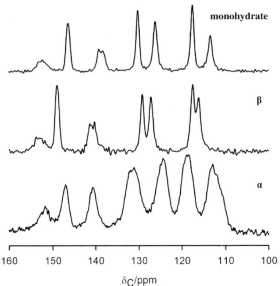

**Figure 3.3.11** Expanded-scale solid state $^{13}C$ NMR spectra for the α, β, and monohydrate forms in the δ = 100 to 160 ppm region. The β and monohydrate spectra were recorded at 75 MHz, whereas the α spectrum was obtained at 125 MHz.

A series of crystallization experiments under a variety of conditions led to the discovery of two new crystal forms, a third anhydrate and a previously unreported monohydrate. Thorough study of the appearance and thermal behavior of the monohydrate led to the conclusion that it, in fact, is the γ form reported by Groth; the third anhydrate form was designated δ. The MNPU molecule adopts four different conformations in the four crystal structures as shown in Fig. 3.3.8; hence, the system exhibits *conformational polymorphism* [30].

Due to the presence of the –NH hydrogen bond donors and the >C=O acceptor one of the main distinguishing structural features of the crystal structures are the hydrogen-bonding patterns (see Fig. 3.3.9). These are shown and summarized by the graph set analysis and notation also developed by Etter [28]. FT-IR spectra on KBr disks are generally quite similar (Fig. 3.3.10), but there are consistent and recognizable differences in some specific regions of the spectral range.

The $^{13}C$ solid state NMR spectra for the α, β and monohydrate forms, shown in Fig. 3.3.11, clearly exhibit distinguishing characteristics for each of the three forms.

The four forms were also studied by thermal methods, both HSM and DSC/TGA. Two HSM sequences are shown in Fig. 3.3.12, demonstrating the wealth of information that visual examinations can yield. It is worth noting that the relative ease and low cost of digital photography, combined with the increasing facility for publishing color in scientific publications make this method for reporting crystal forms and recording the visual manifestation of thermal events even more desirable.

While the HSM experiments often provide dramatic visual evidence for phase transformations, the calorimetric methods (e.g. DSC/TGA) can provide precise thermodynamic data and quantitative information on the composition and nature of solvates and hydrates. The DSC traces for the α, β/δ and monohydrate are shown in Fig. 3.3.13 and 3.3.14. They confirm the thermal events observed in the HSM experiments.

The connection between the *thermal* and *structural* aspects of these events can be followed by carrying out XRPD studies as a function of temperature. Such an experiment is demonstrated in Fig. 3.3.15.

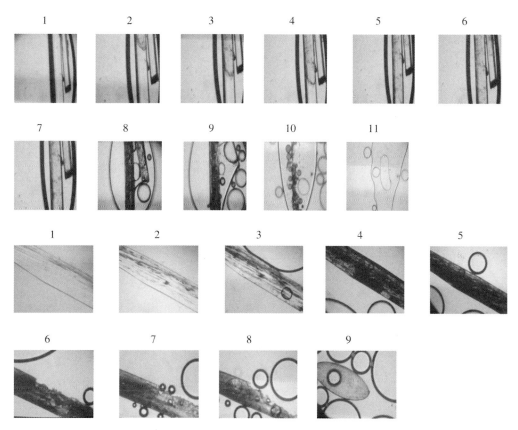

**Figure 3.3.12** Upper: two single crystals of the β/δ forms of MNPU as observed on the hot stage microscope: (1) crystals in paraffin oil before heating; (2–7) when heated, the crystals become yellow due to a phase change of the β/δ form to the α form at the onset temperature range of 160–221 C; (8–11) decomposition of the crystals, taking place in the melt at the onset temperature range of 226–252 C. The bubbles are due to decomposition. Lower: single crystal of the monohydrate form as observed on the hot stage microscope: (1) the crystal in paraffin oil before heating; (2–5) when heated, the crystal becomes darker and bubbles of water vapor evolve from it, at the onset temperature range of 70–101 C; (6–9) decomposition of the crystal, taking place in the melt at onset temperature range of 231–253 C. The bubbles are due to decomposition.

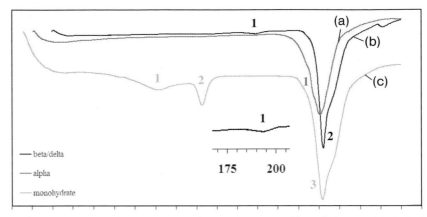

**Figure 3.3.13** DSC thermographs of the α (a), β/δ (b) and monohydrate (c) forms of MNPU. Additional details and interpretation may be found in Ref. [24] from which this figure is reproduced with permission.

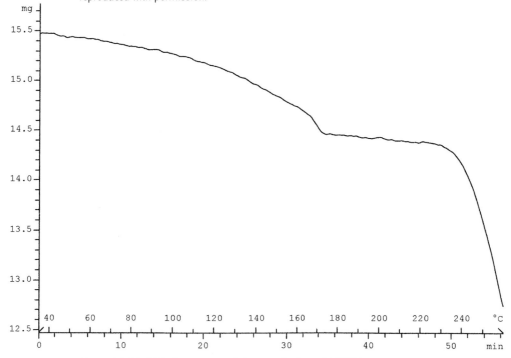

**Figure 3.3.14** TGA thermogram of the monohydrate of MNPU. The shape of the trace gives some indication of how strongly the water is bound and the nature of events following the expulsion of some or all of the water from the crystal lattice. Additional details and interpretation may be found in Ref. [24] from which this figure is reproduced with permission.

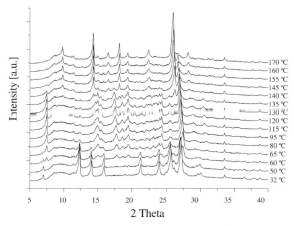

**Figure 3.3.15** Variable temperature XRPD measurements showing the monohydrate thermal behavior. The diffraction pattern of the monohydrate is shown at 32 and 50 C; between 50 and 60 C: dehydration leads to initial formation of the β and δ forms; between 80 and 95 C: the monohydrate goes through complete dehydration leaving only the β and δ forms. The appearance of the β form is indicated at 2 = 17.8 and 23.14 and of the δ form at 2 = 7.48, 15.08, 15.64, 18.66 and 27.36; between 115 and 120 C: shows a solid–solid phase change of the β and δ forms to the α form. β and δ forms totally disappear between 140 and 145 C.

Since different crystal forms have different structures, they can, potentially, exhibit different physical properties and different responses to experimental analytical methods. Some of the more commonly used of these methods have been demonstrated above. A central question for any polymorphic system is the relative stability of the various crystal forms. As noted above, these may be investigated qualitatively by HSM methods, and more quantitatively using thermal analytical techniques. The combined results of these measurements are conveniently summarized on a semi-empirical energy–temperature diagram [31], as shown in Fig. 3.3.16. The thermal

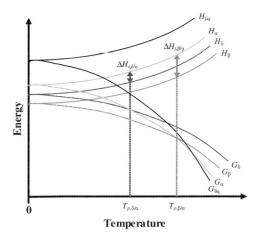

**Figure 3.3.16** Suggested schematic energy vs. temperature diagram for the MNPU trimorphic system (reproduced from Ref. [31], with permission).

data for only true polymorphs may be presented on a such a diagram, but with that *caveat* it is noteworthy how much information can be summarized in such a convenient manner, and the reporting of such a diagram for any polymorphic system for which thermal data have been obtain should be encouraged.

Finally, there is intense interest in being able to computationally accurately account for the relative stability of different crystal forms, and ultimately to be able to predict the relative stability of crystal forms and, with that, the potential existence of polymorphism. These computational methods are in constant development, and are periodically evaluated on a community-wide basis through a blind test at the Cambridge Crystallographic Data Centre [32]. Such computational methods have been applied to the MNPU system as well, in order to determine the differences in conformational energy associated with the conformational polymorphism, as well as the differences in lattice energy associated with the different crystal structures. The comparison of these computational results with the experimentally determined thermal data provides an excellent benchmark for both, and the collaboration of experimentalist and computational chemists in studying polymorphic systems should be greatly encouraged.

### 3.3.5
### Practical Implications and Ramifications of Multiple Crystal Forms – Pharmaceuticals

"Control" is perhaps the guiding principle of chemical endeavors. Control over a reaction means obtaining the expected product in the expected purity with the expected yield. Chemists spend considerable time and effort gaining control over reactions and processes, and much of our measure of success is the level of robustness of that control. The solid state properties of APIs are crucial aspects of the process of both drug development and efficacy. The properties of the various solid forms of a same active principle can affect not only the bioavailability of the drug, but also the procedures of purification, transport, distribution and packaging, let alone the stability of the drug over time. Hence, control over the crystal form obtained along with a thorough knowledge and understanding of the properties of the various possible crystal forms in general and the API in particular are crucial in the pharmaceutical industry.

A dramatic example of the impact of crystal polymorphism on a drug formulation is that of ritonavir (Norvir®), used for the treatment of HIV patients. The problem arose in May of 1998, approximately two years after the launch of the drug, when researchers at the Abbott Laboratories became aware that after 240 production batches it was no longer possible to obtain ritonavir in the crystal form (Form I) approved by the FDA and required for the formulation of Norvir® because of the sudden and unexpected appearance of a more stable and much less soluble crystal form (Form II, Fig. 3.3.17). The loss of control over the production process forced Abbott to withdraw the drug from the market for approximately one year until they learned how to replace the solid formulation with a gel capsule suspension with greater problems of stability and bioavailability. Subsequent investigations have led to the discovery of four other crystalline forms of ritonavir [33].

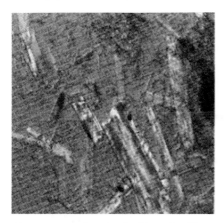

**Figure 3.3.17** Photograph of crystals of forms I and II of
ritonavir (reproduced, with permission, from Ref. [33]).

As a result of this episode the international community and the FDA have begun
to consider very seriously also the consequences of polymorphism: a new crystal
form of an important drug may have dramatic consequences on those patients who
rely on a given therapy. A number of authors have suggested that polymorphism is
essentially universal, and the (late) Walter McCrone is often quoted in this regard:
"It is at least this author's opinion that every compound has different polymorphic
forms and that, in general, the number of forms known for a given compound is
proportional to the time and energy spent in research on that compound". As a
corollary to this rather sweeping, even provocative statement, McCrone noted that
"all the common compounds (and elements) show polymorphism", and he cited
many common organic and inorganic examples [3a].

Other well-known authors have made similar statements. a. F. Findlay, who
authored the classic book *The Phase Rule* noted in that "[polymorphism] is now
recognized as a very frequent occurrence indeed."[34] Buerger and Bloom stated,
"... polymorphism is an inherent property of the solid state and it fails to appear
only under special conditions."[35] Similarly, Sirota in 1982: "[P]olymorphism is
now believed to be characteristic of all substances, its actual non-occurrence arising
from the fact that a polymorphic transition lies above the melting point of the
substance or in the area of as yet unattainable values of external equilibrium factors
or other conditions providing for the transition."[36]

These generalizations are often supported, especially in the field of pharmaceuticals, by citing Kuhnert-Brandstätter's 1965 statistics [37], expanded upon in her
1971 book [38] on three common classes of drugs: 70% of the barbiturates, 60% of
sulfonamides and 23% of steroids exist in various polymorphic or solvate forms.
These statistics were based on HSM studies that Kuhnert-Brandstätter herself
performed. Due to the inherent microscopic nature of this technique and the
metastable nature of many observed phases it is often difficult to translate HMS
observations into fully characterized crystal forms. In fact, as Kuhnert-Brandstätter
noted in her 1965 paper, "...optical crystallographic methods require a more

rigorous training. Without a certain minimum crystallographic knowledge, the use of optical crystallographic techniques in chemistry is not to be recommended." S. R. Byrn, in particular commenting on the Kuhnert-Brandstätter statistics on steroids noted that "This table clearly shows the extent of polymorphism in this important class of compounds [steroids]. It should be noted that these studies are based only on hot-stage results and should be considered unconfirmed until other methods verify the existence of these polymorphs" [39].

A number of statistical analyses of the literature have been carried out in an attempt to estimate the extent of polymorphism. A search of the Cambridge Structural Database on the keywords "polymorph", "form", "modification" and "phase" indicates that about 3.5 % of the ~350 000 entries fall into this category. Approximately 25 % of the entries are either solvates or hydrates. At the other end of the spectrum, Byrn has reported that of the >150 compounds submitted for crystal form screening and analysis to SSCI, Inc. 85 % exhibit more than one crystal form, 37 % are solvates and 31 % are hydrates [40]. Other studies based on different selection criteria reveal results falling somewhere between these two extremes [41]. For instance, Griesser and Burger have collected information on about 600 polymorphic forms and solvates (including hydrates) pharmaceutical compounds that are solid at 25 °C [41c].

The often quoted statements of McCrone et al. tend to give the impression that polymorphism is the rule rather than the exception. The body of literature in fact indicates that considerable caution should be exercised. It appears to be true that instances of polymorphism are not uncommon in those industries where the preparation and characterization of solid materials are integral aspects of the development and manufacturing of products (i.e. those on which a great deal of time and energy are spent): silica, iron, calcium silicate, sulfur, soap, chocolate, pharmaceutical products, dyes and pigments, explosives. Such materials, unlike the vast majority of compounds that are isolated, are prepared not just once, but repeatedly, under conditions that vary slightly (even unintentionally) from time to time. Even with the growing awareness and economic importance of polymorphism, most documented cases have been discovered by serendipity rather than through systematic searches. Some very common materials, such as sucrose and naphthalene, which certainly have been crystallized innumerable times, have not been reported to be polymorphic. The *possibility* of polymorphism may exist for any particular compound, but the conditions required to prepare as yet unknown polymorphs are by no means obvious. There are as yet no comprehensive systematic methods for feasibly determining all of those conditions. Moreover, we are almost totally ignorant about the properties to be expected from any new polymorphs that might be obtained.

It is of worthy of note that the "International Conference on Harmonization (ICH, Guideline Q6A of the October of 1999)" includes under the heading of "polymorphs": "single entity polymorphs; molecular adducts (solvates, hydrates), amorphous forms". The FDA currently requires that pharmaceutical manufacturers investigate the polymorphism of the active ingredients before clinical tests and that polymorphism is continuously monitored during scale-up and production processes [42]. The

European Patent Office (EPO) also demands characterization of solid drugs by means of X-ray diffraction to ensure the integrity of the crystal form [43].

As a matter of fact, the ritonavir example demonstrates that polymorphism may represent a problem if it manifests itself at an advanced stage of the drug evaluation process or (even worse) after market launch, while it can be a useful selection criterion if analyzed at an early stage. Thus, in principle, the exploration of the "crystal space" of an API should be undertaken in the early stages of drug development in order to direct the choice of the crystalline form better suited for subsequent (costly) clinical tests and, in the long run, for the market. As suggested above, a change of crystalline form in the course of clinical testing increases costs and could involve new tests of bioavailability or bioequivalence.

The screening for crystal forms ought to be directed not only to the search for the thermodynamically most stable crystalline form (which will have less tendency to transform spontaneously into a different crystal form) but also to the study of the chemical-physical properties of the various hydrates and solvates, amorphous phase, and, when necessary also to the investigation of the various salts (e.g. hydrochlorides, sulfonates etc.) forms. For instance, in view of the low solubility and bioavailability of the stable form of ritonavir, it is of interest to ask how and when the drug would have been launched if only that form had been known in the early stages of the drug's development.

Besides being a scientific necessity, the investigation of polymorphism has also acquired an important place in the development of strategies of intellectual property protection and patenting in the pharmaceutical industry. Generic companies can enter the market when a patent expires, or prior to patent expiration if the patent can be successfully challenged. As a result of the rather high financial stakes involved, particularly with regard to so-called "blockbuster" drugs, recent years have seen increasing activity in the prosecution, challenge and litigation of patents involving crystal forms of pharmaceutically important compounds.

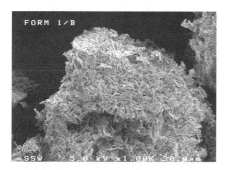

**Figure 3.3.18** Microphotographs of Forms 1 and 2 of the anti-ulcer drug ranitidine hydrochloride produced by Glaxo Wellcome (now Glaxo SmithKline). The differences in physical form of the crystals lead to differences in filtering and drying characteristics, which were among the improvements over Form 1 noted in the Form 2 patent application.

The patent litigation around the Glaxo Wellcome (now GlaxoSmithKline) anti-ulcer drug Zantac® is by now a text book example of the commercial implications of polymorphism. Zantac® is the trademark of ranitidine hydrochloride, an anti-ulcer drug marketed in the US by Glaxo Wellcome since 1984. The compound is known to crystallize in two crystalline forms, called Form 1 and Form 2 (Fig. 3.3.18). Form 2 is the commercial form covered by a Glaxo patent till 2002, while the patent on Form 1, also by Glaxo, expired in 1997, opening the market to the generic manufacturers. However, most attempts to produce ranitidine hydrochloride as pure Form I according to Glaxo's recipe (Example 32 in the Form 1 patent) resulted in the preparation of crystals of Form 2. That issue was one of the major scientific questions in a series of patent litigations between Glaxo and other companies (e.g. Novopharm, Genpharm) [Ref. 3 b, Chapter 10].

A number of additional examples of litigations involving crystal forms (e.g. terazosin hydrochloride, cefadroxyl, aspartame etc.) are described in Ref. [3b]. A more recent high profile case involved the Glaxo SmithKline antidepressant drug Paxil®, paroxetine hydrochloride, which involved an anhydrate and a hemihydrate form of the API [44, 45].

## 3.3.6
**Conclusions**

The scope of this chapter is that of providing to the reader an introductory view of the phenomena of multiple crystal forms in general and polymorphism in particular, and the scientific, commercial, and ethical importance and implications of these phenomena. For reasons of space, many problems have not been addressed, for example, the interaction of drugs with excipients, which themselves can also exhibit polymorphism, the possibility of solid-state reactivity etc. The field will undoubtedly provide many scientific challenges and financial rewards in the coming years. The interested reader will find in the necessarily limited number of references a good starting point for further reading, study and exploration of this fascinating and rapidly developing area of solid state chemistry.

**References**

**1** M.H. Klaproth *Bergmannische J.* I, **1798**, 294.
**2** E. Mitscherlich, *Abhl. Akad. Berlin*, **1823**, 43.
**3** (a) W. C. McCrone, in *Polymorphism in Physics and Chemistry of the Organic Solid State*, D. Fox, M. M. Labes, A. Weissenberg (Eds.), Interscience, New York, 1965, vol. II, p. 726; (b) J. Bernstein, *Polymorphism in Molecular Crystals*, Oxford University Press, Oxford, 2002; (c) D. Braga, *Chem. Commun.*, **2003**, 2751.

**4** (a) G. R. Desiraju, *Crystal Engineering: The Design of Organic Solids*, Elsevier, Amsterdam, 1989; (b) D. Braga, F. Grepioni, A. G. Orpen (Eds.), *Crystal Engineering: from Molecules and Crystals to Materials*, Kluwer Academic Publishers, Dordrecht, 1999; (c) D. Braga, F. Grepioni, G. R. Desiraju, *Chem. Rev.*, **1998**, *98*, 1375; (d) A. J. Blake, N. R. Champness, P. Hubberstey, W. S. Li, M. A. Withersby and M. Schroder, *Coord. Chem. Rev.*, **1999**, *183*, 117; (e) B. Moulton, M. J. Zaworotko, *Chem. Rev.*, **2001**, *101*, 1629.

5 A. Burger, *Topics in Pharmaceutical Sciences*, D. D. Breimer, P. Speiser, Elsevier, Amsterdam, 1983, p. 347; (b) T. L. Threlfall, *Analyst*, **1995**, *120*, 2435.

6 S. Chen, I. Guzei, L. Yu, *J. Am. Chem. Soc.*, **2005**, *127*, 9881; L. Yu, C. A. Stephenson, C. A. Mitchell, C. A. Bunnell, S. V. Snorek, J. J. Bowyer, T. B. Borchardt, J. G. Stowell, S. R. Byrn, *J. Am. Chem. Soc.*, **2000**, *122*, 585.

7 R. K. Merton, E. Barber, *The Travels and Adventures of Serendipity*, Princeton University Press, Princeton, 2004.

8 J. A. Ibers, *Acta Crystalogr., Sect. C*, **2001**, *57*, 641.

9 J. Bernstein, *Chem. Commun.*, **2005**, 5007.

10 A. Grunenberg, J.-O. Henck, H. W. Siesler, *Int. J. Pharm.* **1996**, *129*, 147; J.-O. Henck, M. Kuhnert-Brandstatter, *J. Pharm. Sci.* **1999**, *88*, 103.

11 (a) C. J. Brown, M. Ehrenberg *Acta Crystallogr., Sect. C* **1985**, *41*, 441; (b) G. E. Hardy, W. C. Kaska, B. P. Chandra, J. I. Zink *J. Am. Chem. Soc.* **1981**, *103*, 1074; (c) H. Takazawa, S. Ohba, Y. Saito *Acta Crystallogr., Sect. C*. **1986**, *42*, 1880; (d) Tian-Huey Lu, P. Chattopadhyay, Fen-Liang Liao, Jem-Mau Lo, *Anal. Sci.* **2001**, *17*, 905.

12 (a) T. C. Lewis, D. A. Tocher, S. L. Price, *Cryst. Grow. Des.*, **2004**, *4*, 979, (b) W. Bolton, *Acta Crystallogr.*, **1963**, *16*, 166; (c) G. A. Jeffrey, S. Ghose, J. O. Warwicker, *Acta Crystallogr.*, **1961**, *14*, 881; (d) A. R. Al-Karaghouli, B. Abdul-Wahab, E. Ajaj, S. Al-Asaff, *Acta Crystallogr., Sect. B* **1977**, *33*, 1655; (e) G. S. Nichol, W. Clegg, *Acta Crystallogr., Sect. B* **2005**, *61*, 464.

13 S. L. Morissette, S. Soukasenem, D. Levinson, M. J. Cima, Ö. Almarsson, *Proc. Natl. Acad. Sci. U. S. A.* **2003**, *100*, 2180.

14 G. R. Desiraju (Ed.), *Crystal Design: Structure and Function*, Wiley, Chichester, 2003; D. Braga, F. Grepioni, *Angew. Chem. Int. Ed.*, **2004**, *43*, 4002.

15 J. Bauer, S. Spanton, R. Henry, J. Quick, W. Dziki, W. Porter, J. Morris, *Pharm. Res.*, **2001**, *18*, 859.

16 (a) W. I. F. David, K. Shankland, C. R. Pulham, N. Blagden, R. Davey, M. Song, *Angew. Chem., Int. Ed.* **2005**, *44*, 7032; (b) N. Blagden, R. Davey, G. Dent, M. Song, W. I. F. David, C. R. Pulham, K. Shankland, *Cryst. Growth Des.* **2005**, *5*, 2218.

17 (a) Ö. Almarsson, M. J. Zaworotko, *Chem. Commun.*, **2004**, 1889; (b) C. B. Aakeröy, A. M. Beatty, B. A. Helfrich, *Angew. Chem., Int. Ed.*, **2001**, *40*, 3240.

18 C. P. Price, A. L. Grzesiak, A. J. Matzger, *J. Am. Chem. Soc.*, **2005**, *127*, 5512.

19 D. Braga, F. Grepioni, *Chem. Commun.*, **2005**, 3635.

20 A. V. Trask, N. Shan, W. D. S. Motherwell, W. Jones, S. Feng, R. B. H. Tan, K. J. Carpenter, *Chem. Commun.*, **2005**, 880.

21 E. Y. Cheung, S. J. Kitchin, K. D. M. Harris, Y. Imai, N. Tajima, R. Kuroda, *J. Am. Chem. Soc.*, **2003**, *125*, 14658.

22 A. Burger, U. J. Griesser, *Eur. J. Pharm. Biopharm.*, **1991**, *37*, 118.

23 P. York, *Chem. World*, **2005**, *2*, 50; B. Yu. Shekunov, P. York, *J. Cryst. Growth*, **2001**, *211*, 122.

24 M. Rafilovich, J. Bernstein, R. Harris, D. Apperley, P. Panagiotopolis, S. L. Price *Cryst. Growth Des.* **2005**, *5*, 2197.

25 R. Hiremath, J. A. Basile, S. W. Varney, J. A. Swift, *J. Am. Chem. Soc.* **2005**, *127*, 18321.

26 (a) A. Offret, H. Vittenet, *Bull. Soc. Chim. Fr.*, **1899**, *21*, 152; (b) A. Offret, H. Vittenet, *Bull. Soc. Chim. Fr.*, **1899**, *21*, 788; (c) A. Offret, H. Vittenet, *Bull. Soc. Chim. Fr.* **1899**, *22*, 627.

27 P. H. R. Groth, *An Introduction to Chemical Crystallography* (trans. H. Marshall), Gurnery & Jackson, London, 1906, pp. 28–31.

28 M. C. Etter, Z. Urbañczyk-Lipkowska, M. Zia-Ebrahimi, T. W. M. Panunto, *J. Am. Chem. Soc.* **1990**, *112*, 8415.

29 K.-S.Huang, D. Britton, M. C. Etter, S. R. Byrn, *J. Mater. Chem.* **1995**, *5*, 379.

30 J. Bernstein, in *Organic Solid State Chemistry*, G. R. Desiraju (Ed.), Elsevier, Amsterdam, 1987, p. 471.

31 A. Grunenberg, J.-O. Henck, H. W. Siesler, *Int. J. Pharm.* **1995**, *129*, 147.

32 (a) J. P. M. Lommerse, W. D. S. Motherwell, H. L. Ammon, J. D. Dunitz, A. Gavezzotti, D. W. M. Hofmann, F. J. J.

Leusen, W. T. M. Mooij, S. L. Price, B. Schweizer, M. U. Schmidt, B. P. van Eijck, P. Verwer, D. E. Williams, *Acta Crystallogr. B*, **2000**, *56*, 697; (b) W. D. S. Motherwell, H. L. Ammon, J. D. Dunitz, A. Dzyabchenko, P. Erk, A. Gavezzotti, D. W. M. Hofmann, F. J. J. Leusen, J. P. M. Lommerse, W. T. M. Mooij, S. L. Price, H. Scheraga, B. Schweizer, *Acta Crystallogr. B*, **2002**, *58*, 647.

**33** J. Bauer, S. Spanton, R. Henry. J. Quick, W. Dziki, W. Porter, J. Morris, *J. Pharm. Res.* **2001**, *6*, 59.

**34** A. F. Findlay, in *The Phase Rule and Its Applications*, A. N. Campbell, N. O. Smith (Eds.), Dover Publications, New York, 9th edn., 1951, pp. 7–16.

**35** M. J. Buerger, M. C. Bloom, *Z. Kristallogr.* **1937**, *96*, 182.

**36** N. N. Sirota, *Cryst. Res. Technol.* **1982**, *17*, 661.

**37** M. Kuhnert-Brandstatter, *Pure Appl. Chem.* **1965**, *10*, 133.

**38** M. Kuhnert-Brandstatter, *Thermomicroscopy in the Analysis of Pharmaceuticals*, International Series of Monographs in Analytical Chemistry, R. Belcher, M. Freiser, (Eds.), Pergamon, Oxford, 1971, Vol. 445.

**39** S. R. Byrn, R. R. Pfeiffer, J. G. Stowell, *Solid State Chemistry of Drugs,* SSCI, Inc., West Lafayette, 2nd edn., 1999.

**40** S. R. Byrn, personal communication

**41** (a) J. A. R. P. Sarma, G. R. Desiraju, *Crystal Engineering. The Design and Application of Functional Solids*, Kluwer, 1999, p. 325; (b) U. Griesser, *Acta Crystallogr., Sect. A* **2002**, *58* (Supplement), C241; (c) U. Griesser, A. Burger, IUCr Congress, Glasgow, Abstract P09.013, p. 533; (d) C. H. Gorbitz, H. P. Hersleth, *Acta Crystallogr., Sect. B* **2000**, *56*, 526.

**42** http://www.fda.gov/cder/guidance/6154dft.htm

**43** http://www.european-patent-office.org

**44** J. Bernstein, *Polymorphism in Pharmaceutical Technology*, R. Hilfiker (Ed.), Elsevier, Amsterdam, 2006.

**45** SmithKline Beecham et al. vs. Apotex Corp. et al., United States District Court, Northern District of Illinois 98C3952.

**3.4**
**Nanoporosity, Gas Storage, Gas Sensing**
*Satoshi Takamizawa*

**3.4.1**
**Introduction**

Recently, research into the synthesis of a porous crystalline solid with the capability of gas adsorption has attracted increased attention. Since the physical properties of gas adsorption have primarily been developed in the science of the interface of granular materials, it is difficult for chemists who are used to dealing with objects as single-crystal substances. The gas adsorption state can be regarded as a particular kind of crystal state. Recently, an artificial crystalline porous material within an organic frame has been produced, and this material has pliability, unlike a rigid solid substance. This chapter focuses on the new aspects of nanoporosity.

**3.4.2**
**Description of Porosity**

When unevenness is formed in a solid by various causes, a hole with its concave portion greater than its diameter is called a "pore." Substances without pores are called "nonporous materials" and substances with pores are called "porous materials". Porous materials vary in terms of pore diameter, distribution of pore sizes, and pore volume. Porous solids can be classified according to their pore structure and chemical constitution. Moreover, porous materials are also classified by their structural regularity into ordered and disordered structures. The former include natural and synthetic zeolite crystals, which have a stable frame capable of retaining regular pores. The size of the regular pores in crystalline materials is usually that of a molecule. In contrast, disordered pores are found in materials of agglomerated particles such as silica gel, or solids that have been surface modified by oxidization or corrosion, such as activated carbon and borosilicate glass. The characteristics of their porous properties vary even when they have the same chemical composition.

Classification by pore form is also possible. Typical pore types are shown in Fig. 3.4.1. There are "open pores" into which a molecule can freely enter from the outside and "closed pores" into which a molecule cannot enter. Open pores are divided into penetration pores and non-penetration pores with pore shapes of slits, cylinders, networks, bottlenecks, cones, etc. The volume of pores per unit mass is referred to as the "pore volume" or "porosity".

The types of pores are also classified according to pore diameter (Fig. 3.4.2). A pore having a diameter of 20 Å or less is called a micropore, and this classification can be further divided into supermicropores and ultramicropores bordering on 7 Å in diameter. Pores with diameters of 50 Å or more are called macropores, and pores with diameters between those of macropores and micropores are called mesopores. Macropores usually include the void between particles. The classifications are

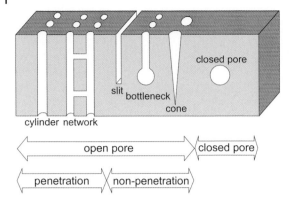

**Figure 3.4.1** Types of pore by shape.

based on the diameter of a nitrogen molecule. Alternatively, the term nanopore has recently appeared as a word for describing small pores. The typical observation techniques for porous materials are shown in Fig. 3.4.2. In the region of micropores and mesopores, the gas probe method is the most useful. Thus, it is natural to consider the relationship between porous crystals and gas adsorption phenomena.

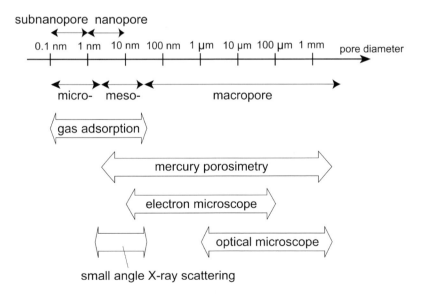

**Figure 3.4.2** Types of pore by pore diameter and typical characterization methods.

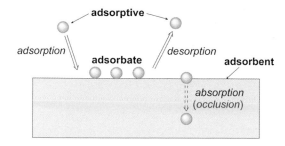

**Figure 3.4.3** Terms for gas adsorption phenomenon.

3.4.3
**Nanoporosity for Gas Adsorption**

Gas adsorption is a phenomenon in which a gas molecule is removed from the gaseous phase by a solid surface. (Fig. 3.4.3) The gas itself is called the adsorptive and the concentrated gas on the surface in high density is called the adsorbate. The substance adsorbing the gas is called the adsorbent and the phenomenon in which the gas adheres to the surface (pores) is called adsorption. The phenomena in which the adsorbents are taken into the solid are called absorption or occlusion. Occlusion is a special case where gas is reversibly adsorbed into a crystal lattice with intrinsic saturated composition. These phenomena are further divided into physisorption and chemisorption based on the strength of the interaction energy between the gas and the solid. When the heat of adsorption is greater than 20 kJ mol$^{-1}$ it is classified as chemical adsorption, but the classification is not strict.

Because the pore size of crystalline materials is usually molecular in size, the materials can exhibit the striking feature that they are able to form deep potential

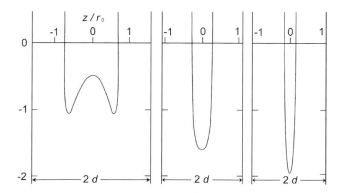

**Figure 3.4.4** Potential field between walls for a slit model.

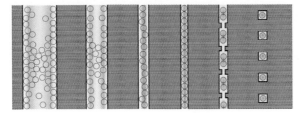

**Figure 3.4.5** Condensed state of molecular adsorption into micropores.(Diameter and length of 1D channel decreases from left to right.)

fields for gas adsorption. For split pores the result of the calculated potential between slits using the Lennard-Jones potential is shown in Fig. 3.4.4 [1]. The slit separation becomes small, in the range of several times the diameter of a molecule, the van der Waals potential from both sides begins to overlap and the pore quickly becomes deep, generating a space with a strong gas adsorption potential field. In cylindrical pores, a stronger potential field can be offered in comparison with slit pores since the space is surrounded by four quarters. In addition, the closed hole, which is completely surrounded by walls, can offer the strongest field since an external guest cannot enter into the conventional solid and the use of enclosed pores is impossible. (Interestingly, adsorption into enclosed pores can be observed in a few metal–organic materials described in Section 3.4.8.) In narrow pores, the state of the adsorbed molecules can change dramatically with the relationship between the host and guest properties. In narrow pores, the states and characteristics of the adsorbate can differ completely from those of its bulk state (Fig. 3.4.5).

### 3.4.4
**Brief Thermodynamic Description of the Gas Adsorption Phenomenon**

Let us briefly look at the fundamentals of gas adsorption.

The amount of adsorbed gas $M$ is a function of pressure $P$, temperature $T$, concentration $c$, and the adsorption interaction potential $E$ between a solid and a gas, which is expressed by:

$$M = f[T, P(c), E] \tag{3.4.1}$$

In adsorption experiments $M$ is generally shown on the vertical axis, and the horizontal axis represents absolute pressure $P$ or relative pressure $P/P_0$ ($P_0$ is the saturation vapor pressure of an adsorptive at an experimental temperature), partial pressure in mixed gases and absolute or relative concentration for a solid-liquid system. Here, since we are focusing on the property of gas adsorption, a solid-gas system is discussed.

The amount adsorbed $M$ can be a function of pressure $P$ at a fixed temperature $T$. This is called an adsorption isotherm, and is most often used for the qualitative analysis of gas adsorption. In contrast, the amount of adsorption as a function of temperature $T$ when the pressure is kept constant is called an adsorption isobar. The relationship between $P$ and $T$ for a constant adsorption amount $M$ is called an adsorption isostere, which is used for calculating the isosteric heat of adsorption from the isotherm curves at various temperatures.

Isotherm:    $M = f(P)_{T,E}$                                          (3.4.2)

Isobar:      $M = f(T)_{P,E}$                                          (3.4.3)

Isostere:    $P = f(T)_{M,E}$                                          (3.4.4)

If an adsorptive is exposed to an adsorbent, the amount of adsorption will increase with time till the amount of adsorption reaches an equilibrium with equilibrium pressure. The temporal response of the amount of adsorption is the adsorption rate. The amount of adsorption is expressed as the amount of adsorbed gas per unit mass of adsorbent $n^a$ (mol g$^{-1}$) or $w^a$ (g g$^{-1}$). If an adsorbate is a gas, it may also be expressed by the converted gas volume of the adsorbate $v^a$ (STP cm$^3$ g$^{-1}$) under a standard condition (1 atm, 0 °C). Generally, the equilibrium amount adsorbed tends to be proportional to the surface area $A_s$ of the adsorbent. Moreover, coverage $\theta (= n^a/n^a{}_m)$ is frequently seen ($n^a{}_m$: amount of monolayer absorption). The equilibrium amount adsorbed $n^a$ is a function of pressure $P$ (or concentration $c$) and temperature $T$.

$$n^a = n^a(P, T) \qquad\qquad (3.4.5)$$

The enthalpy change $\Delta H$, internal energy change $\Delta E$, and entropy change $\Delta S$ accompanying adsorption are treated by the usual thermodynamics if the adsorbed state can be considered as a phase similar to a gaseous or liquid phase. According to the laws of thermodynamics, the Helmholtz free energy $F$ at constant $T$ and $V$, and the Gibbs free energy $G$ at constant $T$ and $P$ become the minimum amounts in a state of equilibrium. Therefore, $\Delta F$ and $\Delta G$ are 0 at the adsorption equilibrium.

$$\Delta F = \Delta E - T\Delta S = 0 \ (T, \ V = \text{const.}) \qquad\qquad (3.4.6)$$

$$\Delta G = \Delta H - T\Delta S = 0 \ (T, \ P = \text{const.}) \qquad\qquad (3.4.7)$$

Since, on adsorption, the system is changed to the adsorption layer in two- or lower dimensional space from the gaseous phase or liquid phase in three-dimensional space, the entropy generally changes and $\Delta S$ accompanying the adsorption becomes negative ($\Delta S < 0$). Therefore, the internal energy change $\Delta H$ is negative at constant $T$ and $V$. Moreover, since the heat of adsorption $Q$ at constant $T$, $V$ and $T$, and $P$ are given by $-\Delta E$ and $-\Delta H$, respectively, adsorption of gas molecules is generally accompanied by the generation of heat. Therefore, the equilibrium amount adsorbed increases as the temperature decreases. In the special case where the adsorbed species can move freely on the surface by dissociation, the change in entropy $\Delta S$ may be positive and an endotherm may be seen in that case.

The thermodynamic variables that can be directly obtained in adsorption experiments are $\Delta E$ or $\Delta H$; they are obtained from the heat of adsorption $Q$ at constant $T$ and $V$, or $T$ and $P$, respectively. $Q$ is the heat of adsorption per 1 mol of substance adsorbed, the integral heat of adsorption $Q_{int}$ and the differential heat of adsorption $Q_{diff}$ are defined by:

$$Q_{int} \equiv \frac{Q}{n^a} \tag{3.4.8}$$

$$Q_{diff} \equiv \left\{ \frac{\partial Q}{\partial n^a} \right\}_T \tag{3.4.9}$$

If the molar internal energy of the adsorptive in the gaseous phase is $E_g$, the internal energy of $n_g$ mole of a gaseous phase is given by $n_g E_g$. In contrast, since the internal energy of the adsorbed phase depends on $n^a$, the internal energy of $n^a$ mole is expressed as $n^a E_a = n^a E_a(n^a)$. Since the heat of adsorption $Q$ is obtained at constant $T$, $V$ is the difference between the internal energy $n^a E_g$ and $n^a E_a(n^a)$, and $Q$ is given by the following formula:

$$Q = n^a \left\{ E_g - E_a(n^a) \right\} \tag{3.4.10}$$

Therefore,

$$Q_{int}(n^a) = E_g - E_a(n^a) \tag{3.4.11}$$

$$Q_{diff}(n^a) = E_g - E'_a(n^a) \tag{3.4.12}$$

Ea' is given by the next expression, and is called the molar differential internal energy of the adsorbate.

$$E_a(n^a) \equiv \left( \frac{\partial(n^a E_a)}{\partial n^a} \right)_T = E_a(n^a) + n^a \left( \frac{\partial E_a}{\partial n^a} \right)_T \tag{3.4.13}$$

Although the heat of adsorption or enthalpy change accompanying adsorption is directly obtained by calorimetry, it can conveniently be evaluated from the adsorption isostere. According to thermodynamics, the relationship between temperature $T$ and pressure $P$ under a state of $\alpha$–$\beta$ phase equilibrium can generally be expressed with the Clausius–Clapeyron equation:

$$\frac{dP}{dT} = \frac{\Delta H_{trans}}{T(V_m^{(\beta)} - V_m^{(\alpha)})} \tag{3.4.14}$$

$\Delta H_{trans}$ is the enthalpy difference between the $\beta$- and $\alpha$-phases, and $V_m^{(\beta)}$ and $V_m^{(\alpha)}$ are the molar volumes in the $\beta$- and $\alpha$-phases. Regarding phases $\alpha$ and $\beta$ as a gaseous phase and an adsorption phase, respectively, the molar volume of the adsorption phase $\beta$ can be disregarded as compared with the molar volume of the gaseous phase α. Since $\Delta H_{trans}$ is a change in enthalpy accompanying adsorption, the heat of adsorption $Q$ (= $-\Delta H$) can be evaluated from the upper equation regarding gaseous phase α as an ideal gas. Usually, the heat of adsorption $Q_{iso}$ is obtained from the relationship (adsorption isostere) between adsorption equilibrium pressure $P$ and temperature $T$, at constant adsorbed amount $n^a$

$$Q_{iso} = -R\left(\frac{\partial \ln P}{\partial (1/T)}\right)_{n^a}$$

(3.4.15)

where $R$ is the gas constant.

The heat of adsorption $Q_{iso}$ at each adsorption level is obtained from the temperature gradient of the pressure in the adsorption isostere. The heat of adsorption $Q_{iso}$ obtained thus is called the isosteric heat of adsorption.

By substituting the averaged molar enthalpy $H_a(n^a)$ of the adsorbate for the averaged molar internal energy $E_a(n^a)$ of the adsorption phase, the differential adsorption molar enthalpy $H_a'(n^a)$ is obtained. The isosteric heat of adsorption $Q_{iso}$ is given as the difference between $H_g$ the molar enthalpy of the gaseous phase and $H_a'(n^a)$ the differential molar enthalpy.

$$Q_{iso} = H_g - H_a'(n^a) = E_g + RT - H_a'(n^a)$$

(3.4.16)

Because the molar volume of adsorbate $V_\alpha$ can be disregarded as compared with that of the gaseous phase, we can approximate $H_a'(n^a)=E_a'(n^a)$. Therefore, Eq. (3.4.16) can be written as:

$$Q_{iso} = Q_{diff} + RT$$

(3.4.17)

In the case of strong adsorption interaction with the surface (when the heat of adsorption is large), it is generally $(\partial E_a / \partial n^a)_T > 0$ in Eq. (3.4.13). This is caused by unevenness from the perspective of adsorption and shows that adsorption arises preferentially from the sites that can handle a large heat of adsorption. Therefore, the integral heat of adsorption $Q_{int}$ and the differential heat of adsorption $Q_{diff}$ decrease with the amount of adsorption $n_a$. (If an adsorption interaction with the surface is weak, the influence of the interaction between adsorbed species must be taken into account.)

## 3.4.5
### Crystalline Organic and Metal–Organic Gas Adsorbents

The history of crystalline gas adsorbents is old and starts with the mineral zeolite. Nowadays, there are successful synthetic techniques for the production of zeolites and zeotypes with controlled diameter pores. Moreover, processes able to produce porosity in pure organic and metal–organic skeletons have been developed. Typical compounds suitable for the topics of this chapter are shown in Fig. 3.4.6. They possess inner accessible spaces for gaseous guests to generate a gas inclusion state

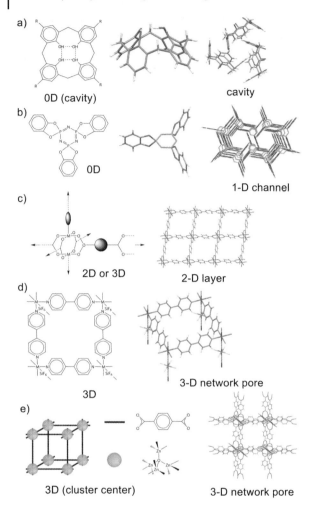

a) 0D (cavity)

cavity

b) 0D

1-D channel

c) 2D or 3D

2-D layer

d) 3D

3-D network pore

e) 3D (cluster center)

3-D network pore

**Figure 3.4.6** Typical substances bearing pores for gas adsorption in the crystalline state.

with individual gas adsorption properties. The types of constituents are discrete molecules of calix[4]arene [2] (a) and tris(*o*-phenylenedioxy)cyclotriphosphazane (TPP) [3] (b), which form pure organic van der Waals cavities and 1D channels in the crystal state by molecular assembly. Interestingly, TPP can only function as a host in the solid state due to its lack of host capability as an isolated molecule. Pore units can be formed in 2-D or 3-D lattices [4–6] (Fig. 3.4.6 (c)–(e)). At the present time, most porous materials capable of adsorbing gases and consisting of organic frames have a neutral skeleton without electric charge.

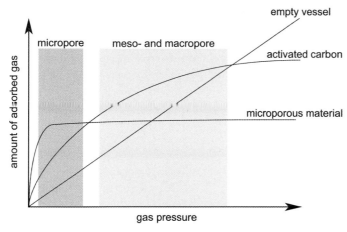

**Figure 3.4.7** Diagram of the correlation between pressure and amount of adsorbed gas for a porous gas adsorbent.

## 3.4.6
**Dawn of Metal–Organic Gas Adsorbents**

The gas adsorption phenomenon involving porous metal complexes was first reported by Kishita at an annual meeting in Japan in the 1970 s, and this was also the earliest report of a crystalline gas adsorbent based on a metal complex [7]. This accidental discovery [8] did not attract much attention, being made at the time of magnetic research on copper (II) terephthalate (1,4-benzenedicarboxylate) (Fig. 3.4.6(c)). It was more than 20 years before that "old" metal complex solid, able to adsorb gases, attracted attention. Mori and Takamizawa at Osaka University and the Osaka Gas company started a joint research program aimed at the exploitation of adsorbing natural gas (methane) as a means of storage (ANG method) [9].

For many gases such as methane the room temperature is well above their critical temperature, therefore they are in a supercritical state, i.e. they cannot be condensed into the liquid state at any pressure. While concentration for the storage and conveyance of natural gas (of which methane is the main constituent) at room temperature requires compression at high pressure (CNG method), the ANG method, which makes use of solid adsorbents, is viable even at low gas pressures (Fig. 3.4.7) Because high efficiency depends on a uniform microporous structure with high porosity, the designable, crystalline metal complex was considered to be a very promising exotic material.

## 3.4.7
**Design of Porosity in Coordination Polymer Systems**

Crystalline solids of coordination compounds can be obtained as a precipitate from solution after mixing the raw materials under mild conditions, usually at room temperature. Thus, the system can be systematically expanded by carefully select-ing the kinds of metal centers and ligands in order to simultaneously design the annular cavity portion and the three-dimensional network structure. However, it was thought that stability of structure was indispensable in porous coordination materials, so that a rigid network frame is preferable, especially if large porosity is required. Network porosity obtained by linking ligands such as dicarboxylic acid and bipyrizine still does represent the main research area in the field of molecular porous materials.

### 3.4.8
### Structural Description and Dimensionality of the Host Component

Porosity is realized by the three-dimensional geometry of the inner space [8, 10] therefore a porous solid may be classified by the dimensionality of its structural element (Fig. 3.4.8). It is clear that a three-dimensional network solid is thermally more stable and, conversely, low stability can be expected for a solid obtained by assembly of lower dimensional units. 1D systems may exhibit conflicting solid state properties of rigidity (stability) and flexibility (instability). There is an example of a one-dimensional chain complex of metal benzoate-pyrazine (rhodium(II) and copper(II) complexes) [11, 12], which has good thermal stability, with a decom-position temperature of 190 °C and 290 °C for the copper(II) and rhodium(II) complexes, respectively, and reversible capability of adsorbing gases while main-taining single-crystal habit (Fig. 3.4.9). As will be explained later, this system is the first single-crystal gas adsorbent, as observed by single crystal X-ray structural analysis. It should be noted that 2D or 3D network porous single crystals usually turn into crystalline powder upon removal of solvent of crystallization from the pores, and sometimes they lose crystallinity and/or gas adsorption capability.

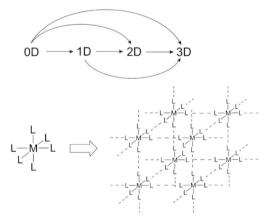

– – – = chemical bonds and/or intermolecular interaction
M = metal center (mono-, multinuclear core, or metal cluster)

**Figure 3.4.8** A dimensional view of the constructing units for a porous crystal. (An example of a coordination compound.)

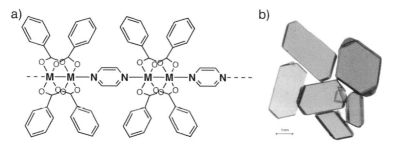

**Figure 3.4.9** The linear component structure of [$M_2$(bza)$_4$(pyz)](bza= benzoate, pyz= pyrazine) (a) and the photograph of a single-crystal gas adsorbent (M= Cu(II))(b).

### 3.4.9
### Transformation of a Single-crystal Gas Adsorbent during Gas Adsorption [13, 14]

An empty crystal of rhodium(II) benzoate pyrazine, [$Rh^{II}_2(O_2CPh)_4$(pyz)]$_n$ consists of a chain skeleton bridged by pyrazine in the axial direction of the lantern-like core of dinuclear rhodium benzoate (Fig. 3.4.9). The polymeric chains are densely gathered together to form the crystal. There are no channel structures sufficient to hold guest molecules, but rather empty cages of size $10 \times 6 \times 4$ Å$^3$ with narrow gaps of size ca. 2 Å at their four corners, which are formed by the benzene rings of benzoate moieties parallel to the chain vector. In a $CO_2$ atmosphere, the crystal undergoes a crystal phase transition from monoclinic to the triclinic space group,

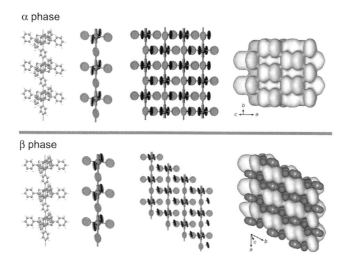

α phase

β phase

**Figure 3.4.10** Crystal structure change on $CO_2$ gas adsorption by [Rh(II)$_2$(bza)$_4$(pyz)]. The slippage of the 1D chains and the tilting of the benzene rings were observed during α–β host crystal phase transition.

generating one-dimensional channels (Fig. 3.4.10). The cages are transformed into channels by a slippage of the chain skeletons along the chain vector, the benzene ring tilting 9° away from the chain vector (Fig. 3.4.10). It is noted that the two modes of chain slippage and offset stacking occur simultaneously in the inclusion β phase. They regulate the formation of the inclusion crystal and shorten the interplanar distances, which may allow the volume of space in the channels to be maximized and the crystal to be stabilized. External $CO_2$ is incorporated into the channels in the form of molecular wires in which the molecular axes of the $CO_2$ molecules are almost parallel and perpendicular to the channel direction at sites A and B, respectively (Fig. 3.4.11). The adsorbed state is stabilized by cooperative interaction between the guest and the hydrocarbon of the channel by intermolecular atomic contact.

The $CO_2$ sorption isobar measurement at ambient pressure yielded a monotonic desorption curve with a saturation point at approximately –80 °C. (Fig. 3.4.12(a)) The saturated quantity of 3.0 molecules per $Rh_2$ unit at –80 °C agrees well with the composition determined from the single-crystal X-ray diffraction data at –180 °C (Figs. 3.4.11 and 3.4.12). The DSC measurements repeatedly revealed exo- and endothermic phase transitions in a $CO_2$ atmosphere. (Fig. 3.4.12(b)) The temperatures where the sets of cyclic DSC peaks occur shift to lower values as the concentration of $CO_2$ (partial pressure) decreases, demonstrating that the observed crystal phase transition is gas-adsorption induced. The estimated value indicates that the α crystal can include no more than one $CO_2$ molecule under the current "α-saturation" condition and subsequently transforms to the inclusion crystal, which can include up to three $CO_2$ molecules per $Rh_2$ unit in the channels (Fig. 3.4.12(c)).

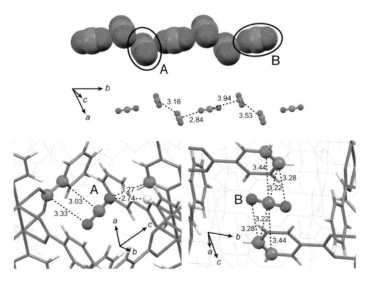

**Figure 3.4.11** The environment of the generated 1D $CO_2$ molecular chain within the 1D channel.

## 3.4.10
## Abnormal Guest Diffusivity Within Pores

The single-crystal adsorbent $[Rh(II)_2(bza)_4(pyz)]_n$ exhibits smooth and reversible gas adsorbing ability as commonly observed in microporous media, and water vapor cannot advance in the crystal due to the hydrophobic nature of the pores (Fig. 3.4.13) They have, however, an insufficient channel structure with a narrow neck diameter of ca. 2 Å, while $O_2$ and $N_2$ gases are plugged in the narrow channels of inorganic materials such as Zeolite 3A and Zeolite 4A under the same condition (77 K) [15]. Some kind of pore structural changes should facilitate the passage of

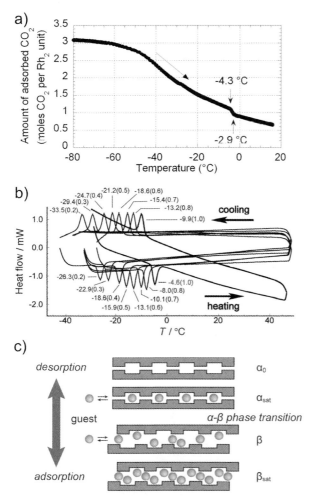

**Figure 3.4.12** $CO_2$ gas desorption isobar curve at ambient pressure (a), DSC curves at various $CO_2$ partial pressures (b), and the deduced crystal phase transition induced by incorporated $CO_2$ gas inside the crystal(c).

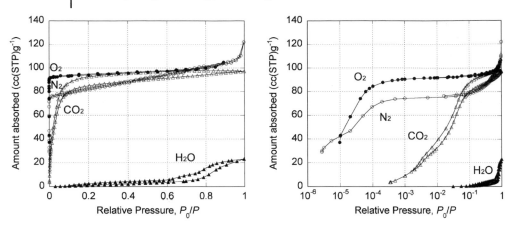

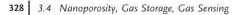

Figure 3.4.13 Gas sorption isotherm curves of [Rh(II)$_2$(bza)$_4$(pyz)] for N$_2$(77 K), O$_2$(77 K), CO$_2$(195 K), and H$_2$O(293 K).

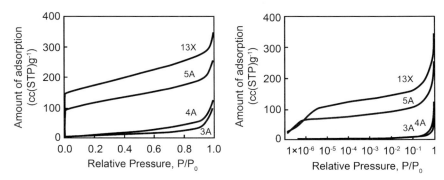

Figure 3.4.14 N$_2$ gas adsorption isotherm of zeolite materials at 77 K.

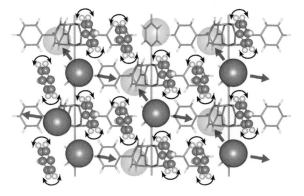

Figure 3.4.15 Schematic drawing of surface motion transfer, which can facilitate guest diffusion within narrow pores surrounded by flexible organic moieties of the host skeleton.

the diffused guests through the necks. "Surface motion transfer" should be considered for the guest diffusion within crystals consisting of a "soft" organic framework [16]. The tilting vibration and/or flip motion of the benzene rings of the necks probably drives the captured guests from the benzene ring to the opposite ring of the same neck by crossing the minimum potential positions sufficient for guest hopping (see Fig. 3.4.14). In comparison with inorganic porous materials, such as zeolite and porous silica, porous metal–organic solids have an inner surface consisting of rich hydrocarbon and $\pi$-conjugated systems, which attracts the guest molecules and enhances surface local motions (Fig. 3.4.15)

## 3.4.11
### Method for the Accurate Detection and Measurement of Gas Adsorbed State

A single-crystal adsorbent is one of the reachable goals of designable porous solids; light gases can then be "co-crystallized" by putting the crystal adsorbent in a gaseous guest atmosphere. Because the crystal remains intact under high pressure and at low temperature, conditions important for the efficiency of the adsorption process, it is possible to analyze the resulting host–guest aggregate in great and accurate detail by conventional single-crystal X-ray diffraction (Fig. 3.4.16). The amount of gas adsorbed can be controlled by varying the gaseous guest pressure (pressure swing) at appropriate temperatures (temperature swing). The convenient experimental method to apply target gas pressure is explained in Fig. 3.4.17. This method gives constant density conditions before gas condensation begins. Figure 3.4.18 shows the effects of such processes on the structures determined via X-ray diffraction. The structures obtained by the adsorption of oxygen and methane are discussed below.

At high pressure (9 MPa at room temperature) and low temperature (90 K) the oxygen inclusion structure is obtained [17, 18]. Oxygen molecules are coherently aligned in a linear fashion and the chain structure is most likely regulated by the channel structure. Since the observed temperature is higher than the melting point

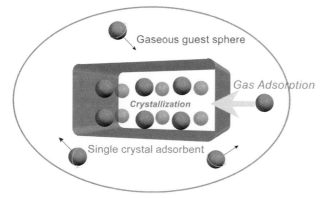

**Figure 3.4.16** Observation Method of observation of co-crystallization of gases using a single-crystal adsorbent.

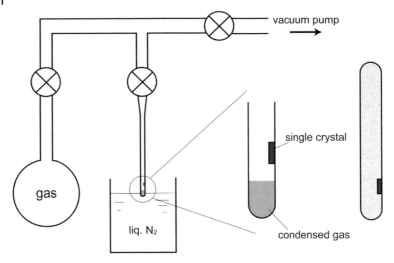

**Figure 3.4.17** Convenient pressurizing method for single-crystal X-ray diffraction measurement.

(55 K) of bulk oxygen, this result shows the strong co-crystallizing ability for guest gases. In contrast, the O–O atomic distances in the oxygen molecules become reasonable at 1.147(17) Å and 1.143(16) Å at 10 K. These are the first exact bond lengths of molecular oxygen measured by single-crystal X-ray analysis, while the O–O bond length in α-phase solid oxygen was ambiguously evaluated to be in the range 1.15–1.22 Å by powder X-ray diffraction analysis at 23 K [19]. The periodic arrangement and configuration of the oxygen in the chain becomes more crowded at 10 K.

On the other hand, the structure of adsorbed supercritical methane ($CH_4$) in an adsorption–desorption equilibrium state gives a methane inclusion crystal even at

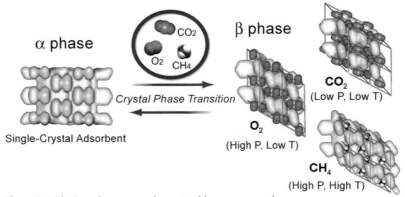

**Figure 3.4.18** Crystal structure, determined by pressure and temperature swinging method, for [Rh(II)$_2$(bza)$_4$(pyz)] with $CO_2$, $O_2$ and $CH_4$.

298 K by the forcible pressure swing adsorption method (ca. 13 MPa). The adsorbed methane molecules are located in the pocket-like narrow corners of the necks of the 1D channel [20]. Because the thermal motion of the pseudo-spherical methane molecules seems to be effectively suppressed in its translation mode but rotation is allowed, the forcible adsorption of methane gas produces an "inclusion plastic crystal" [20], which can be regarded as a mesophase between the fluid and solid state of the phase of a guest incorporated in a crystal host: the guest molecules are randomly oriented, but their alignment follows the crystal periodicity.

## 3.4.12
### Hydrogen Storage [21]

Nowadays, with the shift to a hydrogen energy society, there is an urgent need for the basic technology in connection with the manufacture of hydrogen gas, refining, storage, conveyance, and supply. Since the energy density of hydrogen gas is extremely low, exploitation of a high concentration storage technique becomes much important.

The hydrogen gas sorption isotherm measurement revealed a certain sorption ability, even at 77 K. Ad- and desorption occurred reversibly without hysteresis. (Fig. 3.4.19(a)) The property of two-step adsorption was observed at around 300 mmHg after rapid adsorption in the form of micropore filling, indicating the formation of the first of the stable adsorbed states. The hydrogen molecules are paired and form parallel dimers that face the benzoate planes, and are regularly distributed in the 1D channel.(Fig. 3.4.19(b)) The interatomic intradimer distances are 2.44 Å and 2.35 Å, which indicate strong contact to achieve the condition of highest density (the van der Waals diameter of the H atom is 2.4 Å). Since the

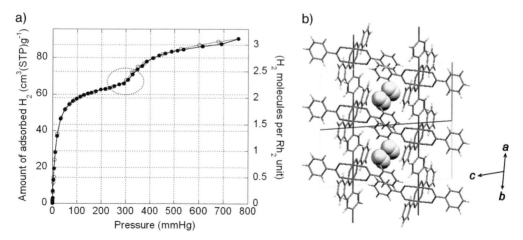

a)                                                    b)

**Figure 3.4.19** $H_2$ adsorption isotherm at 77 K(a) and crystal structure of $[Rh(II)_2(bza)_4(pyz)]\cdot2(H_2)$ at 90 K. (The circle in (a) indicates the first stable state with two $H_2$ molecules adsorbed per $Rh_2$ unit.)

hydrogen molecule, $H_2$, is very light, a quantum effect may show up. The observed slipped-parallel configuration might suggest a specific anisotropic interaction such as quadrupole–quadrupole interaction through ortho-$H_2$ dimer formation, concentrated during the formation of the aggregated pair, since it is known that ortho-$H_2$ has greater adsorption capability than does para-$H_2$, due to the difference in nuclear spin symmetry [22]. It may be necessary to consider the quantum effects in using extremely small regular spaces for hydrogen adsorption to develop methods for molecular hydrogen storage.

### 3.4.13
### Phase Transition of the Adsorbed Guest Sublattice in the Gas Inclusion Co-crystal State [23]

The state of adsorbed guest aggregates inside a pore should become unusual as the space is decreased. Since oxygen is the simplest and the most stable paramagnetic gas molecule with a $^3\Sigma_g^-$ ground state, with $S=1$ spin configuration, observation of the magnetic properties of adsorbed 1D oxygen molecules can give information on the state of guest aggregates inside regular pores. As the applied field is increased, a different magnetic state grows in the middle temperature region, with boundaries at 54 K and 104 K, while the other temperature region is only slightly influenced by the magnitude of the magnetic field (Fig. 3.4.20). On comparing the boundary temperatures of $[\text{Rh(II)}_2(\text{bza})_4(\text{mpyz})]_n \cdot 3(O_2)$ (mpyz = 2-methylpyrazine) [ 54 K ($T_{\text{low–middle}}$) and 104 K ($T_{\text{middle–high}}$)] with the phase transition temperatures of bulk oxygen (mp 54.8 K, bp 90.2 K), it can be seen that the structural phase transition of the included oxygen system is temperature-induced, and that

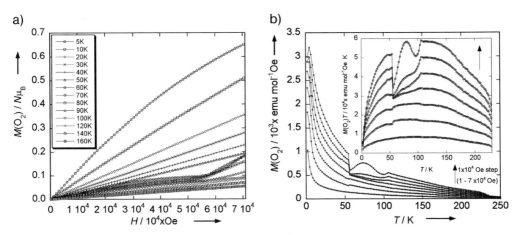

**Figure 3.4.20** Magnetic behavior of adsorbed 1D molecular oxygen within $[\text{Rh}_2(\text{bza})_4(\text{mpyz})]$ (bza = benzoate, mpyz = 2-methylpyrazine).Magnetization curves at various temperatures (a), and temperature dependence at various magnetic fields(b).The three temperature phases of adsorbed oxygen were clearly observed in (b).

the middle temperature phase is a mesophase between the solid and fluid states. Consequently, the abnormal magnetic behavior of the middle temperature phase can be explained by the fluctuation, caused by the external magnetic field, of its "soft" and flexible arrangement to reach the most stable configuration.

This method has the potential to advance the experimental study of low-dimensional physicochemical properties in generating real low-dimensional molecular/atomic aggregates/clusters with various degrees of structural freedom, under the control of arrangement dimensionality. This would improve the observation of specific quantum effects by signal enhancement to perceptible bulk levels.

## 3.4.14
### Dynamic Change in Pore Topology by Design of Host Flexibility [24]

In this section, it is demonstrated that the addition of higher-order flexibility to the host skeleton can realize unprecedented dynamic structural change. Designable dynamic porous single-solid devices may generate new applications by the active controllability of anisotropic guest diffusivity based on sorption and diffusion phenomena (see Fig. 3.4.21(e)). By introducing a "leverage" mechanism in a host component of a metal–organic crystal, a channel switching property driven by incorporated guest gas stress was developed in a porous single-crystal adsorbent of $[Rh^{II}_2(bza)_4(2\text{-epyz})]_n$ (2-epyz = 2-ethylpyrazine), where the chain skeleton was bent by the steric hindrance of the ethyl group on a pyrazine linker against a dinuclear rhodium benzoate core; small vacant cavities of 35(1) Å$^3$ in volume were

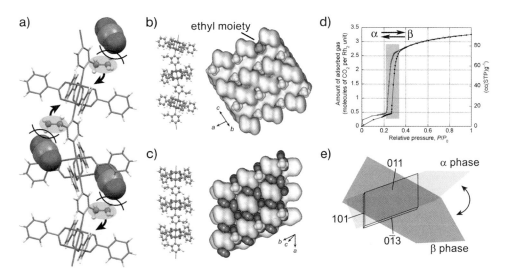

**Figure 3.4.21** Channel switching property by the molecular leverage mechanism of $[Rh_2(bza)_4(2\text{-epyz})](2\text{-epyz}= 2\text{-ethyl-pyrazine})$. Contact of $CO_2$ at ethyl lever on the channel surface (a), topological change of inner space by changing from a bent 1D chain skeleton to a straight one during α–β crystal phase transition (b, c), $CO_2$ isotherm curves with large leap during phase transition (d), and channel direction change in a single-crystal.

generated, sandwiched in between the ethyl groups exposed on the crystal cavity (Fig. 3.4.21(b)). Via modification of the host skeleton the inner surface can also be modified. The crystal can absorb a large amount of gas, with α–β bulk phase transition when a critical amount of gas is adsorbed (Fig. 3.4.21(d)) From single-crystal X-ray structures for $CO_2$ gas adsorption states in the α- and β-phases, it can be seen that the incorporated gas molecules destabilize the α-phase crystal lattice by acting as a stress on the ethyl groups, and this initiates a local structural change leading to the bulk phase transition. This can be considered a "molecular leverage mechanism" where the pushing force of the incorporated guest concentrated on the ethyl moieties (the point of force) forces the bent M–pyrazine–N link (the point of application) to stretch into the straight 1D skeleton (Fig. 3.4.21(a)). The fulcrum force is generated by the reaction of the packing force to position the 1D chain skeleton at a fixed location in the crystal.

### 3.3.15
#### Mass-induced Phase Transition [24]

In order to clarify the nature of the gas adsorption induced transition, vapor adsorption isotherm measurements were carried out over a narrow temperature region. The ethanol vapor sorption isotherm measurements showed the ad- and desorption jumps which are caused by the crystal phase transition (Fig. 3.4.22(a)). The natural logarithm of the vapor pressure ($\ln P_{EtOH}$) at the starting points of the ad- and desorption jumps and the reciprocal of the temperature ($T^{-1}$) show a good linear correlation, coincident with the *Clausius–Clapeyron* equation (Fig. 3.4.22(b)) This relationship clearly shows that the observed jump area has the same $\Delta H$ of 53.9 kJ mol$^{-1}$ over a wide range of temperature. This correlation gives the isosteric heat of adsorption ($\Delta H_{iso}$) using two isotherm curves at different temperatures for the same adsorption quantity. Thus, A (adsorption amount)–$\Delta H_{iso}$ curve was numerically estimated by the equation using two adsorption curves at 10 and 20 °C. The enthalpy leaps were clearly observed around the initial and terminal regions of the adsorption jump, and a plateau exists between them (Fig. 3.4.22(c)). The discontinuous leap in enthalpy indicates the first-order phase transition of the crystal adsorbent. The plateau indicates the coexistence of the α- and β-crystal phases, where the propagation of the β-crystal phase continues as the adsorption amount increases, displaying the same situation for gas adsorption (Fig. 3.4.22(d)). At the phase transition, the degree of freedom becomes $F = 1$, by Gibbs' phase rule ($F = C$ (number of components)– $P$(number of phases) +2. Since, in this case, $C = 2$ (gas and host crystal) and $P = 3$ (gas, α-phase and β-phase crystals), the pressure will not change at constant temperature, even if composition changes. It should be noted that the ratio of α–β phases should be determined by the adsorption fraction in the range of $A_{pro}$, which can be controlled by guest supply. Thus, the current gas adsorption induced phase transition can be called "mass induced phase transition." In the α–β phase equilibrium, the adsorption actually occurs at the α–β phase boundary, which should have a fluctuating and changeable structure. The abnormal adsorption enthalpy during phase equilibrium may be

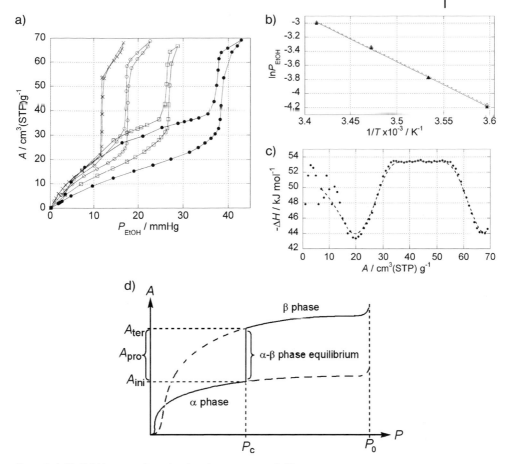

**Figure 3.4.22** EtOH vapor adsorption istotherm curves at 5, 10, 15, 20 °C for [Cu$_2$(bza)$_4$(pyz)](a), Clausius–Clapeyron plot from ad- and desorption jumps (b), temperature difference between the curves at 10 and 20 °C (c), and the deduced "mass-induced phase transition" profile during α–β host crystal phase transition induced by the gas adsorption (d).

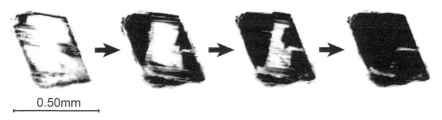

0.50mm

**Figure 3.4.23** Observation with a polarization microscope of α–β crystal phase boundaries in a single-crystal [Cu$_2$(bza)$_4$(pyz)] during ethanol vapor adsorption at room temperature. (Dark areas are the generated β-crystal phase.)

correlated with the dynamic property of the phase boundary. Actually, the α-β phase equilibrium state can be observed by visual inspection (Fig. 3.4.23).

### 3.4.16
### Sensing Gas by Porous Crystals

The exploitation of color changes when a crystal enters into contact with a particular gas is an extremely useful technique for gas sensing.The so-called phenomenon of "vapochromism" [26] is much studied for this reason. An example is given by the crystalline double salt [Pt($p$-CNC$_6$H$_4$R)$_4$][Pd(CN)$_4$] (R=alkyl group), which shows different color changes in response to various organic vapors; the color change originates from the kind of assembly of metal-complex units in the crystal structure, and is induced by inclusion of gas. Although, at present, both gas sensing and selective inclusion of gas can be regarded as independent phenomena, a material able to show both properties might be attractive in the near future. Recently, the important role that channel flexibility can play in vapochromism has been pointed out [27].

Selective gas inclusion is an alternative route to gas sensing. We have been able to observe the possibility of thermodynamic gas inclusion and release action of a mass-induced crystal phase transition; this phase transition is essentially triggered by gas clathration. The transformation property through the adsorption process is characterized by a combination of phase boundary transfer and guest diffusion methods, which provide specific adsorption behavior. This characterization can be viewed in terms of the ease of β-phase (or α-phase in the reverse reaction) propagation. Large structural differences between the α and β crystals sometimes require a superpressure condition and activating energy to pass the transition state between the α- and β-phases, which would show a curve with specific hysteresis or

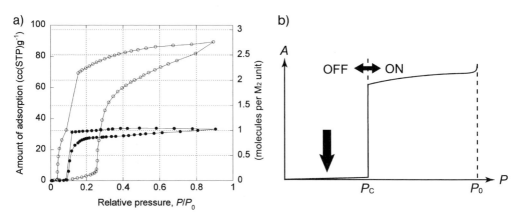

**Figure 3.4.24** Vapor sorption isotherms for benzene(open circle) and n-hexane(solid circle) at 20 °C (a) and the pressure selective on–off adsorption (b). The difficulty of fitting these large guests into the α-crystal phase is illustrated by the negligible adsorption before crystal phase transition into the β-crystal phase.

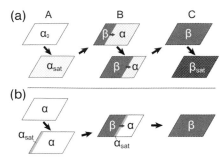

**Figure 3.4.25** Generation and diffusion of crystal phase boundary in mass-induced phase transition in the single crystal. The cases of smooth guest diffusion (a) and not smooth (b) are shown. This can explain the pressure selective on–off adsorption shown in Fig. 3.4.24.

irreversibility. For non-ideal systems that lack the pathway between the $\alpha$ and the $\beta$ phases without recombination of the host fragments, transformation may stop halfway with a greatly fluctuating structure, apparently representing the crystalline–amorphous transition. Moreover, the discussed transition would be influenced by the type of guest spreading inside the solid host, which corresponds to the distribution of gas–clathrate domains in the adsorbent crystal. The amount of adsorption is much less below the critical pressure because the gas–clathrate formation only occurs at the near-surface of the adsorbent solid necessary for $\beta$-phase nucleation. The adsorption then suddenly starts to expand in the $\beta$-phase from the localized domain with $\alpha$-saturation boundary (Fig. 3.4.25(b)). The pressure should correspond to the vapor pressure of the clathrate domains. This specific adsorption behavior was observed for the relatively large gases of benzene and $n$-hexane for $[Rh(II)_2(bza)_4(pyz)]_n$, (Fig. 3.4.24) while a reversible sharp transition through the $\alpha$–$\beta$ phase equilibrium was observed for small guests such as ethanol.(Fig. 3.4.25 (a)).

## 3.4.17
## Concluding Remarks

It should be emphasized that the crystalline material consisting of an organic frame is not very "solid" but is flexible, with a potential degree of freedom even in its crystalline state. Therefore, by a weak physical adsorption interaction, a change in the solid structure (local structure, pore structure, crystal structure) arises easily, and can stabilize various gas adsorption states. Here, the relationship between incorporated gaseous guest and the walls of the pores cannot be compared to the contact between a ball and a flat surface, but to the atomic contact of a point (atom) to a point (atom), raising the necessity of regarding a gas adsorption state as a cocrystal state. Furthermore, it was demonstrated that the gas adsorption behavior of single-crystal adsorbents is related to whether they have crystal phases that can change depending on the amount and type of guest molecules and on the process of guest inclusion, via control of the host structural flexibility; the host frame can be seen as a large "hand" that can actively grasp and release gaseous guest molecules. Since the action is correlated to the gas adsorption phenomenon with controllability by external stimuli, novel dynamic selectivity for gas condensation, storage/

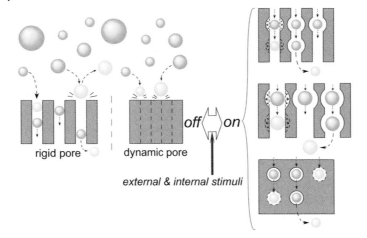

**Figure 3.4.26** The expected dynamic gas selectivity attainable by the control of thermal factors and crystal phases, which is much different from those of the well-known rigid porous materials.

release, separation, etc. will be expected (Fig. 3.4.26). The homogeneity, anisotropy, integrity, transparency, etc. of crystal adsorbents with precise molecular design contributes to the advancement of new techniques for future society.

## References

1 D. H. Everett, J. C. Powl, *J. Chem. Soc., Faraday Trans. 1* **1976**, *72*(3), 619.
2 J. L. Atwood, L. J. Barbour, A. Jerga, *Science* **2002**, *296*, 2367.
3 P. Sozzani, S. Bracco, A. Comotti, L. Ferretti, R. Simonutti, *Angew. Chem. Int. Ed.* **2005**, *44*(12), 1816.
4 W. Mori, F. Inoue, K. Yoshida, H. Nakayama, S. Takamizawa, M. Kishita, *Chem. Lett.* **1997**, 1219.
5 S.-i. Noro, S. Kitagawa, M. Kondo, K. Seki, *Angew. Chem. Int. Ed.* **2000**, *39*(12), 2082.
6 M. Eddaoudi, J. Kim, N. Rosi, D. Vodak, J. Wachter, M. O'Keeffe, O. M. Yaghi, *Science* **2002**, *295*, 469.
7 F. Inoue, W. Mori, M. Kishita, *26th Annual Meeting of the Chemical Society of Japan* **1972**, p.265 (Japanese).
8 W. Mori, S. Takamizawa, in *Organometallic Conjugation*, A. Nakamura, N.

Ueyama, K. Yamaguchi (Eds.), Kodansha Springer, Tokyo **2002**, Ch. 6, pp. 179–213.
9 M. Fujiwara, K. Seki, W. Mori, S. Takamizawa, JP 09132580, **1995**; EP 0727608, **1996**.
10 D. R. Turner, A. Pastor, M. Alajarin, J. W. Steed, Molecular containers: design approaches and applications, *Struct. Bond.*, **2004**, *108* (Supramolecular Assembly via Hydrogen Bonds I), 97–168.
11 S. Takamizawa, T. Hiroki, E. Nakata, K. Mochizuki, W. Mori, *Chem. Lett.* **2002**, 1208.
12 S. Takamizawa, E. Nakata, H. Yokoyama, *Inorg. Chem. Commun.* **2003**, *6*, 763.
13 S. Takamizawa, E. Nakata, H. Yokoyama, K. Mochizuki, and W. Mori, *Angew. Chem. Int. Ed.* **2003**, *42*, 4331.
14 S. Takamizawa, E. Nakata, T. Saito, *Inorg. Chem. Commun.* **2004**, *7*(1), 1.

**15** D. W. Breck, *Zeolite Molecular Sieves*, Krieger Publishing Company, **1984**, pp. 593–724.

**16** S. Takamizawa, E. Nakata, T. Saito, *CrystEngComm* **2004**, *6*(9), 39.

**17** S. Takamizawa, E. Nakata, T. Saito, *Angew. Chem. Int. Ed.* **2004**, *43*, 1368.

**18** S. Takamizawa, E. Nakata, T. Saito, T. Akatsuka, K. Kojima, *CrystEngComm* **2004**, *6*(34), 197.

**19** C. S. Barrett, L. Meyer, J. Wasserman, *J. Chem. Phys.* **1967**, *47*, 592–597.

**20** S. Takamizawa, E. Nakata, T. Saito, T. Akatsuka, *Inorg. Chem.* **2005**, *44*(5), 1362.

**21** S. Takamizawa, E. Nakata, *CrystEngComm*, **2005**, *7*(79), 476.

**22** For example: (a) I. F. Silvera, *Rev. Mod. Phys.* **1980**, *52*, 393; (b) V. Buch, J. P. Devlin, *J. Chem. Phys.* **1993**, *98*, 4195.

**23** S. Takamizawa, E. Nakata, T. Akatsuka, *Angew. Chem. Int. Ed.* **2006**, *45*(14), 2216.

**24** S. Takamizawa, K. Kojima, T. Akatsuka, *Inorg. Chem.* **2006**, *45*(12), 4580.

**25** S. Takamizawa, T. Saito, T. Akatsuka, E. Nakata, *Inorg. Chem.* **2005**, *44*(5), 1421.

**26** C. C. Nagel, U. S. Pat. No. 4826774,1989.

**27** T. J. Wadas, *J. Am. Chem. Soc.*, **2004**, *126*, 6841.

# Index

*Making Crystals by Design.* Edited by Dario Braga and Fabrizia Grepioni
Copyright © 2007 WILEY-VCH Verlag GmbH & Co. KGaA, Weinheim
ISBN: 978-3-527-31506-2

## Related Titles

Scheel, H. J., Capper, P.

### Crystal Growth Technology
**From Fundamentals and Simulation to Large-Scale Production**

2007
ISBN 978-3-527-31762-2

Hilfiker, R. (Ed.)

### Polymorphism
**in the Pharmaceutical Industry**

2006
ISBN 978-3-527-31146-0

Tilley, Richard J. D.

### Crystals and Crystal Structures

2006
ISBN 978-0-470-01821-7

Tiekink, E. R. T., Vittal, J. (Eds.)

### Frontiers in Crystal Engineering

2006
ISBN 978-0-470-02258-0

Tanaka, K.

### Solvent-free Organic Synthesis

2003
ISBN 978-3-527-30612-1